Introducing Biological Energetics

Introducing Biological Energetics

How energy and information control the living world

'Energy is the queen of the world, and entropy is her shadow'
German physical chemist W. Ostwald (1853–1932)

'Information flow, not energy per se, is the prime mover of life'
US cell biologist Werner R. Lowenstein

NORMAN W. H. CHEETHAM

Adjunct Professor
University of the Sunshine Coast, Queensland, Australia

OXFORD
UNIVERSITY PRESS

Great Clarendon Street, Oxford OX2 6DP

Oxford University Press is a department of the University of Oxford.
It furthers the University's objective of excellence in research, scholarship,
and education by publishing worldwide in

Oxford New York

Auckland Cape Town Dar es Salaam Hong Kong Karachi
Kuala Lumpur Madrid Melbourne Mexico City Nairobi
New Delhi Shanghai Taipei Toronto

With offices in

Argentina Austria Brazil Chile Czech Republic France Greece
Guatemala Hungary Italy Japan Poland Portugal Singapore
South Korea Switzerland Thailand Turkey Ukraine Vietnam

Oxford is a registered trade mark of Oxford University Press
in the UK and in certain other countries

Published in the United States
by Oxford University Press Inc., New York

© Norman W. H. Cheetham 2011

The moral rights of the author have been asserted
Database right Oxford University Press (maker)

First published 2011

All rights reserved. No part of this publication may be reproduced,
stored in a retrieval system, or transmitted, in any form or by any means,
without the prior permission in writing of Oxford University Press,
or as expressly permitted by law, or under terms agreed with the appropriate
reprographics rights organization. Enquiries concerning reproduction
outside the scope of the above should be sent to the Rights Department,
Oxford University Press, at the address above

You must not circulate this book in any other binding or cover
and you must impose the same condition on any acquirer

British Library Cataloguing in Publication Data
Data available

Library of Congress Cataloging in Publication Data
Data available

Typeset by SPI Publisher Services, Pondicherry, India
Printed in Great Britain
on acid-free paper by
CPI Antony Rowe, Chippenham, Wiltshire

ISBN 978–0–19–957593–0 (Pbk)
   978–0–19–959371–2 (Hbk)

1 3 5 7 9 10 8 6 4 2

To Yuki, for her love, patience, and sense of humour

Preface

There are many ways to look at the complex phenomenon we call life. The more viewpoints we take, the greater the probability we are likely to 'understand' it—whatever that means. The same is true of any challenging topic, for example genetics. No one can appreciate its full implications from a single lecture course, a single text, or a single approach.

Above all I have tried to make this book accessible. I hope anyone who reads any of it will read all of it. Look on it perhaps as a friendly textbook. To this end I have kept the story as brief as possible without compromising the discussion to the extent of triviality. I have tried to ensure that no specialist scientific terms or concepts have been introduced without at least some explanation, and have tried to minimize the formal side of each science, while retaining much of its rigour. This approach holds inherent dangers. Readers with expertise in a particular area could be affronted that a pet theme has been omitted or dismissed in a few sentences. The more pedantic readership might deplore a perceived lack of depth or background, or the order of presentation. In anticipation of such responses I say, *Introducing Biological Energetics* is selective rather than definitive, in that I could not possibly cover the range of topics chosen in any great depth.

Why choose to look at life through the window of energy? The answer probably lies in my background as a professional scientist and teacher. As a student of chemistry, I was interested in things at the molecular level. Given this interest, it was not much of a step to wondering about what actually drives the chemical reactions I studied. Formally, the answer lies largely in the subject of thermodynamics. The science of thermodynamics was born during the development of steam engines and has been called 'the child of the age of steam'. It began as a narrow, very applied area, and even now is sometimes mistakenly thought to be largely about steam engines.

My serious introduction to this rather daunting area of physical chemistry was not spectacular. One day at university, I was preparing for a laboratory class. I said to my lecturer in physical chemistry:

'I didn't realize there was any thermodynamics involved in the stretching of rubber'.

He looked down his long nose and said quietly, as was his way, 'There's thermodynamics in *everything*'.

I almost burst out laughing, he was so serious. Of course he was right, but I didn't really appreciate this until the relevance struck home when I began to study biochemistry. The ways in which organisms manage to access their chemical and physical needs for survival are wonderful to contemplate. The pathways involved in extracting energy from a complex foodstuff, directly from sunlight, or from 'raw' chemicals in deep-ocean vents, all obey the laws of physics and chemistry. What makes the study so fascinating is that all these processes are carried out under the constraints imposed by being part of something alive. This usually means that chemical reactions need to take place in the presence of an excess of water under quite mild conditions of temperature, pressure, and pH. Most living organisms can't cope with the high temperatures, strong acids, oxidizing agents, and exotic solvents often used by organic chemists. The fact that these constraints have been overcome by natural selection over 3.5 billion years of evolutionary time makes life all the more intriguing. If they hadn't been overcome, none of us would be here.

Contents

Introduction	1
1. The Flow of Energy in Living Systems	3
References	10
2. Origins: The Early Earth	11
General references	29
References	29
3. Force and Energy Explored	31
3.1 Gravitational force	36
3.2 The electromagnetic force	40
3.3 The strong nuclear force	43
3.4 The weak nuclear force	43
References	47
4. Which Way? An Introduction to Thermodynamics	48
General references	72
References	72
5. The Building Blocks	73
General references	108
References	108
6. How Fast, How Far? Chemical Kinetics and Equilibrium	109
General references	138
References	138
7. The Strange Story of Water and Oil	139
General references	161
References	161
8. Size Matters: Proteins and Enzymes	162
8.1 Principles of protein structure	162
8.2 Some very special molecules: enzymes	178
8.3 Regulation of enzyme activity	189
8.4 Coenzymes, vitamins, and enzyme classification	192
References	199
9. Molecular Genetics—the Chemical Basis of Heredity	200
General references	222
10. Electron Gymnastics: Energy Revisited	223
10.1 How many ATP molecules are produced when electrons traverse the entire ETC?	233
10.2 What about the rotary engine?	234
References	249

11. Cells and Metabolism: Putting it all Together — 250
  11.1 General aspects of metabolism — 250
  11.2 Glycolysis — 254
  11.3 The reactions of glycolysis — 257
  11.4 The pentose phosphate pathway — 265
  11.5 The citric acid or tricarboxylic acid cycle — 266
  11.6 Regulation of the citric acid cycle — 271
  General references — 273
  References — 273

12. From Prokaryotes to Eukaryotes: Getting Ready for Multicellular Life — 274
  General references — 286
  References — 286

13. Multicellular Life: The Last Hurdle? — 287
  13.1 Multicellularity arrives — 287
  13.2 The evolution of plants — 290
  13.3 The evolution of fungi — 294
  13.4 The evolution of animals — 296
  13.5 The organization of the mammalian body — 300
  General references — 302
  References — 302

Appendix A Electromagnetic Radiation — 303

Appendix B Glycolysis: Chemical Structures of Intermediates — 311

Appendix C TCA Cycle: Chemical Structures of Intermediates — 312

Appendix D The Calvin Cycle in Photosynthesis, Showing Chemical Structures of the Intermediates, and the Enzymes Involved in each Step — 313

Appendix E Amino Acid Structures — 315

Index — 319

Introduction

All too often, science is presented and taught in bits that are seldom brought together in the minds of the student or the general reader. Some may be inspiring, others may appear dull or of little relevance. Chunks of chemistry, pieces of physics, bits of biology, with few or no unifying themes, are offered to the student or the general reader alike. This has always seemed to me a great failing because the human mind craves unifying, simplifying explanations for almost everything.

Why should I think it important to integrate aspects of several 'separate' sciences in a book on biology? Firstly, some people experience a great sense of satisfaction when such an integration is achieved, and the light dawns. Secondly, in reality all science is related via the fundamental laws, although this is not immediately obvious. With these points in mind, I have chosen the all-encompassing thread of energy flow, which unites, indeed dominates, much of science, as the theme of this book. The specific aim of my approach is to increase the reader's enjoyment and understanding of biology by pointing out some unifying aspects of its workings. I attempt to introduce some scientific rationale into what we see around us. There are sound scientific reasons for everything we observe in the natural world, but complex social, historical, and political reasons behind the way we interpret them. The continuity of life over the millennia is a prime reason for the world being as we see it. Paradoxically, energy has at once been the driving force and a major limiting factor for the life which thrives on our planet.

To combine in a short book the aspects of physics, chemistry, geology, molecular biology, and more has been quite a challenge. I hope you find that it has been a worthwhile challenge. I believe that a familiarity with the principles of many of the sciences, and with their interrelationships, is desirable for the professional scientist.

Multidisciplinary research in science is becoming commonplace. Researchers are being encouraged to collaborate across traditional discipline boundaries, and indeed many granting bodies explicitly favour proposals of such a nature. In general, the training of researchers is still according to long-established disciplinary structures. Even experienced researchers in one scientific discipline find it difficult to appreciate the differences in approach employed by their colleagues in other areas. This becomes a problem when researchers from two or more distinct disciplines find it necessary to collaborate.

Sub-discipline titles such as biological chemistry, biochemistry, physical biochemistry, medicinal chemistry, nutritional chemistry, biophysics, and molecular biology, to name a few based on the molecular sciences, have been around for years. The list for biology is enormous. Despite the apparent overlap of material between various groups, it is likely that there will be many conceptual gaps in the biological knowledge of the physical scientists, and vice versa for their biologically trained colleagues.

Introducing Biological Energetics attempts to redress the reciprocal imbalance in knowledge experienced by physical and biological scientists. To achieve this successfully, I believe, the depth of treatment should not be so great as to discourage one or the other group from reading it. Accordingly, the text is written in an informal style; a major aim has been to make the book accessible.

When traditional discipline boundaries break down, as happens frequently now, researchers are becoming increasingly aware of the deficiencies in their knowledge of related disciplines. Good researchers realize that even a basic knowledge and appreciation of these related disciplines will be of help in their creative thinking and discussions with colleagues. Today, multidisciplinary knowledge is an essential professional tool.

Although this volume covers a range of scientific topics, it is not a daunting tome. It has been kept to a modest size deliberately to avoid such a perception. The chemical basis of biology is strongly featured. The book presents an 'integrated' approach to energy in biology, in that it treats energy as the overarching theme and emphasizes the all-pervading influence of energy transformation in every process, living and non-living.

An understanding of basic scientific concepts and thinking is important for everyone in the community. I say this for science, but it may be argued similarly for any other discipline. No one ignorant of science and technology is capable of making informed judgements about many of the issues in modern life that concern us all. Increasingly, individuals and various lobby groups are insisting that their views be heard by politicians, and rightly so. One barrier to serious debate on science-based matters is the lack of basic scientific understanding in the community. The same can be said about discussion on the resulting developments in technology and engineering. How is it possible for politicians, company leaders, finance providers, or others in executive positions to make rational decisions, often involving billions of dollars or influencing the lives of thousands, in areas where they do not even speak the language? I do not advocate loads of compulsory science for everyone up to school-leaving age, nor do I imply that all students should be subjected to the same levels of science. I do believe there should be more people capable of recognizing nonsense wrapped up in scientific jargon. Perhaps we should start broadening the science educational process with journalists, closely followed—despite recent significant advances in the teaching of science in primary schools—by primary school teachers and their students.

Unfortunately, many people never proceed past the bare understanding stage of scientific concepts. The reasons are varied, but one of the least important is because science is too difficult, while one of the most important is because science is perceived as being so. Devising means to achieve a change in this widely held perception, in young people especially, is perhaps the greatest challenge faced by science educators.

1
The Flow of Energy in Living Systems

With few exceptions, life on Earth depends on energy we receive from the Sun—solar energy. A tiny fraction of this electromagnetic energy, radiated into space in all directions, is captured by plants and specialized bacteria, and used to drive chemical reactions essential for life. Other wavelengths of solar radiation that we know as infrared are converted to heat when they contact the Earth and its atmosphere, and this heat is a major energy source that powers the world's weather. Most of us know these basic facts, but let's not be dismissive of them for that. The energy from the Sun plus a unique combination of chemicals on Earth were here long before any life existed. The young Earth had to wait about a billion years, during which period something both momentous and unique occurred. Somewhere, no one is sure exactly where, there began a series of events which, over the next 3.5 billion years, has led to the words you are reading now. Some scientists speculate that, given the combination of chemicals and energy available at the time, perhaps the emergence of life was inevitable (de Duve 1991, 2002; Morowitz 1992; Luisi 2006; Morowitz & Smith 2007). We may never know. We can't be sure of those conditions, nor can we, in the foreseeable future, reproduce them in an experiment to start life again. What we can be quite confident in stating is that there was and is no guarantee that life will survive indefinitely. As we will see later, there have been several mass extinctions of life over the history of this planet, and any one could have wiped it out before we humans evolved.

Let's start by discussing the living system with which you are most familiar—your own body. Consider what happens, in broad terms, to the energy flow through your body over a 24-h period. Immediately we strike a problem. What is meant by the term 'energy'? We all have a general idea about the meaning of such terms as 'children are full of energy' or 'I feel low in energy', and that someone 'has lots of sexual energy'. There are dozens if not hundreds of individual concepts, from the strictly scientific to the downright erroneous, so let me define what I mean.

Energy is a measure of the capacity to do work

Like so many formal definitions, the above statement does not tell us anywhere near the whole story.

This means that to get a clear idea of what energy means, we need to go into quite a discussion. I do so here for energy, and later for many other concepts, without apology. I am (almost) ready to guarantee that by doing so I will expose a few misconceptions many readers have picked up.

Energy is a measure of the capacity of an animal, a person, or an inanimate object to carry out work. Again the term 'work' has a number of connotations. Work as a scientific term has a particular meaning. Work comes in a number of forms, most of which will be familiar.

Very broadly, work is the ability to cause changes in a controlled manner.

Again, this does not, by itself, tell us much. Let's restrict work for the moment to the ability to move something. This is usually called mechanical work. We do mechanical work when we push a cart or lift a weight against the force of gravity, or merely move ourselves along as we walk. Wind does work in moving a sailing boat or bending a tree, and water does work in moving soil down a river or driving a hydroelectric generator. The greater the amount of energy available from a particular source, the greater the amount of work that source is able to perform. A river with a large water flow can deliver more energy and thus do more work than the trickle of a small stream. Even the simplest process involves a flow of energy. What do I mean by a process? Dictionary definitions of the word 'process' include

'A continuous action, operation, or series of changes taking place in a definite manner'.

Notice the words 'definite manner' and 'controlled manner' in the definitions. These are important in the definition of processes involving work. The involvement of energy changes in living organisms is the overall theme of this book. Formally, the study of energy changes is called thermodynamics, derived from the Greek *therme* (heat) and *dynamis* (power). Early studies in the area were concerned with converting heat into work in heat engines, for example steam engines. The amount of energy in a system or process is measured in joules (J) or, more practically, kilojoules (kJ, a thousand joules). Work, which is one readily measurable manifestation of energy, is also measured in kilojoules, as the two are intimately related. Energy is a measure of the capacity to do work. These and other relevant units will be defined in Chapter 3.

Each process involving work needs its own source of energy, and there are many such sources. Indeed for convenience, physicists and chemists have classified, and studied in detail, most of these sources, which they usually term 'forms' of energy. We will mention most energy forms in due course, but our particular concern is with the conversion of one form of energy to other forms because this is what occurs in most of the processes which will be discussed. When work is done, there is always a certain amount of energy converted from one form into other forms.

The absolute (total) amount of energy in an object or system does not always concern us, rather it is the change in energy during some process that is of greater interest. For example, how much energy is required to heat a kettle containing a litre of water from room temperature to 100°C?

The water at room temperature already has a certain energy content, but we don't really need to know that. All we need to know for our calculation is the quantity (mass) of water involved, the temperature increase in degrees from room temperature to boiling, and the quantity of energy in kilojoules required to raise the temperature of a kilogram of water by 1°C. If we want to be really accurate, allowance must be made for the heat taken to warm the kettle itself, and a factor for heat loss to the surroundings. The appropriate calculation will allow the quantity of energy in kilojoules to be determined, and from that we may calculate the cost of boiling the kettle of water, provided we know the cost of a kilojoule of electrical energy from our electricity supplier.

Another example is the conversion of some of the kinetic energy in flowing water into electrical energy in a hydroelectric generator. At the same time, some is converted to sound energy (actually a form of kinetic energy involving movement of air molecules) as the water strikes the pipes and turbine blades, and some is converted to heat energy, although this is not so obvious in this example, as it would be impossible in practice to measure the change in temperature of our hydro system. All types of energy conversion can be measured quantitatively in kilojoule units.

Here it is convenient to introduce one of the fundamental laws of physics, the law of conservation of energy, a simple form of which states:

Energy cannot be created or destroyed, but it may be transformed (converted) from one form to others, without overall loss.

This will do for now. The energy conservation law will be discussed more rigorously under its other name—the First Law of Thermodynamics—in a later chapter. Those who are not familiar with the energy conservation law, and have not seen any of the evidence to support it, might have some difficulty in accepting the above statement. It has tremendous implications, but the one that concerns us initially is the relationship between energy intake, its utilization, and its output in living systems during a 24-h period.

I'm sure by now you will have noticed my tendency to define, to specify conditions, and restrict certain statements. Scientists will recognize this as a necessary part of science. Ill-defined or undefined statements can often become meaningless in the scientific sense, as they may be inconsistent or, more likely, incomplete. All this can lead to confusion at a later date if not dealt with upfront. The need to be explicit in these matters partly explains the wariness of scientists to give statements to the media. The media typically look for short, pithy, unqualified statements. Scientists know that such statements are likely to be misleading.

No doubt you will have also noticed that having started with a statement about your own biology, I moved swiftly into physics and one of its laws. This does not show a bias towards physics, but serves to underline the theme of this book—the essential role of energy and its transformations in the processes of life. The Russian-born geneticist and naturalist Theodosius Dobzhansky (1900–1975) boldly proclaimed 'Nothing in biology makes sense except in the light of evolution'. This has become a widely quoted statement, especially in recent years.

It is equally valid to say that nothing in biology (or in any other science for that matter) makes sense except in the context of energy and its transformations.

This approach of course leads us towards a more 'fundamental' (but not necessarily more important) level of biology. I hope to convince you that it is a valid and useful way to look at life. Scientists, particularly physical scientists, love to discover simple, unifying laws of wide applicability. They speak of the 'beauty' and 'elegance' of such laws in almost reverent terms. The energy conservation law is one of them.

Let us return to energy flow in the human body. Most people will be familiar with the concept of flow of energy, and indeed it is a very useful way to look at things. Flow invokes thoughts of the movement of a river, and many of the properties of a flowing river are analogous to those of the flow of energy. The water in a non-tidal river flows from higher ground to lower. It does so spontaneously, or without any outside assistance to start it moving. As it flows, the potential energy of the water (due to its height above sea level) is converted to kinetic energy (due to its motion). The kinetic energy possessed by the moving water can do mechanical work, in that it may carry boats or silt from one place to another, or turn a waterwheel connected to some kind of machinery. The kinetic energy of the moving water may be partially converted to electrical energy by means of a hydroelectric turbine. Everyone is happy with the statement that a non-tidal river flows downhill, and with the idea that by the time a raging mountain torrent calms down as it runs more slowly across a coastal plain, it will have lost (transformed actually) some of its kinetic energy. No longer is the river able to move large rocks, its water moves more slowly, and suspended soil will be deposited as silt.

Heat energy is related to the movement, or thermal motion, of the atoms and molecules in a substance. As, for instance, a solid body is heated, its atoms and/or molecules absorb energy and vibrate more vigorously. They move about their average positions in the solid to a greater extent. They jiggle about more. This increased molecular movement is an indication of the amount of kinetic energy the molecules have absorbed. It results in an increase in temperature of the substance. Temperature is formally a measure of the average kinetic energy of the molecules in a body (or a volume of liquid or gas) at thermal equilibrium with its surroundings. Temperature is conventionally measured by some kind of thermometer. There is in principle a temperature at which the kinetic energy of molecules is zero, their motion ceases, and their heat content is zero. This is called absolute zero or thermodynamic zero, and is very close to minus 273.15°C. The absolute temperature scale/Kelvin scale/thermodynamic scale is used in thermodynamics calculations instead of the Celsius scale, in which, many years ago, the freezing point of water was arbitrarily set at zero, and the boiling point at 100, merely for the convenience of humans. Absolute temperature is measured in degrees Kelvin (K). The magnitude of a degree Celsius is the same as that of a degree Kelvin, and one may convert from Celsius scale to the absolute (Kelvin) scale simply by adding 273.15, and vice versa. Thus, zero degrees Celsius is the same as 273.15 degrees absolute. This is written as

$$0°C = 273.15 \text{ K}$$

Note there is no degree superscript for degrees Kelvin. Although actual achievement of the absolute zero remains elusive, the concept is of theoretical and indeed practical interest. Matter begins to behave strangely as the temperature approaches absolute zero. Some materials become superconducting, that is they conduct electricity almost without resistance, and a small electric current can run for an extraordinarily long time. Such materials are used routinely to produce very powerful superconducting electromagnets for use in modern scientific instruments.

Strictly speaking we do not 'get cold', rather we lose heat, as heat from our warm body flows spontaneously to the cooler surroundings. Similarly, heat will flow from a hot cup of tea to our colder hands. Our bodies are able to regulate themselves to maintain our internal temperature at about 37°C. Humans, indeed most living organisms, are able to live comfortably within a very restricted temperature range. By adjusting clothing and living conditions, this range has been extended as humans moved from the warmer African climates where human evolution began, to colder regions of the Earth. Modern technology, from protective clothing and air-conditioning to space suits and submarines, makes it possible for us to live, protected in our narrow temperature range, almost anywhere on Earth, or indeed in space. Other organisms do not have such internal control, and have evolved a variety of protective measures to cope with extremes of temperature.

Already in the preceding paragraphs we have introduced several types of energy—mechanical, sound, and heat (which are forms of kinetic energy) and electrical, together with everyday examples involving transformation of these forms. The chemical reactions that occur in the body involve a further form, chemical energy. Energy is stored in chemical bonds, and energy changes take place when chemical bonds are made and/or broken during a reaction. Chemical energy is a vital source of work for living organisms. The totality of chemical reactions carried out by a living organism is called its metabolism. The details of metabolism vary from animal to animal, from plant to plant. Plants and animals differ in the way they gather and process energy, and this is reflected in the different metabolic pathways they employ. A metabolic pathway is a series of chemical reactions leading to a particular end product. The product may be a vitamin, a bodily structural feature, or even energy itself stored in a particular form, such as starch in a plant, or glycogen and fat in an animal. All energy flow in living systems must obey the fundamental laws of chemistry and physics.

Over the period of a day, our body usually undergoes very little observable change. This is why I have chosen such a length of time to consider. Depending on our age and state of health, daily changes are in fact taking place, but for the sake of discussion, our body today is essentially the same as it will be tomorrow. We will be the same weight, size, colour, and shape, but as we all know many processes will have been occurring constantly. Among other things, we have worked, eaten, breathed air, drunk liquid, slept, and eliminated waste products, yet have little to show for it. An approximate description of such a state is that of dynamic energy balance or steady state, that is although apparently we have not changed during our selected day, much energy conversion has been going on merely to maintain the status quo. This is one vital characteristic of living systems. In contrast, a brick sitting on a table could be said to be in a state of static energy balance. The force of gravity is trying to move it down through the table, but there is no flow of energy through the brick as there is in a living organism.

To maintain the state of dynamic energy balance, our bodies release and largely utilize the available energy in our food, or its equivalent from bodily energy stores. It is important to recognize that the energy in the food must be in a usable, available form. With the assistance of water, and of oxygen in the air, the available chemical energy in food is converted (metabolized) into other forms, which are used to 'drive' processes in the body. In its simplest form, metabolism of simple sugars may be considered as a type of controlled burning. Much, but by no means all, of the energy resulting from this controlled burning of sugars is trapped in forms that may be utilized, enabling the body to do work and maintain itself. This is a major objective of metabolism. Some energy is released as heat, part of which helps to keep our internal temperature fairly constant. Too much energy in the form of heat may be harmful and 'denature' some of our delicate bodily systems, so any excess heat, such as that produced by the large muscles during exercise, must be removed. This is partly achieved unconsciously by producing perspiration, which on evaporation produces a cooling effect. In addition, we usually modify our behaviour by removing heavy clothing, consuming drinks, taking a swim, or turning on the air conditioning. Similarly, we are able to adapt to lower temperatures, initially by subconscious movements of our muscles (shivering), which produces heat, or again by modifying our behaviour and clothing to reduce heat loss.

Most human tissues can increase their oxygen consumption and heat production in response to thyroid hormones, which modify the metabolism of carbohydrates and fats. This is one example of a biological control mechanism, many of which are required to maintain our dynamic energy balance. Complex animals such as mammals have evolved a range of such survival mechanisms.

The ability of the human body to adapt to extreme conditions, such as mental and physiological shock in times of war, and restore itself to the steady state is legendary. It compensates for blood loss by withdrawing fluid from the tissues and from the lymphatic system. Body temperature can be controlled not only by the methods mentioned above, but also by appropriate diversion of the blood flow to different areas of the body. All these control mechanisms are necessary to maintain the body in a state suitable for the efficient functioning of its many systems.

So much for a few examples of the ways evolution has 'solved' some of life's problems. As we shall see in later chapters, the adaptability made available to living organisms through the process of evolution by natural selection is truly impressive in its dimensions. Even natural selection, however, must obey the laws of physics. This is not to imply that the totality of natural selection can be explained by simple application of the laws of classical physics—far from it. The physical scientist deals with factors that have remained essentially constant since the formation of the solar system, as summarized in the laws of chemistry and physics. Biologists, especially evolutionary biologists, think and work in ways that are different from those of physical scientists because they must deal with and try to explain the myriad factors that account for all aspects of living organisms, past and present.

Maintaining our state of dynamic energy balance involves a complex series of processes. The amount of energy we take in must be balanced by that which we expend—a personal example of the energy conservation law, with the realization, however, that in thermodynamic terms our bodies are examples of what are called open systems. They exchange matter and energy with the surroundings. Excess weight in humans is a result of excess uptake of energy in the form of food. The 'average' 70 kg man needs about 10,000–11,000 kJ per average day to maintain constant weight, that is to maintain a daily energy balance and thus constant weight, he must 'burn off' as much energy as he extracts from his food. This is often expressed as an equation

$$\text{energy in} = \text{energy out}$$

This is a simple equation, and a simple concept, but its execution is found to be almost impossible by millions of humans. Whole industries have evolved to combat the problems of the overweight, and the economies of nations stagger under the burden of the resulting health problems. Modern societies suffer from a number of preventable diseases caused by high-fat high-energy diets. Heart attacks and strokes are usually the result of arteries clogged with fat (atherosclerosis), cancer rates are increased in individuals with high-fat diets, while certain diabetes types may develop in people who are overweight from an excessive consumption of food. Broadly interpreted, all these problems arise as a consequence of the law of conservation of energy.

Consider again our 24-h body study, this time for someone who eats just a little more than is really necessary. Not much, but just enough such that the energy balance is positive, rather than zero. What happens to the excess energy extracted from that little extra food? The digestive system will still process it, and the net effect will usually be the storing of the extra energy in body cells as lipid (fat). About 85% of the energy in excess of our needs ends up stored as body fat. Fat cells (adipocytes) possess an enormous capacity to enlarge if food intake is high, and may also increase in number.

The effects of excess energy consumption are cumulative. Every biscuit at afternoon tea, every spoonful of sugar in coffee, every scoop of ice cream will count in the daily balance-of-energy equation. There is no escape, no matter how fervently we may wish there to be. The ways by which the body controls the storage and the use of energy make a complex but fascinating story, which we will explore further.

The above discussion serves as a brief and restricted introduction to the importance of energy in our daily lives. There is much more to discuss other than mere energy balance. The science of the underlying mechanisms by which this balance is achieved, and other control mechanisms, are fascinating topics and will be examined later.

The flow of energy through most living organisms is similar in principle to that which occurs in humans. Chemical energy in the form of some kind of food is consumed, absorbed during some form of digestion, and metabolized with the assistance of oxygen to produce water and carbon dioxide. The chemical energy released during metabolism is used in a number of complex biochemical pathways to drive essential chemical reactions, synthesize new structures (i.e. grow), move muscles, control body temperature, and reproduce. Such organisms, known as heterotrophs, rely on ready-made, organic compounds of plant or animal origin for their supplies of chemical energy and carbon, and can be considered the consumers of the living world.

Among the consumers, some biologists give the classification of decomposers to certain bacteria, fungi, and other organisms that help in the breakdown and recycling of useful materials from the tissues of dead animals and plants.

On the other hand, there are the producers of most of these ready-made foods. The major producers, the autotrophs, are green plants, the photosynthetic bacteria, and some non-photosynthetic bacteria and Archaea. Such organisms are the primary biological sources of energy for the food chains that sustain life on Earth. The photosynthetic life forms are capable of directly taking some, estimated to be only about 10% of the total, of the radiant energy received from the Sun and, together with carbon dioxide and water as ingredients, converting it to chemical energy, initially in the form of simple sugars. In addition, and importantly for us, oxygen is released during the process. There also remain to this day many types of anaerobic organisms, that is those that can survive only in the absence of oxygen. Most biologists are of the opinion that the first forms of life on Earth were in fact anaerobic, as the primordial atmosphere was essentially devoid of oxygen. Only with the evolution of photosynthesizing bacteria and plants was oxygen released into the atmosphere in significant amounts. Hundreds of millions of years of adaptation were then to elapse before much of the living world became aerobic, or oxygen dependent.

Almost certainly we owe our very existence as air-breathing, complex, thinking beings to those lowly, anaerobic, single-celled, incredibly numerous transducers of accessible energy.

REFERENCES

de Duve, C. (1991) *Blueprint for a Cell: The Nature and Origin of Life*. Neil Patterson Publishers, Burlington.
—— (2002) *Life Evolving: Molecules, Mind and Meaning*. Oxford University Press, Oxford.
Luisi, P.L. (2006) *The Emergence of Life*. Cambridge University Press, Cambridge.
Morowitz, H.J. (1992) *Beginnings of Cellular Life*. Yale University Press, New Haven.
—— and Smith, E. (2007) Energy flow and the organization of life. *Complexity* 13, 51–9.

2
Origins: The Early Earth

Before we can seriously contemplate the origins and development of life on Earth, we must look at the origin of Earth itself, at least briefly. This is entirely reasonable, as it was in the wafer-thin surface layer of the young Earth we now call the biosphere that the life we know almost certainly originated. All the physical and chemical ingredients needed to be together, at the appropriate time and in appropriate amounts, for this (presumably) unique series of events to have occurred. Whether or not there were important ingredients that arrived from space remains a matter for debate. Certainly, a significant number of chemical precursors to biomolecules has been found in meteorites and identified in interstellar space (de Duve 2005; Luisi 2006).

Whatever their origins, the ingredients were there as a result of interactions between matter and energy under the influence of the fundamental laws of the Universe, the same laws that allowed life to begin and that have allowed it to continue and flourish ever since. Scientists seem to agree that as far as can be determined, the laws we refer to as 'universal' really do apply across the universe, and atoms that constitute stars millions of light years away from us have properties identical to those of the respective atom types we study here on Earth. If we allow for their various forms (isotopes—see Chapter 5), hydrogen is hydrogen, and carbon is carbon wherever their atoms may occur. The light elements, hydrogen, helium, and lithium, were produced early in the history of the universe, whereas the heavier ones evolved later, inside stars. I use the word 'evolved' here in the cosmological sense, a different context from that of biological evolution, but the general meaning is the same.

The entire universe is changing with time, as it always has since the Big Bang and always will, with the various parts of it, separated by enormous distances, often being at different stages of stellar evolution. Stars are still evolving in galaxies throughout the universe, and it is more than likely that new planetary systems are being formed around them as well, at this very instant. New stars are also coming into existence, mainly from the vast quantities of interstellar gas and dust, and old ones are dying. Our own Sun is about half-way through its estimated life of 10 billion years.

Our Sun will pass through the same dying stages experienced by stars of similar mass. Very briefly; the core of our Sun will become richer and richer in helium than the outer layers; the temperature will rise, causing the helium to fuse into heavier matter. The heat formed will then cause the Sun to expand enormously to become a red giant, and destroy any life left on Earth. Later, it will pass through further, relatively quiet, stages and become a white dwarf, an intensely hot, very dense object about the size of a smallish planet. Eventually, unrecognizable as our life-giving Sun, it will cease to emit visible light, grow slowly cooler, and come to the end of its life as a star. More massive stars die more violently, exploding in perhaps the most spectacular stellar event observed by astronomers, a supernova. Again depending on the size, these larger stars pass through events to become either a neutron star, so dense that a thimblefull would have a mass of about 100 million tonnes, or the real giants, which will become that most intriguing 'singularity' a Black Hole, where the force of gravity is so immense that even light cannot escape.

How do we know all this?

Astronomers believe that the Sun will pass through the predicted stages because of information built up over the past 200 years or so, plus the development of powerful theories which have given them confidence. Part of that confidence arises from the notion that the same laws of physics apply across the universe.

Imagine our newly formed planet, some 4.55 billion years ago, freshly aggregated from the gas and dust of the solar nebula. This process of aggregation and accretion was not uniform across the solar system, or even within each planet that eventually formed. Heterogeneous accretion led to the denser, gas-depleted inner planets Mercury, Venus, Earth, and Mars, and to the less dense, largely gaseous outer planets—the gigantic Jupiter, ringed Saturn, 'twin' Uranus and Neptune, and the distant, hardly visible Pluto (although the latter has recently lost its planet status). The Earth formed a dense, predominantly solid-iron core of some 1300 km radius, over which settled a thicker (2180 km) outer core of liquid iron plus some lighter elements—sulphur, oxygen, and silicon. Above this again there formed a yet-lighter mantle (2880 km) high in magnesium silicates (Figure 2.1).

A period of long, slow cooling led to solidification of the still-lighter crust, averaging about 17 km (range 8–70 km) deep and rich in aluminium silicates. The crust was initially far too hot to allow the steam that burst forth from fissures and volcanoes to condense as liquid water. Frequently, for a quite short period around 4.55 billion years ago, meteorites tore their fiery way through the heavy atmosphere, some occasionally so huge as to blast craters up to 150 km diameter on the land. The laws of physics and chemistry dictated the moulding and shaping of the new landscape—a harsh, lifeless, alien world. Gradually, over millions of years, vast quantities of energy were transformed, largely to heat, and radiated into space, dissipated into the far reaches of the Universe as the Earth cooled.

About 4.50–4.52 billion years ago a momentous event in Earth's history took place; a meteorite about the size of Mars apparently collided with the Earth and was vaporized. Its remnants went into close orbit around the Earth, and eventually coalesced to form the moon (Hartmann and Davis 1975). The early moon was probably as close as 25,000 km, but has gradually receded to its present distance of about 400,000 km. Thanks to accurate laser measurements taken during the Apollo missions, we know it is still moving away at about 3–4 cm per year. When the moon was close, enormous tidal forces were generated, distorting the shapes of both the Earth and the moon. The Earth's thin, malleable crust

Origins: The Early Earth 13

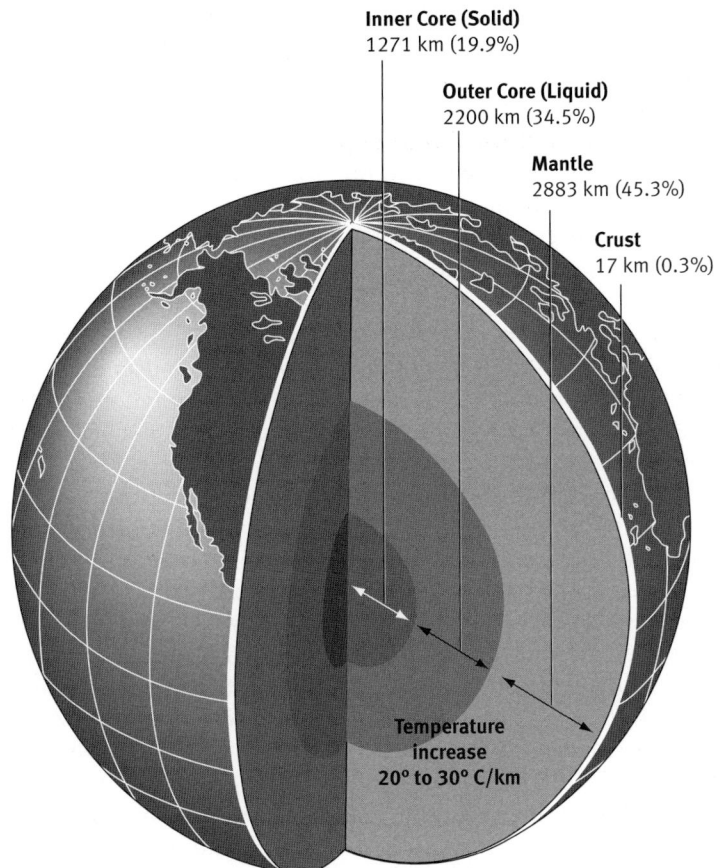

Figure 2.1. Structure of the Earth. (Illustration from Schopf, J. William, *Cradle of Life* © 1999 Princeton University Press. Reprinted by permission of Princeton University Press).

rose and fell as much as 60 m twice a day. These distortions and tidal effects dissipated the rotational energy of the earth–moon system, slowing the rotation of both bodies, and commencing the recession of the moon that continues today (Wills and Bada 2000).

Gravitational effects of the moon (and the Sun) have always exerted a great influence on life on our planet. Combined with the tilting of the Earth's axis which produces the seasons, the gravitationally generated tides provide large variations in conditions around continental and island coastlines. These variations have provided challenges and opportunities for life ever since it commenced.

At last, about 4 billion years ago, cooling reached the stage when liquid water began to fall on the jagged landscape. It probably rained for centuries—vast, unimaginable downpours, washing, scouring, and sculpting. What size, shape, and form the land took in those early days we can only speculate. Temperatures were high, evaporation was rapid, and water began its cycle of condensation, evaporation, and recondensation that has continued ever since. Volcanoes were violently active, pouring huge quantities of lava, steam, and carbon dioxide from below, together with some methane, carbon monoxide, and hydrogen sulphide. Some scientists believe that a significant amount of our water originated from comets colliding with the Earth. Large numbers of comets are known to have been present in the early years of the solar system. Comets have been described as 'dirty snowballs' reflecting their composition—largely water, contaminated by small amounts of organic matter (Ferris 1997). Other research has shown that deuterium, an isotope of hydrogen, is present in a significant number of comets at about twice the levels found in earthly seawater (Wills and Bada 2000). This seems to dispel the proposal that the majority of water arrived here from space.

Searing ultraviolet rays from the Sun pierced the clouds that shrouded the land and the sea, while solar infrared radiation was continuously converted to heat that helped keep in motion the great energy cycle of evaporation, rainfall, erosion, and deposition that continues to this day. The lower regions of Earth filled with water of ever-increasing salinity from soluble salts leached from the surface layers and rocks, and finely ground particles were carried down from higher regions as the cycles of weathering and erosion commenced. The rocks which had done the grinding were eventually washed and tumbled along and ground in turn to finer and finer granules. Chemical reactions between water-soluble bicarbonates and silicate minerals liberated silica (SiO_2), some of which became the sand that flanks our continents and forms our deserts today. Carbonaceous precipitates also began to form, and settled to supplement the ever-deepening sediments. Enormous amounts of carbon dioxide gas were 'fixed' chemically at this time as sedimentary carbonate, much of which remains today. The weathering cycle rolled on. By 3.8 billion years ago, oceans of liquid water prevailed, and an atmosphere containing perhaps 98% carbon dioxide blanketed the Earth. The 'skin' of the cooling Earth became sufficiently thick and rigid at about this time to allow the formation of the first true continental crust, and perhaps the first of a long series of plate tectonic processes (Figure 2.2).

The notion of these drifting plates, introduced by geophysicist J.T. Wilson in 1965, involves huge segments of the Earth's crust, called plates, that travel on the pliable mantle that underlies them. Plate tectonics revolutionized thinking in the Earth sciences in the 1960s, extending the formal concept of continental drift propounded first in 1912 by German meteorologist and geophysicist Alfred Wegener (1880–1930) following centuries of vague speculation about large-scale movements of the continents. Wegener knew the continents were made of material less dense than the ocean floors. Similarities in shape between the Atlantic coastlines of Africa and South America commenced him thinking, but only with the publication of a scientific paper describing finds of identical fossils on both continents was he convinced. For the rest of his life, Wegener fought long and hard for acceptance of his proposals. Solid evidence for the movement of continents came in the 1950s with the discovery of mid-ocean ridges. This submarine mountain chain of some 66,000 km virtually circles the Earth. It has been shown to have a deep, narrow rift along the centre line through which hot, molten rock called magma wells up from below. As the magma spreads and solidifies, new crust is formed in a process known as sea-floor spreading.

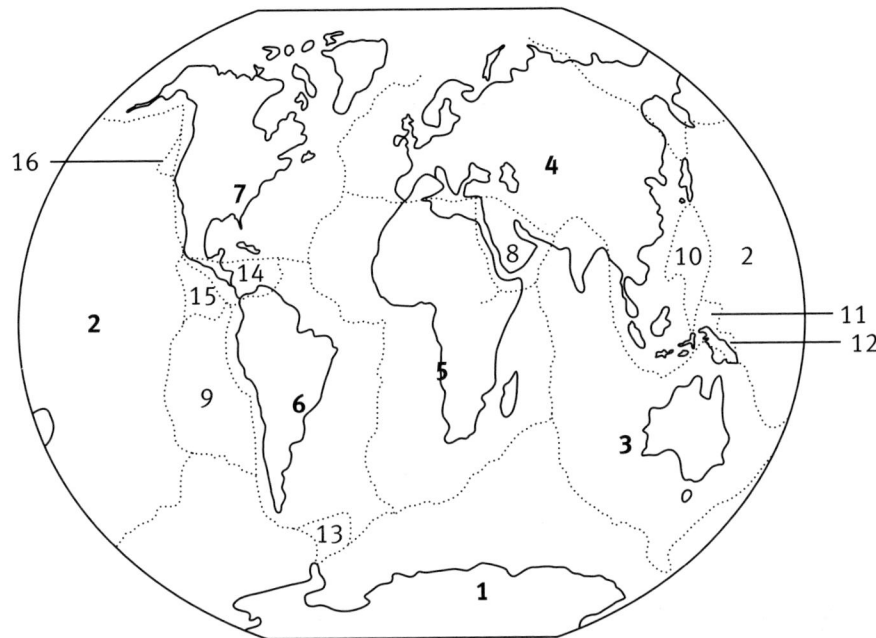

Figure 2.2. The Earth's tectonic plates. 1, Antarctic; 2, Pacific; 3, Indo-Australian; 4, Eurasian; 5, African; 6, South American; 7, North American; 8, Arabian; 9, Nazca; 10, Philippina; 11, Carolina; 12, Bismarck; 13, Scotia; 14, Caribbean; 15, Cocos; 16, Juan de Fuca.

Studies of the direction of the permanent magnetic fields that form when new rocks first solidify helped to verify the process. Cores from the Mid-Atlantic ridge extracted in the late 1960s showed that the rocks of the ocean floor were youngest near the ridge, and became progressively older farther away on either side. The theory of plate tectonics proposes that the Earth's crust is divided into huge segments, thousands of kilometres across and up to 130 km thick. They move slowly over the mantle, but what is the energy source for such a feat? The energy requirements must be incomprehensibly enormous. Quite early in the argument, British geologist Arthur Holmes proposed that radioactive decay deep underground might provide enough heat to drive convection currents that circulated the molten magma to provide sea-floor spreading. Even Wegener (not having the sea-floor spreading information at the time) found this hard to accept, but according to modern opinion it appears to be so.

So continental drift is driven by heat energy from the interior of the Earth. When the plates move away from each other, they form cracks or fissures, the mid-ocean ridges, through which new crust can well up. When such fissures occur on the above-sea continents the land splits. An example is the East African Rift Valley, parts of which were to play such an important role in the very early history of human evolution on our planet. When plates collide, one may be forced under the other, a process called subduction, which created the areas of volcanic activity around the edge of the Pacific Ocean known as the Rim of Fire. Some of the forced-down material melts under the heat and tremendous

16 Introducing Biological Energetics

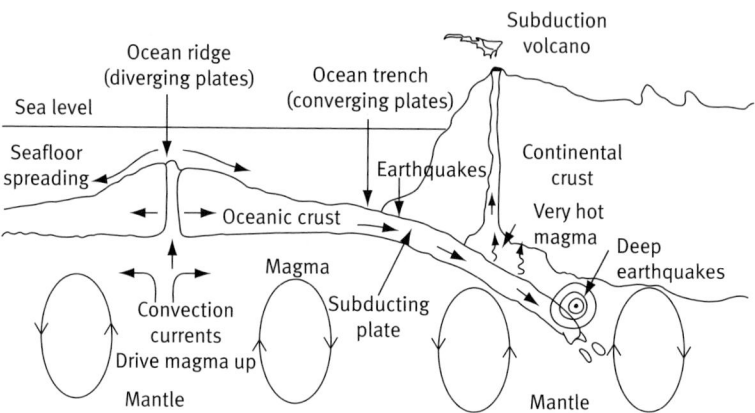

Figure 2.3. Ocean floor spreading from ocean ridges sometimes forces the ocean plate under a continental plate, causing the former to melt and force its way upwards to form volcanic mountain chains such as the Andes.

pressure to produce very hot magma, which in some instances melts the rocks above it. The molten rock forces its way through faults to form volcanoes, and even volcanic mountain chains (Figure 2.3).

Again, when plates of continents collide they may also be thrust upwards to form mountain ranges. When the Indian Plate crashed (slowly) into Asia, the ultimate result was the squeezing up of the Himalaya mountains. Mount Everest is still being raised at the rate of about 2 cm a year, although the Himalayas are being constantly eroded away by extreme weather conditions. Plate boundaries are usually regions of frequent earthquake activity, as experienced along the San Andreas fault in California, across the long sweep of Indonesia, and in the Philippines and Japan in the western Pacific. Studies based on the theory of plate tectonics show us that the surface of the Earth is constantly evolving. Continents and supercontinents have come and gone, but I will mention only the youngest ones here.

About 240 million years ago, the supercontinent Pangaea, began to split into two blocks, Laurasia and Gondwanaland.

This was followed by the break up of Gondwanaland, commencing about 160 million years ago. Gondwanaland was a mega-continent, which included the areas known today as Africa, South America, Australia, Antarctica, Madagascar, and India. These split apart as a result of continued plate mobility, and we have seen what happened to a wayward India. Its drifting and collision with Asia resulted in the formation of the Himalayas, the highest mountain range above sea level on the Earth today.

Some of the evidence for the proposed relationships in Laurasia and Gondwanaland comes from the fossil record in the continents as they are today. Other confirming evidence is geological, with the history being preserved in the chemical composition of rocks and minerals, and in their magnetic properties. Figure 2.4 outlines some of the major changes which led to the continents we know today.

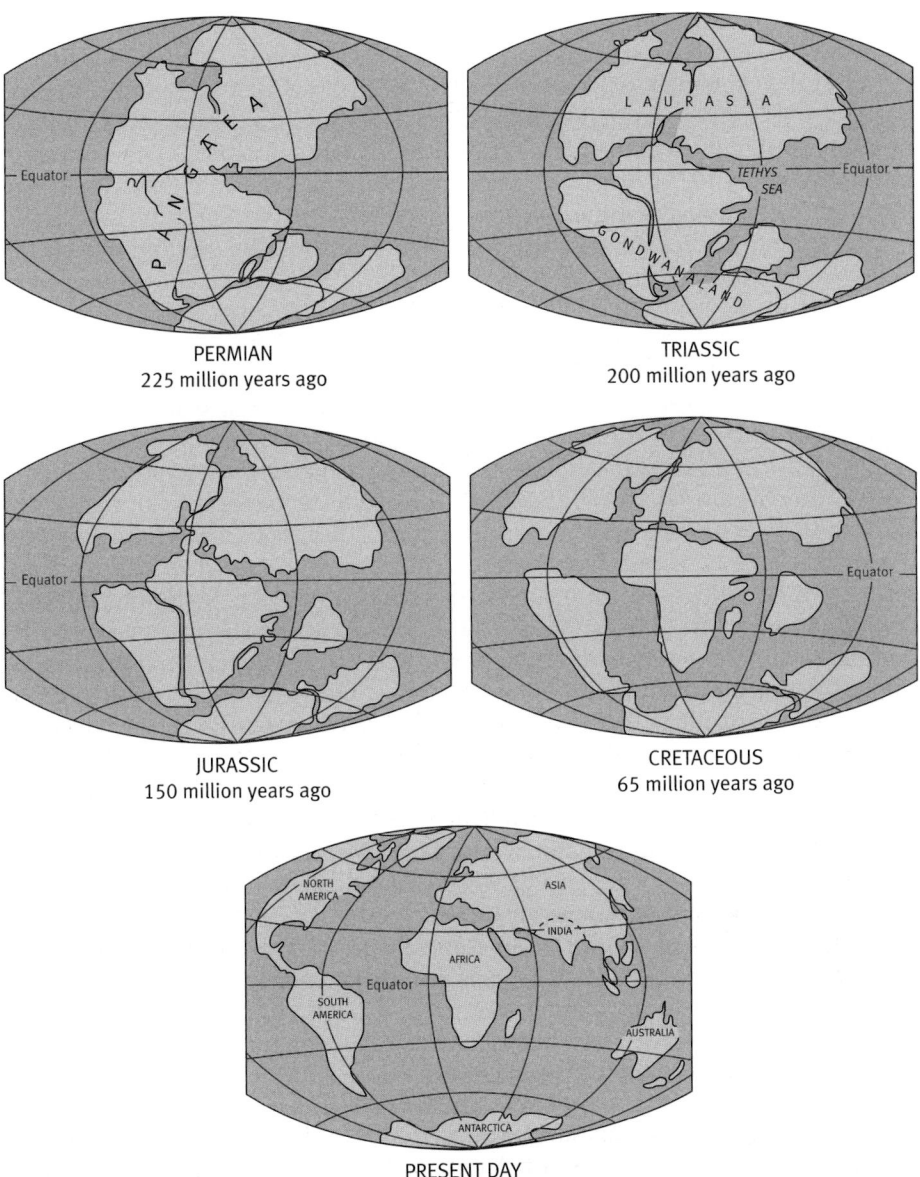

Figure 2.4. Origin of the present continents. Used courtesy of the US Geological Survey.

Why have I gone into the above description of the geological history of the Earth?

It is important to bear in mind that the Earth is, and has always been, in a state of change. Exchanges of energy on a vast scale between its various layers have occurred since its very formation. When the large amount of carbon dioxide originally present in the atmosphere became depleted by plants and by deposition as carbonates, its blanketing effect was greatly reduced, and as a result the

Earth should have become much colder. This did not happen. The compensating factor was an increase in the amount of solar radiation, which has increased by some 30% in the last 4 billion years, because of the increased rate of nuclear reaction in the Sun. Fortunately for life on Earth, the extra warmth compensated for the carbon dioxide taken up in the growth of plants. This is a vastly simplified account of the many changes in atmospheric carbon dioxide levels that have occurred over geological time.

Apart from the occasional impact of large meteorites, large volcanic events, etc. the Earth underwent its normal convulsions. If we could view a greatly speeded-up videotape of the Earth's history, we would readily notice the changes, in contrast to the normal rate of such events. We can easily pass a lifetime without noticing any geological changes at all. Only those who live through an earthquake, volcanic eruption, a tsunami, or a major flood are likely to have any appreciation of the scale of changes the Earth has undergone. Our videotape would show a dynamic Earth, the surface rippling, mountains being thrust upwards and being eroded away by rain, snow, ice, and wind. Continents would be seen drifting about the oceans, driven by the power of moving plates. Sea levels would rise and flood vast areas, later to recede once more during the expansion of the Ice Ages, exposing land bridges across which plants and animals, including humans, would migrate to new lands, eventually to the farthest corners of the Earth. We know from our own experience that climate and life are closely linked. Everyday, one intimately and profoundly influences the other. This has probably always been so, and a good argument has been made that the two in fact co-evolved, and that climate continues to play a role in both the evolution of new species and the extinction of existing ones (Schneider and Londer 1984; Schneider 1996). The vast forests on land in particular can affect climate. Freshwater and particularly marine algae absorb and transform enormous amounts of solar energy, and by absorbing vast quantities of carbon dioxide help to keep the temperature of the Earth cooler than it might otherwise have become.

I want to emphasize that the Earth and its atmosphere have changed continually (but not evenly) over time, and that these changes have had profound effects on the origins and the development of life. On Earth over the last 500 million years there have been five major mass extinctions of life. It has been estimated that, depending on which extinction we are looking at, between 60 and 95% of species living at the time may have been extinguished.

The most familiar is probably the extinction, about 65 million years ago, of a large variety of species, including the dinosaurs and other large reptiles. On the basis of the discovery and compositional analysis of a layer of clay rich in the element iridium, Luis Alvarez (1911–1988) proposed that a large body some 10 km in diameter had collided with the Earth at Chicxulub, Mexico, at that time. Iridium, having a strong affinity for iron, is mostly found in the iron-rich core of the Earth. Its appearance so close to the surface is therefore mysterious. Other chemical analyses suggested the body was probably a meteorite, and some meteorites were known to contain iridium. The position of the thin, iridium-rich layer, sandwiched between rocks of the Cretaceous period and those of the Tertiary period, set the time of collision. The collision would have generated an enormous cloud of dust, which would have completely enveloped the Earth for a number of years. Imagine the effect of cutting off a significant proportion of the Sun's energy. Photosynthesis was inhibited, plants died, and the climate either cooled or, blanketed by the dust, was heated as in the greenhouse effect. Acid rain fell, possibly for years.

Whatever the details, the balance of chemistry and/or energy was distorted too rapidly for many species to adapt, and they perished. The fossil record shows a shift from a prevalence of dinosaurs with a few mammals before impact, to no dinosaurs, and a great increase in mammals after impact. The mammals and other species, fortunate at the time to be adaptable enough in food habits or temperature tolerance, survived—perhaps for some of their descendents to perish as the result of a later collision or violent volcanic eruption. This is perhaps the best-known hypothesis. The meteorite impact proposal is by no means universally accepted by geologists. Plimer argues that more accurate dating shows that the meteorite impact took place some 300,000 years before the dinosaur extinction (Plimer 2009). The extinctions occurred during an 800,000-year eruption period of basalts that formed the Deccan traps in India, during which period an enormous volume of toxic, acid sulphur dioxide was emitted into the atmosphere. Iridium also arises from volcanoes. Mass extinctions probably result from a combination of events.

A study by two astronomers from Caltech, Eric Leach and Gautam Vasisht, resulted in an even more intriguing proposal. Each century, one to two supernovae occur in our own spiral galaxy, the milky way. The milky way rotates in such a way that every 50 million years, the solar system passes through the spiral arms, which happen to be where most of the supernovae are likely to occur. Using computer modelling techniques, Leitch and Vasisht have calculated back in time to determine the dates when our solar system would have been 'close' to regions known to have supernova activity. High-energy cosmic radiation is emitted in vast amounts during the height of supernova activity, and it is well known that such bursts of radiation can be damaging to living organisms. Amazingly, the calculations of Leitch and Vasisht show very close correspondence between the known periods of massive life extinction on Earth and the times of its proximity to the supernova belt. The researchers have estimated that there is about a 50% probability that the radiation from such a supernova could cause massive extinctions of life on Earth (Leach and Vasisht 1998). The 'final' answer is still awaited.

The region in which life exists on Earth, the biosphere, is a physically insignificant layer (in terms of the dimensions of the Earth) extending from perhaps 1 km below ground level to about 8 km above it. The depth estimate might well have to be revised, should the 'deep hot biosphere' hypothesis proposed by Thomas Gold become generally accepted (Gold 1999). Gold presents evidence that supports the existence of a subterranean biosphere, where bioenergy may be produced without photosynthesis. Recent discoveries of strange ecosystems at hydrothermal vents adjacent to mid-ocean ridges lend credence to his deep hot biosphere arguments, first proposed in 1992. Gold's controversial ideas also include the proposal that the hydrocarbons that make up petroleum are largely of non-biological origin, and that they are being constantly replenished from deep within the Earth (Gold 1999).

Life has been found at all levels in the oceans, from the shallow continental shelf, the continental slope, and the abyssal plains found near the margins of ocean basins at depths between 3000 and 5000 m. Even the deep oceanic regions such as the Mariannas Trench in the western part of the north Pacific Ocean near the Philippines, which descends to just over 11,000 m, has recently been shown to support life. In sheer biomass terms, the oceans have been estimated to support more life than all the terrestrial regions combined. The biosphere, including all its life, makes up a mere one ten-billionth of the mass of the Earth.

Enormous amounts of energy have been and continue to be involved in movements of the Earth and its atmosphere. These movements, whether fast on our human timescale, as in weather patterns or volcanic eruptions, on the modest geological timescale of thousands of years, or on the vaster scale of hundreds of millions of years, will continue to have an important effect on terrestrial life. It is a 'mere' 18–20,000 years since the peak of the last ice-age gripped the Earth, a short time geologically, but most of us think of the Earth in terms much shorter, often revealing the obsession with our human past to the exclusion of all else. The Egyptian pyramids were built during the Old Kingdom, between about 2700 and 2200 BC. This is about as far back as most people bother to be interested.

To understand the full story of the past Earth, and to have some appreciation of what is likely to happen in the future, a much longer perspective is essential. For example, between 35 and 65 million years ago, during the first-half of the Tertiary period, both deciduous and evergreen trees grew in Antarctica. Fifty-five million years ago Australia began to separate from Antarctica, and certainly by 35 million years ago the latter was isolated. The long-term results are still with us. A cold, circular-flowing current developed, and isolated Antarctica from warmer waters to the north. By some 4 million years ago, Antarctica was heavily glaciated, the trees gone. Today the ice on that inhospitable continent has immobilized water corresponding to an estimated 60 m fall in sea level. This, together with all the other water estimated to be present as ice, means that the Earth would experience about a 100–150 m increase in sea level should *all* the ice melt. This is a most unlikely event, especially in the short term, but even a several metre rise in sea level would be disastrous for the hundreds of millions of humans who live in low-lying areas such as some Pacific islands and areas of Bangladesh and China. London and parts of Europe would also be affected.

The West Antarctic Ice Sheet (WAIS) is currently in the news as the political rhetoric on climate change and sea level rises approaches the hysterical. The WAIS is a major factor, and its complete collapse, the timing of which is the subject of intense debate, could add a further 6 m to sea levels by 2130 (Nichols *et al.* 2008). A more recent estimate (Bamber *et al.* 2009) reduces this to 3.2 m for an immediate collapse for the WAIS, or 81 cm in the next 100 years at an assumed melting rate of 6.4 mm per year. On the other hand, at the time of writing (mid-2009) there have been reports that the overall area of Antarctic sea ice has increased over the last 30 years. This has been linked to changing weather patterns caused by the ozone layer hole (Turner *et al.* 2009). By the time you read this, who knows where the debate will have led? My advice: read all reports on emotive issues such as climate change very carefully, and with a healthy degree of scepticism.

The atmosphere at present appears to be cooling, at least in the short term, no warming having been recorded since 1998. (Plimer 2009). The IPCC continues to warn of impending global warming. Should we worry now, or let things roll on for a century, or perhaps for a millennium or two? How many years are likely to pass before the onset of serious results of the so-called greenhouse effect? This effect, caused by the build up in the atmosphere of excessive energy in the form of heat, is an immediate reminder of the importance of energy balance, this time potentially on a vast scale that could affect much of humanity. Bearing in mind that water vapour is the predominant greenhouse gas, others such as carbon dioxide and methane also trap heat in much the same way as a greenhouse does, selectively reducing its re-radiation away into space.

The greenhouse effect is something over which humanity has, in principle, some control. Greenhouse gas emissions, such as carbon dioxide from burning fuels, methane from cattle and swamp gases, and other emissions such as those resulting in acid rain, are not beyond our ability to handle in the scientific/technological sense. Whether we are mature enough, or become desperate enough, to handle them politically is entirely another matter. The very least we can do, as a matter of urgency, is to try to understand what is happening. An appreciation of the geological timescale, and more reliable estimates of the rate at which changes in the Earth and the biosphere occur, are pre-requisites to such an understanding.

For the first billion years or so after its formation some 4500 million years ago, the Earth was apparently lifeless. It is useful to bear in mind that even the other inner planets had, and still have, atmospheres, surface temperatures, and amounts of solar radiation significantly different from those on Earth. Lack of oxygen and water, temperatures too high or too low, make Mercury, Venus, and perhaps Mars unlikely places to find life as we know it. We are 'fortunate' on Earth in that its mass, and hence the force of gravity, is such that oxygen and water have remained. Earth's distance from the Sun is such that we are not too hot for the molecules of life to be unstable, or too cold for all water to freeze. Most experts agree that the outer planets, in chemical composition and distance from the life-giving Sun, are even less likely candidates.

Recent evidence puts the first life forms appearing on Earth some 3.5–3.8 billion years ago. Chemical evidence lies in the oldest sedimentary rocks known, the 3.75 billion year-old-Isua series in Greenland. Although these rocks have been too greatly altered (metamorphosed) by heat and pressure to preserve the remains of living organisms, it has been proposed that their ratio of the two common isotopes of carbon, ^{12}C and ^{13}C, closely reflects that in living matter. Photosynthesis uses the two isotopes differentially, and raises the $^{12}C/^{13}C$ ratio. The Isua rocks possess carbon isotope ratios close to those associated with living organisms.

The oldest confirmed remains of living organisms have been found in unmetamorphosed (i.e. unchanged by heat and pressure into metamorphic rocks) sediments dating back 3.465 billion years, in the Apex cherts in the Pilbara region of Western Australia. In one study they were shown to consist of 11 groups (taxa) of filamentous microbes, described as cellular cyanobacterium-like organisms (Schopf 1999).

Since its beginnings, life has not only survived, but has changed and multiplied its forms and its complexity enormously. During that staggering, unimaginable time of 3500 million years, the process of natural selection has directed the formation of all the myriad types of bacteria, archaea, fungi, viruses, plants, and animals we observe today. One estimate of the total number of living species puts it at about 10 million (Eldredge 1998) while another (Wilson 1992) predicts that 100 million species inhabit this planet, most of which have not been discovered, let alone studied. Humans are among the most recent developments of natural selection. This does not automatically make us in any way superior to other species. We may be much more intelligent, in the literal sense, but is it really very bright to slaughter our own kind for no sensible reason, or to threaten our environment to the extent that our long-term survival is by no means certain? Slowly, humanity as a whole is beginning to realize that a viable future requires a holistic understanding of our position in the biosphere. The problem will be agreeing what to do about it.

To reach its current state of incredible diversity, natural selection needed time—lots of it. To begin to comprehend a timescale of billions of years, our so-called 'advanced' human brain struggles. From our human perspective, we are used to thinking in terms of perhaps a lifetime, a mere 70 or 80 years. Young children are incapable even of that. I remember my mother asking a neighbour's daughter, a youngster of five, the age of her new teacher. 'Oh, in her early sixties or seventies,' was the confident reply. My mother was staggered when she met the teacher, an attractive young woman in her mid-twenties!

The use of analogy is one of the best ways to handle the very large and the very small: the distance to the stars, the speed of light, the size of atoms, and a timescale of billions of years. A common way of representing the geologic timescale is shown in Figure 2.5. Such tabular presentations are informative, but give us no real idea of the vast timescale over which the events occurred.

To try another approach, let us shrink the 3.5 billion years of life on Earth to the timescale of 1 year. This is a reduction such that a year becomes about nine one-thousandths (0.009) of a second!

On this timescale, suppose life begins on the stroke of midnight that commences January 1st. For about the first 2 billion years, that is until some 1.5 billion years ago (or to about July 27 on our year scale), the only living things were the various types of simple, single-celled life forms, the prokaryotes (from *pro* meaning 'prior to' and *karyote* meaning 'kernel' or 'nucleus'). The very early life forms almost certainly drew their essential energy from such sources as hot springs and suitable chemical reactions. These were anaerobic (i.e. they carried out their metabolism in the absence of oxygen) organisms, which despite their 'simplicity' were and still are very successful forms of life.

Anaerobic bacteria still thrive in the Earth's biosphere, and their fossils have been found throughout the full span of life on Earth. This vast array of bacteria continues to have an enormous influence on the rest of the living world. Part of the human digestive tract swarms with *Escherichia coli*, or colon bacilli, which are themselves anaerobic bacteria. There is very little free oxygen in the lower

GEOLOGIC TIME SCALE					
Time Units of the Geologic Time Scale				Development of Plants and Animals
Eon	Era	Period	Epoch	
Phanerozoic	Cenozoic	Quaternary	Holocene —0.01—	Earliest *Homo sapiens*
			Pleistocene —1.6—	
		Tertiary	Pliocene —5.3—	Earliest hominids
			Miocene —23.8—	'Age of Mammals'
			Oligocene —33.7—	
			Eocene —55—	
			Palaeocene —65—	Extinction of dinosaurs and many other species
	Mesozoic	Cretaceous —145—	'Age of Reptiles'	First flowering plants First birds Dinosaurs dominant First mammals
		Jurassic —208—		
		Triassic —248—		
	Palaeozoic	Permian —286—		Extinction of trilobites and many other marine animals
		Carboniferous: Pennsylvanian —320—	'Age of Amphibians'	First reptiles Large coal swamps
		Carboniferous: Mississippian —360—		Amphibians abundant
		Devonian —410—	'Age of Fishes'	First amphibians First insect fossils Fishes dominant
		Silurian —438—		
		Ordovician —505—	'Age of Invertebrates'	First land plants First fishes Trilobites dominant
		Cambrian —545—		First organisms with shells Abundant Ediacaran faunas
		Vendian —650—	'Soft-bodied faunas'	
Proterozoic				First multicelled organisms
	2500	Collectively called Precambrian Comprises about 87% of the geological time scale			
Archean	3800			First one-celled organisms Age of oldest rocks
Hadean	4600 Ma			Origin of the earth

Figure 2.5. The geological timescale.

Source: Campbell, A., Cooke, P., Otrel-Cass, K., and Earl, K. (2004) Evolution for Teaching website. www.sci.waikato.ac.nz/evolution.

parts of the mammalian digestive tract. Some early anaerobic bacteria produced oxygen as a waste product. For some hundreds of millions of years, however, most of the released oxygen was 'captured' by reaction with ferrous iron (Fe^{2+}) that had collected in the early seas, forming hydrated iron oxides, or 'rust'. This settled on large areas of the sea floor as sediments, and eventually formed banded rocks such as those forming the Hammersley Ranges in Western Australia, which provide modern industry with vast amounts of iron ore. No plants or higher animals existed at this time, and most organisms were obliged to remain anaerobic.

These early creatures have survived, and to have done so in a world with steadily increasing oxygen levels they must have been incredibly adaptable. Indeed, bacteria are so adaptable today that they have medical experts greatly concerned with the resistance that many of the common pathogenic bacteria have developed to antibiotics. We are running out of effective means of controlling such bacteria as some strains of *Staphylococcus aureus* in hospitals and the tuberculosis bacillus, once believed to be under permanent control, are re-emerging as a serious problem even in 'advanced' countries, for example in some poor and crowded areas of New York.

Some time between 2 and 1.5 billion years ago, the simple single-celled cyanobacteria (previously called blue-green algae), themselves prokaryotes, developed the ability to photosynthesize. Free oxygen, that is uncombined, molecular oxygen, which today constitutes about 21% of the Earth's atmosphere, was essentially absent until the advent of photosynthesis. During photosynthesis, the cyanobacteria (and today green plants as well) take water and split it into hydrogen (which they utilize with carbon dioxide to make sugars and other chemical compounds) and oxygen, much of which is released into the water and thence to the atmosphere. Gradually, the level of free oxygen in the early seas and atmosphere increased. This was a critical step towards the development of life as we see much of it today—aerobic, or dependent on oxygen. The diversity of life also increased as oxygen levels in the atmosphere built up (Figure 2.6).

About July 27 (1.5 billion years ago) the first single-celled organisms having a nucleus enclosed by a membrane appeared. These are the eukaryotes (*eu* means 'true' or 'good', so eukaryote means 'having a true nucleus'), which are microbes having discrete chromosomes in a discrete nucleus and various organelles, such as chloroplasts and mitochondria, within their cells.

This event, the arrival of the eukaryotes, began a process that led ultimately to multicellular life and thus to ourselves. It is surely no coincidence that the emergence of the eukaryotes came soon after we believe free oxygen reached significant levels in the atmosphere. Bacteria were the only inhabitants of the Earth for some 2 billion years, but they could not be accused of 'doing nothing' over this period of time. They transformed the Earth's surface layers and atmosphere, and developed all the biochemical systems, which led directly to the emergence of multicellular life as we know it. Not until about October 20 (670 million years ago) do the first fossil records of multicellular organisms appear (Mayr 1997).

Somewhere around October 31 (570 million years ago) there was a sudden (in geological terms) and enormous increase in the number of living species on

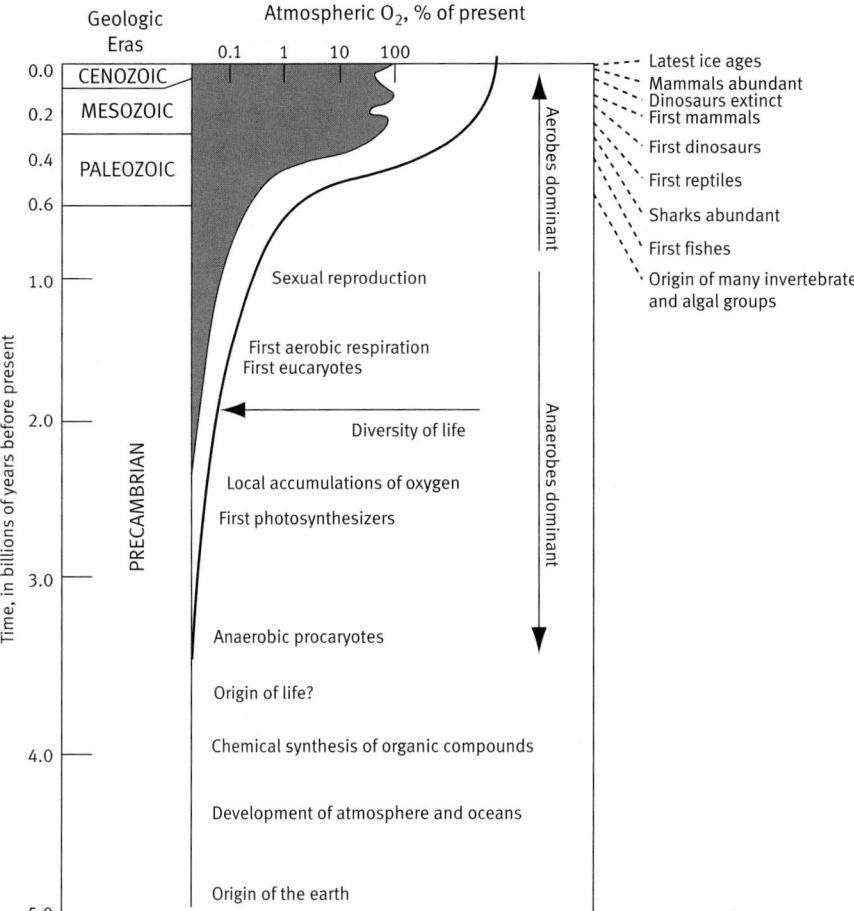

Figure 2.6. The diversity of life relative to geological time and levels of atmospheric oxygen. (Illustration from Sumich, J.L. (1996) *An Introduction to the Biology of Marine Life*, 6th edn, p. 5. © Times Mirror Higher Education Group, Inc.)

Earth, known as the Cambrian Explosion. By about November 15 (430 million years ago) the oxygen content of the atmosphere had increased to about 10%, and another profound change took place. The land began to be colonized by plants. About this time (440 million years ago) there occurred the first of five recorded mass extinctions of life on Earth, with an estimated 85% of multicellular organisms being eliminated. This extinction particularly affected animal life in the oceans. Soon after, plants began to colonize the warmer, moister parts of the continents. With plants established as a food source, animals soon followed, as the land provided a whole new range of ecological niches ripe for exploitation. In the state of Victoria, Australia, an amphibian left, in muds dating back 360 million years and which later became covered with silt, the oldest four-legged tracks known on Earth.

The second major mass extinction of life took place about 368 million years ago, possibly the result of a large meteorite which struck Sweden.

Further mass extinctions took place at approximately 245 and 208 million years ago. The former was not associated with a meteorite impact, and was the most disastrous in that an estimated 77–95% of species on Earth were eliminated. Strangely, very few of these were land plants, while almost all shallow water marine animals disappeared, together with considerable numbers of land animals. What happened? No one knows for certain, but possibly a nearby star about the size of our Sun exploded as a supernova, and showered the Earth with the equivalent of 100,000 years of cosmic radiation within a day or so. This explanation is considered unlikely, as no one can rationalize why the land plants would be unaffected.

The 208 million years ago extinction has been attributed to a huge meteorite which hit Quebec, Canada, leaving a crater 100 km wide.

Around December 25 the dinosaurs became extinct (at about the end of the Cretaceous Period, 65 million years ago) in the last of the great extinctions. This well-known event was possibly caused by a meteorite or comet impact, in the Gulf of Mexico. The crater is filled with sediment, but its shape has been determined by measurements of the Earth's gravity and magnetic field. The presence of the element iridium also indicates a meteorite origin. As mentioned above, enormous basalt flows have recently been proposed as the prime cause of the dinosaur extinction.

All the great extinctions were followed by a surge of evolutionary activity, as new species emerged to fill the ecological wastelands that had resulted from the massive changes in energy as dust clouds cut off the solar radiation and temperatures (presumably) plunged.

All the explanations proposed for the above mass extinctions reflect the widely held view that external, catastrophic events were responsible. As I have mentioned briefly, it is difficult to envisage just how some of the proposed events could have caused the observed extinctions. Scientists have recently developed theories that involve alternative processes. One of these involves what is called self-organized criticality (SOC). In his book, *How Nature Works*, physicist Per Bak, from the Niels Bohr Institute in Denmark, proposes that the inherent properties of complex systems may explain catastrophic events in biology, geophysics, and indeed economics (Bak 1996). He believes that the processes of evolution may involve highly non-linear systems, in which it is impossible to predict 'emergent' behaviour such as catastrophies.

Taking the appearance of early humans at about 2 million years ago in real time, they would not appear until December 31 at 7 pm in our reduced timescale—very recent arrivals indeed. The pyramids of Egypt would have been completed at about 24 s before midnight, and the industrialized world came into being less than 2 s before the end of our year.

There is a danger in my use of the above year analogy. It might de-emphasize the very aspect of biological/geological time I am trying to convey. Treat it just as a means of putting the major evolutionary events mentioned in perspective.

Another useful way of looking at the development of life is to relate it to a 24-h clock (Figure 2.7).

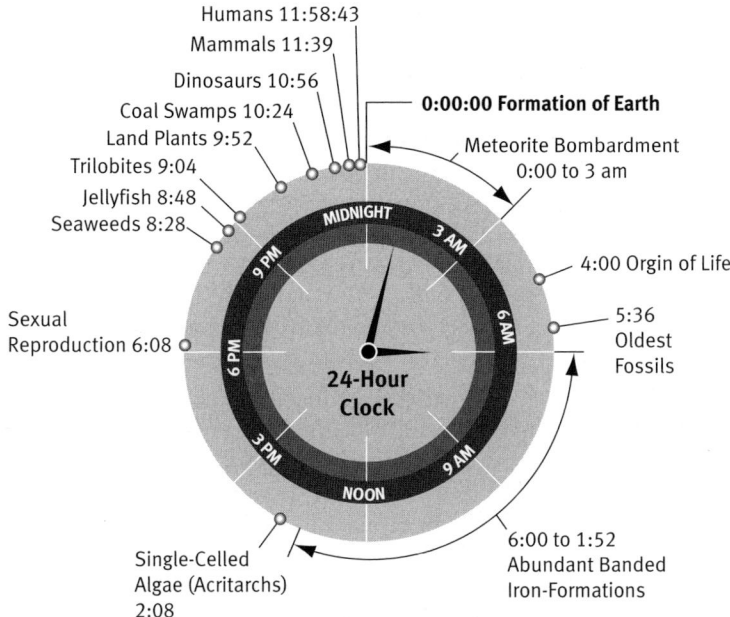

Figure 2.7. The development of life on Earth as a 24-hour clock. (Illustration from Schopf, J. William, *Cradle of Life*. © 1999 Princeton University Press. Reprinted by permission of Princeton University Press.)

'Deep Time' is probably as good a term as any to use when contemplating evolutionary or geological time. The term 'Deep Time' was coined by John McPhee, who wanted to emphasize the difference between the geological timescale and the vastly shorter one that dominates our own lives (McPhee 1981).

Consider the various ways of representing the geological timescale, and wonder. Try to keep them in mind when we speak of evolutionary events. It will not be easy, but it should give you a greater appreciation of what was involved.

Any number of questions can arise from consideration of the timescale of life on Earth.

- Why did a billion years, as far as we know, elapse between the formation of the solid Earth and the emergence of life?
- Why did eukaryotic cells appear when they did, not before or later?
- How did multicellular organisms evolve?
- Was the emergence of life inevitable, given the conditions existing on the prebiotic Earth?
- If life were to begin again under exactly the same conditions, would it necessarily follow the same course and roughly the same timescale? (Gould 1989).
- If all life on Earth were to be extinguished tomorrow, would it be likely to emerge again?

28 Introducing Biological Energetics

We shall return to some of these questions, and others, in later chapters.

Finally, let us look at some other aspects of the rise in oxygen levels brought about by photosynthesis. Photosynthesis is an example of the transformation of energy, which is of the utmost importance to us. The energy in sunlight 'drives' the reactions in which carbon dioxide and water are combined. Thus, sunlight (part of the electromagnetic energy spectrum) is converted by the processes of photosynthesis into chemical energy in the form of simple sugars. In its simplest form this can be written:

$$12H_2O + 6CO_2 \xrightarrow{h\nu} C_6H_{12}O_6 + 6O_2 + 6H_2O$$

where $h\nu$ represents the solar energy for photosynthesis.

This transformation of solar energy is assisted by chlorophyll, the green pigment in plants. The whole process is vastly more complex than this, involving many biochemical reactions and specialized pieces of cell machinery. We shall discuss many of these in a later chapter. Photosynthetic organisms, such as green plants and photosynthetic bacteria, are the primary source of the chemical energy in molecules such as sugars that constitute the basis of food for most life forms on Earth. They are at the bottom of many food chains, including those of mammals, the group which includes ourselves. As such, the photosynthesizers are a precious resource, whose diversity and survival we must ensure to preserve our own.

The presence of oxygen had another major effect beneficial to the development of terrestrial life. High-energy ultraviolet (UV) radiation from the sun is damaging to many of life's essential molecules and is capable of causing skin cancer in humans. In other reactions, oxygen molecules, O_2, are split by UV energy into oxygen radicals, $O\cdot$. These are so reactive that they combine with oxygen to form ozone, O_3. A layer of ozone in the atmosphere has the fortunate property of absorbing most of the harmful solar UV radiation, thus protecting life at sea level.

Most people have heard of the effects of chloroflourocarbons (CFCs) on depleting the atmospheric ozone, causing a worrying 'hole' to form in the ozone layer. These refrigerant gases inevitably escape from their 'sealed' units in refrigerators and air conditioners, and eventually find their way into the upper atmosphere, where they react with free ozone. Such large quantities have built up in recent years, causing a worrying depletion of ozone, for example in the Antarctic regions. First reported in 1985, the so-called 'hole in the ozone layer' has been increasing in size since about the mid-1970s and is monitored constantly. A major concern is the likelihood of increased skin cancer in humans, as a result of the extra UV radiation reaching ground level. More recently, the ozone layer hole has been linked to changing weather patterns in Antarctica, which seem to have contributed to an increase in the area of sea ice in Eastern Antarctica since the late 1970s (Turner *et al.* 2009). CFCs have been banned as refrigerants in most countries as a result of the 1997 Montreal Protocol. Despite this, the stability and concentrations of the contaminating compounds are such that the ozone layer will exist each spring for about the next 50 years (BAS 2008).

So far we have seen a just brief glimpse of what was to become the cradle of life as we humans know it. It is a pity no one was there to record the actual birth. Perhaps H.G. Wells' time-traveller would have revelled in it, and what a fascinating experience it would have been. As it is, what we have managed to glean of our origins so far is intriguing enough. What is yet to come is probably beyond our imagination.

There is a saying beloved by some historians which goes something like:

'We can only learn about where we are going by knowing about where we have been.'

This saying applies to the whole history of life on Earth, and is important to keep in mind when we are trying to understand life as it appears now. The distribution and forms of life at any time in history depended on the distribution of life and the conditions which came before, and so on back through the 3.5 billion years of life's existence. This is an expression of the principle of common descent, forcefully espoused by Darwin. All life is linked, ultimately, to all other life. If all life on Earth were to be extinguished tomorrow, would it commence again? I use this highly unlikely scenario merely to emphasize a point. If life *were* to begin again tomorrow, it would be vastly different in its origins and history. The conditions on Earth today differ greatly from those 3.5 billion years ago, from the composition of the atmosphere and the oceans, to the positions of the continents. Humans evolved in Africa last time around, but in another 3 billion years or so, presuming that evolution should toss us up again, who can say? The development of the Earth in the future must follow pathways that will be constrained and limited by those which we are experiencing now, and any new wave of life we contemplate will interact with and modify those pathways just as the old wave has done.

A final sobering thought, though not of concern to ourselves or to our immediate descendants, is that the Earth is about half-way through its own time as a carrier of life. In about another 5 billion years, the Sun, itself undergoing evolution as do all stars, will expand to a vast red giant and extinguish whatever life has survived here, before collapsing to a white dwarf star on its way to the stellar graveyard. Even the time-traveller would be unlikely to survive events such as those.

GENERAL REFERENCES

Johnson, D.P. (2004) *The Geology of Australia*. Cambridge University Press, Cambridge.
Plimer, I. (2001) *A Short History of Planet Earth*. ABC Books, Sydney.

REFERENCES

Bak, P. (1996) *How Nature Works*. Copernicus/Springer-Verlag, New York.
Bamber, J.L., Riva, R.E.M., Vermeersen, B.L.A., and LeBrocq, A.M. (2009) Reassessment of the potential sea-level rise from a collapse of the West Antarctic Ice Sheet. *Science* **324**, 901–2.
BAS (2008) The Ozone Hole. *British Antarctic Survey, Science Briefing*. http://www.antarctica.ac.uk/press/journalists/resources/science/the_ozone_hole_2008.pdf.

de Duve, C. (2005) *Singularities.* Cambridge University Press, New York, pp. 7–9.
Eldredge, N. (1998) *Life in the Balance.* Princeton University Press, Princeton, pp. 67–8.
Ferris, T. (1997) *The Whole Shebang.* Weidenfeld & Nicolson, London, pp. 177–9.
Gold, T. (1999) *The Deep Hot Biosphere.* Springer, New York, pp. 117–8.
Gould, S.J. (1989) *Wonderful Life.* Norton, New York, pp. 45–8.
Hartmann, W.K. and Davis, D.R. (1975) *Icarus* **24**, 505.
Leach, E.M. and Vasisht, G. (1998) Mass extinctions and the sun's encounters with spiral arms. New Astronomy, http://arxiv.org/pdf/astro-ph/9802174.
Luisi, P.L. (2006) *The Emergence of Life.* Cambridge University Press, Cambridge, pp. 47–50.
Mayr, E. (1997) *This is Biology.* The Belknap Press of Harvard University Press, Cambridge, MA, p. 149.
Nichols, R.J., Tol, R.S.J., and Vafeidis, A.T. (2008) Global estimates of the impact of a collapse of the West Antarctic ice sheet: An application of *FUND. Climatic Change* **91**, 171–91.
McPhee, J. (1981) *Basin and Range.* Farrer Strauss & Giroux, New York.
Plimer, I. (2009) *Heaven and Earth: Global Warming, the Missing Science.* Connor Court Publishing, Ballan, pp. 178–9, 490.
Schopf, J.W. (1999) *The Cradle of Life.* Princeton University Press, Princeton, p. 100.
Schneider, S.H. (1996) *Laboratory Earth.* Weidenfeld & Nicolson, London, p. 60.
Schneider, S.H., and Londer, R. (1984) *The Coevolution of Climate and Life.* Sierra Club Books, San Francisco, Ch. 6.
Turner, J., Comiso, J.C., Marshall, G.J., Lachlan-Cope, T.A., Bracegirdle, T.J., Maksym, T. *et al.* (2009) Non-annular atmospheric circulation change induced by stratospheric ozone depletion and its role in the recent increase of Antarctic sea ice extent. *Geophys. Res. Lett.* **36**, L08502, doi:10.1029/2009GL037524.
Wills, C. and Bada, J. (2000) *The Spark of Life.* Perseus Publishing, Cambridge, MA, pp. 67–71, 74.
Wilson, E.O. (1992) *The Diversity of Life.* The Belknap Press of Harvard University Press, Cambridge, MA, p. 132.

3
Force and Energy Explored

Since the times of Galileo, Descartes, and Newton in the sixteenth to eighteenth centuries, many great minds have grappled with the concepts of force and energy, and with their applications. The full story would take several large books to do it justice. Sometimes in this book it will be possible to outline the evidence for and against certain statements made and conclusions reached. Mostly, however, it will not be feasible to convey the full extent of the argument, or to give more than the barest notion of the time and effort involved in both the thoughts and the experiments which led to the conclusions.

It is sufficient to say here that the reader may readily find any basic information of interest in standard physics or physical chemistry texts, while examples of the many applications of energy may be found in a good encyclopaedia of science and technology.

At some stage all branches of science must become quantitative. All manner of characteristics need to be measured with varying degrees of accuracy—the velocity of light, the weight of the space shuttle, the wingspan of a bird, the rate of a chemical reaction. All these measurements should be made in internationally recognized units, as science is communicated world-wide. Even now there is not complete standardization of units, but international committees have been set up to improve the situation. Units and standards have been established, and means of calculation of derived properties and characteristics from the basic units have been developed. The system of units adopted by the international scientific community is the Systeme Internationale d'unites, or SI.

One of the most famous lines in the history of science was coined by Galileo in his description of the Universe: 'This grand book is written in the language of mathematics, and its characters are triangles, circles, and other geometrical figures', (Gribbin 2002).

Much more recently, astronomer, science writer, and Pulitzer Prize winner Carl Sagan eloquently expressed the essential nature of quantification:

'If you know a thing only qualitatively, you know it no more than vaguely... Being afraid of quantification is tantamount to disenfranchising yourself, giving up on one of the most potent prospects for understanding and changing the world.' (Sagan 1997)

It is unfortunately true that the essence of most science is mathematically based. I say unfortunately only because most of us have only the slightest smattering of this most useful language. This need not be the case, at least for the mathematics that would enable more people to appreciate the basics of science. Attitudes towards, and poor teaching of, early mathematics are problems that could be overcome, given the political will—but that is another story.

Energy is an abstract concept not easily explained in words. Except when it brings about some kind of change that can be seen and/or measured, the idea of energy may be difficult to grasp, and was something of a controversy until the middle of the nineteenth century. The nature of force was actually formalized before that of energy, so let us first take a brief look at this important concept.

Galileo, Descartes, and Newton formulated their much-celebrated mechanics in terms of time, distance, velocity, acceleration, and force. Mechanics is the section of physics concerned with the action of forces on bodies and with the motions they produce.

Galileo Galilei (1564–1642) was an Italian scientist skilled in mathematics who carried out critical experiments in mechanics. For this he has been called the founder of modern scientific mechanics. Among many other contributions, notably to astronomy in the development of the telescope, he deduced the laws of falling and sliding bodies, and developed mathematical equations to interpret their motion.

Frenchman Rene Descartes (1596–1650) was a distinguished mathematician who, while emphasizing the importance of getting rid of principles that are not based on experimental evidence, at times himself arrived at such principles based on theological or metaphysical arguments. His ideas were influential, especially in France, well into the eighteenth century (Laidler 1993).

Sir Isaac Newton (1642–1727) formulated his three famous laws of motion while at Trinity College, Cambridge. As stated in Newton's second Law of Motion:

A force is something which changes, or tends to change, the state of motion of a body.

Newton's three laws were based on experimental and mathematical analysis, and indeed the first had already been expressed by both Galileo and Descartes. On the basis of the three laws and other work, Newton made great contributions to what is now usually known as classical mechanics (or Newtonian mechanics), which consists of the application of Newton's three laws of motion, first stated in 1687.

The idea of energy was implicit in some of the studies of Galileo, Descartes, and Newton, and any one of them might well have developed the relationships discussed below. History records that this was not the case. The formal definitions of energy and work were introduced much later after much confusion over the meaning of terms such as power, energy, and momentum (Laidler 1993). Their mathematical definitions show that the relationships are exact. Energy changes can be calculated, and experimentally verified, for almost any process. This is obviously enormously important for practising scientists and engineers.

Cutting the story very short, we now have the following very useful definitions for classical mechanics:

(1) velocity (*v*) = distance travelled (*s*) ÷ time taken (*t*)

$$v = s/t$$

unit: metres/second (m s^{-1})

(2) acceleration (*a*) = velocity change ($v_2 - v_1$) ÷ time taken (*t*)

$$a = (v_2 - v_1)/t$$

unit: metres/second/second (m s^{-2})

(3) force (*f*) = mass (*m*) × acceleration (*a*)

$$f = ma$$

unit: newton (N)

(4) work (*w*) = force × distance moved

$$w = f\Delta x$$

unit: joule (J)

(5) power (*p*) = work done ÷ time taken (*t*)

$$p = w/t$$

unit: watt (W)

All the above are SI units. Note that the unit of work, the joule, is expressed here in terms of force and distance. This is convenient for mechanical work such as that carried out by pushing, pulling, or lifting objects, and by simple machines such as a lever. Later we will deal with other forms of work, each having its own source of energy.

Here, it is convenient to introduce the two forms of mechanical energy. I do so without any explanation of their derivation, as this does not really concern us:

$$\text{kinetic energy (KE)} = \tfrac{1}{2}(\text{mass} \times (\text{velocity})^2)$$

$$\text{or KE} = \tfrac{1}{2}mv^2$$

unit: joule (J)

If we let a moving body at any instant have a velocity (*v*) and mass (*m*) then its kinetic energy at that instant will be equal to $\tfrac{1}{2}mv^2$. This quantitative mathematical expression is within our own experience. Objects with more mass possess greater amounts of kinetic energy than lighter ones. Just try catching a cannonball rather than a football going at the same velocity! You will certainly be moved by the experience or, put more scientifically, much more work will be done on you. Similarly, the difference in kinetic energy between a football kicked with low velocity compared with that kicked with high velocity by a professional will be obvious to anyone who tries to stop it. A bullet weighing a few grams is relatively harmless until it is propelled at several hundred metres per second. Work can also be done by each of the other sources of energy mentioned. For each case, mathematical expressions have been derived to allow the calculation, in kilojoules, of the quantity of energy involved in doing a particular amount of work.

Note that the kinetic energy is proportional to the square of the velocity. If the velocity of a body doubles from say 2 m s^{-1} to 4 m s^{-1}, its kinetic energy will be four times as great. For this reason, the energy involved in an impact during an accident at 4 kilometres per hour (kph) will be four times that at 2 kph, and so on. The kinetic energy of a car at 100 kph compared with that at 25 kph will be 16 times larger. Speed certainly kills.

As well as kinetic energy, which is the energy an object has a result of its motion, there is also the very useful concept of potential energy. This term was originally applied to the energy possessed by an object because of its height above sea level. The higher the object, the greater is its gravitational potential energy. If the object were to be allowed to fall freely, it would be able to respond to the force of gravity, and undergo the acceleration due to gravity (g). Expressed quantitatively:

$$\text{gravitational potential energy (PE)} = \text{mass} \times \text{acceleration due to gravity} \times \text{height above sea level}$$

$$\text{or PE} = mgh$$

$$\text{unit: joule (J)}$$

Once it starts to move, a dropped object begins to acquire kinetic energy, which increases as the velocity becomes greater and greater. Its potential energy is continuously decreasing as it is converted to kinetic energy as the velocity increases during the fall, that is its potential energy becomes real, or actual, energy. Although to ride a bicycle to the top of a hill might be quite exhausting, you will at least build up some energy credit. This can be spent in an exhilarating ride down the other side, as much of your potential energy gets converted to kinetic energy. Things dropped from a great height are able to do a large amount of damage. Large hailstones are an example—the savage hailstorm in Sydney in April 1999 caused damage in excess of A$ 1,000,000,000, mainly to cars and homes. On the other hand, people falling from great heights (or from speeding bicycles) mainly damage themselves.

In accordance with the energy conservation law, is energy being conserved during the falling process or the ride downhill?

Yes, there are no exceptions. During *any* change, the total energy of the universe is conserved. The potential energy lost (the objects are losing height, and as h in our formula decreases, so must potential energy) would at any stage during the fall or ride have been converted to the same amount of kinetic energy gained (except for a loss as heat energy, due to friction with the air and the road, and perhaps a little sound energy).

What happens to the kinetic energy when the object strikes the ground, and comes to rest? Is energy conserved here?

Yes. Some energy is lost (converted to mechanical energy) to move aside a certain amount of soil; some is lost as (converted to) sound energy (we can hear a thud); some is lost as heat to the soil (not easy to measure) and some is lost due to air-friction (converted to heat) on the way down. If we could measure all these bits of energy and add them up, it should be possible to come exactly to the value of the original gravitational potential energy before the object was dropped. Similarly, when the cyclist reaches level ground, his speed will soon slow as friction takes over, and soon he will need to pedal to maintain some forward motion.

The concept of potential energy is so useful that it has been applied to other sources of energy waiting to be tapped. The untapped energy in chemicals such as petrol, oil and other fuels, in foods, and in explosives is called chemical potential energy. Chemical potential energy may be converted to mechanical energy in cars or human muscles, in the case of explosives to kinetic energy in a cannon or rifle, or to electrical energy when a reaction is used to pump electrons around a circuit. All these energy sources can be harnessed to do work. In later chapters, much will be said about how living organisms access chemical potential energy.

Similarly we can conceive of elastic potential energy, as in a stretched spring, rubber band, or bow, which can be converted to kinetic energy if the device is arranged to propel a projectile. Initially, some form of mechanical energy must be used to stretch or load the device.

The rather long discussion above largely concerns mechanical energy, the energy involved in the working of machines, both simple (a lever or a pair of scissors) or complex (a steam engine or internal combustion engine). Mechanical energy was the first type to be studied scientifically, and these studies helped to clarify the energy concept. Studies of the conversion of heat energy to mechanical energy in heat engines such as the steam engine contributed greatly to the development of the science of thermodynamics, with its laws that have such important and universal implications.

There are four fundamental forces 'at work' in our universe: gravitational force, electromagnetic force, and the strong and weak nuclear forces. These forces (Table 3.1) are the basis of all forms of energy and work, so it is important to be familiar with them.

Table 3.1. Characteristics of the four fundamental forces.

Force	Relative strength	Range
Gravitational	10^{-39}	Infinite
Electromagnetic	7.299×10^{-3}	Infinite
Strong nuclear	1	10^{-15} m
Weak nuclear	10^{-5}	10^{-17} m

Adapted from Trefil, J. and Hazen, R.M. (2001) *The Sciences, an Integrated Approach*, 3rd edn, p. 295. © 2001 John Wiley & Sons Inc.

3.1 Gravitational force

Gravitational force was the first to be studied formally, but its fundamental nature is still not understood. It is the force behind much of classical mechanics and includes studies of planetary motion, projectiles, and satellites. Gravitational force exists between all matter in the universe. In principle, everything in the universe is attracted to everything else via gravitational force. Quantitatively expressed:

$$F = Gm_1m_2/r^2$$

where F is the force between two bodies of mass m_1 and m_2, r is the distance between them, and G is the universal gravitational constant. In everyday practice, we notice gravitational force via the weight of objects, including our own. Weight is a force. Quantitatively:

$$\text{force} = \text{mass} \times \text{acceleration}$$

In the case of weight:

$$\text{wt} = mg$$

where wt is the weight of a object of mass m and g is the acceleration due to gravity, 9.81 m s^{-2} at the surface of the Earth.

Mechanical work was the first type of work to be formalized and studied by physicists.

a) Mechanical work is done by exerting a force f on a stationary object to move it a distance Δx, as stated above. Quantitatively:

$$\text{work} = \text{force} \times \text{distance moved}$$

For work done when an object falls through a height h due to the force of gravity,

$$\text{work} = \text{force} \times \text{distance} = \text{weight} \times \text{height fallen} = mgh$$

where m is the mass of the object (kg), g is the acceleration due to gravity and h is the height of the body above mean sea level (m). The SI unit is the joule. A stationary object situated h m above sea level has the potential to do work amounting to mgh, if it were allowed to fall. Thus, the gravitational potential energy of such an object is equal to mgh (see below).

b) Mechanical work is also done when a moving body is acted on by a force to either increase or decrease its velocity. If its velocity changes, up or down, there will be a corresponding increase or decrease in its kinetic energy of $\frac{1}{2}mv^2$. Some kind of force is always necessary to change the velocity of a moving body.

For completeness, we also need to clarify the meanings of the terms mass and velocity.

Mass is a measure of the quantity of matter in an object. The SI unit is the kilogram.

A particular object will have the same mass in any part of the Universe. The same cannot be said for the weight of an object. As we have seen above, weight is a force and the acceleration referred to is the acceleration due to gravity (g). The value of g varies, depending on where an object is in the universe. Thus, the force of gravity at the surface of a particular planet is proportional to the mass of the planet. If m_1 is the mass of a person and m_2 is the mass of the planet he inhabits, the force between them, the force of gravity, F, will be

$$F = Gm_1m_2/r^2$$

As the moon (m_2) is much less massive than Earth (about one-sixth), the acceleration due to gravity is approx one-sixth that on Earth, that is $g/6$.

Thus,

weight of our object on the Moon = mass × acceleration due to Moon's gravity = $mg/6$

that is, the weight of our object on the moon is only about a sixth of its weight on Earth. Similarly, on Jupiter, which is much more massive than Earth, our object would have a weight about 2.53 times its weight on Earth. In all of the examples, the mass, or quantity of matter, of our chosen object remains the same, say 1 kg. A person on the moon will still have the same muscular strength as he does on Earth (at least for a while). If, however, he were to exert the same force to walk on the moon, if he were to push down with his feet just as hard as he did on Earth, what would be the result? His step would become huge, and he would probably sail 30 cm or so above the ground, as he only needs to overcome one-sixth of the gravity he is used to on Earth. His weight actually falls to one-sixth that on Earth, so he must adjust his walking style dramatically—most of us have seen film of the first awkward steps/hops by astronauts on the moon. On the other hand, the same person on Jupiter would feel as though his weight had increased 2.53-fold, as indeed it would have done. Assuming he could actually walk on the surface of Jupiter (unlikely) our space walker would soon become quite exhausted from merely 'dragging' himself around. In both cases, his mass would have remained exactly the same.

Why have I gone into such detail about the force of gravity? What has this to do with life on Earth? Quite a lot, in fact. Firstly, the force of gravity determines the weight of an object, and the weight of a land animal in particular is important in determining its maximum size and its shape. Size and shape largely determine the rate of heat loss from animal bodies; this in turn has important effects on their rate of consumption of food, which is related to the manner in which the food is obtained. Size and shape also affect the movement capabilities of animals. All of the above aspects are clearly important in the lives of animals and plants. They must also be considered when we are thinking about the possibility of life on other planets. Compared with the values on Earth, vastly different forces of gravity on other planets will give rise to vastly different planetary atmospheres, different problems for evolution to solve, and probably vastly different life forms.

Velocity is commonly regarded as the same as speed. It is measured in the same units. However, to be strictly correct, velocity must incorporate direction as well as speed. Force and velocity are called vector quantities, that is they incorporate both magnitude and direction. Thus a car may have a speed of 50 km per hour, but a velocity of 50 km per hour in, say, a southerly direction—a reading of 180° by compass.

Power is the rate of doing work. From the expression

$$\text{power} = \text{work done} \div \text{time taken}$$

$$\text{or } p = w/t$$

we can see that the faster the work is done (i.e. the smaller is t) the larger is the power consumed. This is within our experience also. The faster you move a tonne of soil with a shovel, the harder will be your breathing. Take it slowly and steadily, and the job will get done with apparently less effort. Essentially, the same amount of work will be done to move the tonne of soil, but the power exerted will be less as the work will be spread over a longer time period. Similar reasoning can be applied to walking 100 m compared with sprinting it. In this case, however, the sprinter will actually do more work than the walker. Part of the reason is that when a sprinter pushes off powerfully, he raises himself completely off the ground and higher into the air than the walker, who always has one foot on earth. The sprinter thus does more work against the force of gravity than the walker over the same distance, as well as doing it faster.

Another example is the way railway tracks are laid to avoid steep slopes. As we know, railway engines are designed to pull enormous loads. Even on flat ground, trains start quite slowly compared with cars, as the power needed to reach high speed quickly is beyond their designed capacity. Similarly, too great a slope means a lot of work has to be done quickly against gravity to maintain speed, that is more power is needed. With limited power available, the result is that the train must slow down (i.e. use a larger t in the expression above to lower the p required). To avoid the train coming to a stop, the slope or grade of a railway line is limited by railway engineers and railway lines wind up mountains kilometre after tortuous kilometre. Fortunately for passengers, the long haul is often more than compensated for by spectacular scenery. The principles of climbing a slope gradually (i.e. overcoming the force of gravity) also apply to animals on the ground, and obviously to birds when flying. You won't see many birds, particularly large ones, shooting up almost vertically like a rocket. Their muscles are simply incapable of delivering the necessary power.

Now let us look at Newton's three laws of motion.

Newton's first law:

Every body continues in a state of rest or in uniform motion in a straight line (uniform velocity) unless acted on by an external force.

In everyday experience, we observe that many objects, such as stones or pieces of gravel, remain at rest, merely lying around. As we know, they are being acted on by the force of gravity, but unless they are on quite a slope or we pick one up and drop it the force of gravity has little visible effect. We are reminded that the force of gravity is present when shifting furniture that has been resting on a carpet. The flattened area is mute testimony to the force of gravity, as is the exertion we experience. Try to move a piece of heavy furniture too quickly and you will be reminded of the limits to the amount of power you are able to generate. There are few observable examples of straight line uniform motion of bodies on Earth, or indeed anywhere as far as we know. The force of gravity usually ensures that bodies move in a curve, which means their velocity changes, which means they experience acceleration. It follows that uniform motion does not occur. Examples of moving in a curve include throwing a stone or kicking a football into the air. The force of gravity makes these objects move in a curve as it accelerates them back to Earth. Similarly, the force of gravity keeps the planets moving around the Sun, their orbits being ellipses. Man-made satellites zip around the Earth in prescribed orbits, but even they are 'kept up' by the balance between the force of gravity and their tendency to fly off into space. Actually, satellites are falling towards the Earth all the time, but never reach the ground as the curvature of the Earth beneath ensures they remain at the same altitude above it, provided that their speed remains sufficient. Ground control monitors them carefully to ensure that all is well.

Comets that visit our solar system also have curved, elliptical orbits, often with what seem to us long periods before they return to circle the Sun and be flung off once more into outer space by the slingshot of solar gravity. Perhaps the best known is Halley's comet, last seen in 1986, which has a period of 76 years. It is remarkable that the periods of comets, and the more familiar periods of the planets, are so predictable. The same is true for eclipses of the Sun and Moon, and is generally the case with large moving objects such as planets, stars, and galaxies, for which Newtonian mechanics works brilliantly. It is not generally the case with very small moving objects such as atoms, which individually behave unpredictably. The bulk properties of very large collections of molecules (such as a glass of water or a satellite) are usually predictable, however. This is fortunate because we as living organisms consist of just such collections of molecules. What am I on about here? I can assure you that it is important and the significance will become evident in Chapter 5.

Newton's second law:

A force is something which changes, or tends to change, the state of motion of a body. Force = mass × acceleration.

We need to insert 'or tends to change' to cover situations such as pushing on a brick wall without moving it. The tendency to change the state of motion of the wall is there, but we are not strong enough to do so. Should we be able to push hard enough, or should the wall be weak, over it would go. A similar situation arises when we try to lift a block of stone that is too heavy for us to move. The tendency to change its state of motion is there. Certainly something has happened, judging by our perspiration and heavy breathing, even if the block of stone remains unmoved. Most of our effort ends up as heat, rather than useful work. By definition, work is not done as there is no actual movement of the wall (no Δx). In addition, when we push unsuccessfully on the wall, it pushes back on *us* (resists) with the same amount of force which we apply to *it*. If this were not the case, either the wall would fall over or we would. This experience leads to

40 Introducing Biological Energetics

Newton's third law of motion:

To every action there is an equal and opposite reaction.

As an example of the third law, imagine two people on ice-skates face-to-face on a rink. If they should push one another, both will move backwards. If they happen to be equal in weight (and thus equal in mass) each will move backwards with the same acceleration and will come to rest after having moved about the same distance. This is because they both experience the same mutual force pushing them apart. As force = mass × acceleration, and their mass is the same they will move apart with the same acceleration, and of course be slowed down to a stop by about the same amount of friction between their skates and the ice. If the two happen to be greatly different in weight, the lighter one will move backwards with the greater acceleration, and end up farther from the starting point than the heavier one. A similar argument can be made about the collision of any two objects. Most of us have seen astronauts floating about in space, and perhaps pushing apart in a similar manner to the skaters, that is they were free moving. What was the reason for using the example of two skaters? Two people standing normally on the pavement would merely push one another over, as friction would anchor their feet to the ground. This would be a more amusing but less instructive example.

3.2 The electromagnetic force

Electromagnetic energy (also known as radiant energy) is energy possessed by electromagnetic radiation, such as visible light, infrared radiation, microwaves, radio and television waves, ultraviolet (UV) rays, X-rays and gamma rays (Figure 3.1).

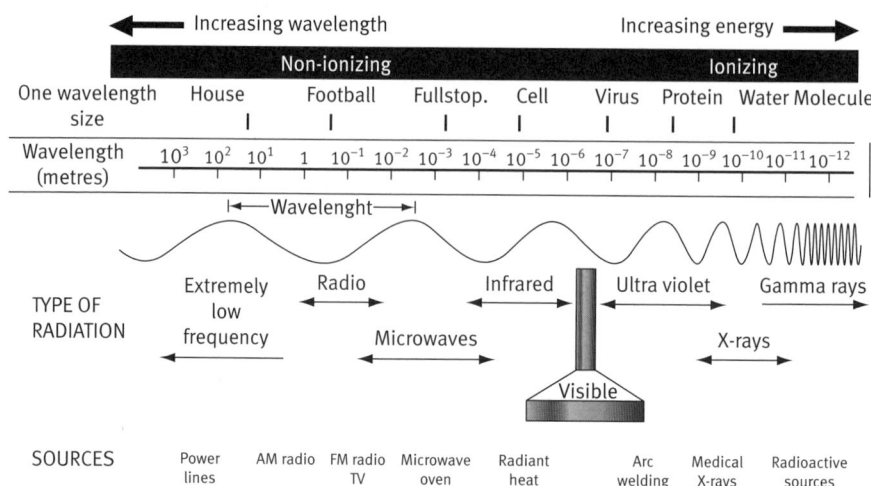

Figure 3.1. The electromagnetic spectrum. (Adapted with permission from http://www.arpansa.gov.au/images/basics/emr.jpg.)

Here I deal with the more general features of electromagnetic energy, most of which will be familiar to readers with a basic science background. The discussion is qualitative and highly selective. Electromagnetic energy is so important to an understanding of the structure of matter, chemical reactivity, and the functioning of scientific instruments that I have treated it more formally in Appendix A.

That many of the sources illustrated in Figure 3.1 possess energy are within our direct experience. Visible light allows us to see our surroundings in colour and to take photographs. Both these processes involve the transformation of radiant energy to other forms of energy. The visual process in mammals such as ourselves involves as an early step, rearrangement of the shape of a particular type of molecule. This occurs as a result of the molecule absorbing radiant energy from the visible region of the spectrum. Further steps involve electrical transmission of nerve impulses to the brain, which correlates and interprets the impulses so that we see an image of a tree, the ocean, or another person.

In (pre-digital) photography, light energy initiates chemical reactions that produce a negative. In a complex series of chemical reactions, the rest of the process leads to the final colour print. The conversion of infrared radiation from the Sun to heat energy, also known as thermal radiation, is one of the most familiar examples of energy transformation. We feel it ourselves every day. Certain wavelengths of infrared energy are selectively absorbed by the molecules it falls on, increasing their average bond vibrations, rotations, and translational movements. This absorption of thermal energy increases the average kinetic energy of the molecules, and results in an increase in temperature. Temperature by definition is a measure of the average kinetic energy of the atoms and molecules in the body whose temperature is being measured.

We are frequently reminded, particularly during the summer months, to avoid or minimize exposure of our skin to radiation energy from the Sun. The culprit is part of the UV region of the solar spectrum. UV rays are found towards the higher energy, higher frequency/lower wavelength end of the solar spectrum, and are damaging to sensitive cells in our skin. UVA (315–400 nm) contributes to premature ageing, wrinkling of the skin, and possibly cancer. UVB (280–315 nm) is more harmful and is probably the main contributor to skin cancers and eye damage (cataracts). UVC (100–280 nm) is highly dangerous, but is absorbed in the atmosphere by ozone and does not reach the surface of the Earth (ARPANSA 2008).

Unfortunately, UV rays arrive from the Sun along with warming but less harmful infrared rays, and we cannot avoid them entirely. It has been known for a long time that dense, white, opaque chemicals such as zinc oxide and titanium dioxide may be mixed with a cream base and used simply to block out some of the UV rays, rather like a layer of white paint (some paints do in fact contain these compounds). It occurred to scientists that if DNA and other organic molecules could absorb UV, perhaps it would be possible to design special chemical compounds to put into a cream to absorb UV rays specifically and thus protect the skin beneath. A range of organic compounds capable of absorbing the dangerous UV energy and mainly transforming it to harmless heat has been synthesized. This approach has been highly successful and sunscreens that can protect skin, at least partially, from UV for half an hour or more have been available for some time. Similar compounds have been chemically attached to clothing material, as research has shown that quite serious sunburn could occur through much of the light fabrics worn in summer. It is possible that the increased UV radiation reaching ground level as a result of the human-induced, increased size of the hole in the ozone layer reduces the effectiveness of our immune systems (Health Thematic Guide 1996).

Fortunately, exposure of our skin to sunlight is not all bad news, as a certain amount of exposure to UVB is necessary to avoid vitamin D deficiency. Such a deficiency leads to rickets, and affected children develop deformities in their weight-bearing bones. The chemical precursor to vitamin D (7-dehydrocholesterol) needs radiant energy from sunlight to initiate the reaction for its conversion to vitamin D_3 (cholecalciferol). The active hormone forms are then made by enzymic hydroxylation of vitamin D_3. The active forms stimulate the absorption of ingested calcium and are involved in the calcification of bone.

There was a report some years ago in the UK of vitamin D deficiency in a group of Pakistani immigrants, and a later one in Denmark. The colder weather and lower sunlight intensity was too much for these people from a climate where they had led much more outdoor lives, and who in the Europe covered themselves and remained indoors to the extent that they suffered vitamin D deficiency. This was remedied after the cause had been established. Interestingly, the same problem has been identified recently in much sunnier Australia. Some immigrants have been affected, especially females who cover their bodies for cutural/religious reasons. The problem is exacerbated when the person has dark skin. Darker-skinned people require more exposure to sunlight to activate pro-vitamin D to vitamin D (Holick 2004).

The energy of UV rays is used to initiate some useful chemical reactions, for example modern dentistry uses a small UV lamp to initiate the hardening of tooth fillings.

There are many industrial examples of the use of the energy in UV and X-rays, such as curing of surface coatings. X-rays and gamma rays are in general to be avoided by living tissue. These are high-energy, skin- and bone-penetrating forms of radiation. X-rays for medical purposes are strictly controlled in terms of dose-rate and area of exposure. Their energy is damaging to living tissue at any level, and even with the latest technology there is a finite risk of chemical changes in DNA.

Gamma rays and X-rays are produced (along with enormous amounts of light, UV, and infrared radiation) in nuclear explosions. Gamma rays are even more energetic and damaging to living tissue than are X-rays. The tragedies of Hiroshima, Nagasaki, and Chernobyl attest to this.

Electrical energy is familiar to us all, and details of its generation will not be discussed here. Electrical energy has its origins in electromagnetic force. It is sufficient to say that the electricity we are familiar with in the home consists of a flow of electrons along a wire. The analogy to electricity flow is the flow of water in a pipe. The characteristics of interest are the pressure (the intensity factor) of water and the quantity (the extensive factor) of water flowing in a given time. In electrical energy flow, the voltage is a measure of the potential for the electricity to flow (analogous to water pressure in the pipe). The direction of flow is from higher voltage (intensive; analogous to the flow of water from higher to lower pressure) to lower voltage. The current (measured in amperes, or amps) is a measure of the number of electrons that flow per unit time (an extensive property; analogous to the quantity of water flowing in a pipe).

The primary sources of heat energy that drives steam turbines to produce electricity are coal, gas, and oil, with lesser contributions from hydroelectric, solar, nuclear, wind, tidal, and geothermal energy. Electrical energy is also produced and utilized by biological systems, as we shall see later, although too much electrical energy, indeed too much energy of any kind, may be dangerous to living organisms.

3.3 The strong nuclear force

This is involved in maintaining the stability of atomic nuclei by holding protons and neutrons together. The electromagnetic force tends to be repulsive in the nucleus—the positive charges on protons mutually repel. The interplay between the repulsive electromagnetic force and the attractive strong nuclear force largely determines the stability of atomic nuclei.

3.4 The weak nuclear force

This is involved in what is termed beta decay. A nucleus that does not have the optimum ratio of protons to neutrons may undergo beta decay, during which, depending on whether neutrons or protons are in excess, (i) a neutron will change to a proton plus an electron or (ii) a proton will change into a neutron plus a positron (a positively charged electron). The electron or positron created is ejected from the nucleus.

In general, unstable nuclei undergo radioactive decay, an example of a nuclear reaction, to form nuclei of more stable, lighter atoms, with the emission of electromagnetic radiation and/or subatomic particles. Other nuclear reactions can be brought about by bombardment of atoms with particles such as neutrons in a machine made for such experiments. An example is the neutron bombardment of cobalt-59 to form cobalt-60. The isotope cobalt-60 is radioactive with a half-life of 5.3 years. It is used as a source of gamma rays for research, cancer therapy, and industrial radiography.

Some of the naturally occurring radioactive isotopes such as carbon-14, uranium-238, and potassium-40 have useful dating applications in geology, biology, chemistry, and archaeology (Plimer 2001). Samples consisting of such nuclei decay at a constant rate, characterized by their half-life ($t_{1/2}$). For carbon-14, $t_{1/2} = 5.73 \times 10^3$ years. This means that in 5730 years half the original carbon-14 will remain, and after another 5730 years ½ × ½ = 1/4 will remain, and so on. This constant decay rate, plus a knowledge of carbon isotopes and their distribution in living and dead organisms, allows the time at which an organism ceased living to be determined.

Potassium-40, which captures an electron to become argon-40 ($t_{1/2} = 1.25 \times 10^9$ years) and uranium-238, which decays to lead-206 ($t_{1/2} = 4.47 \times 10^9$ years) are important for dating in geochemistry (Chang 1994).

The decay of naturally occurring radioactive nuclei is responsible for a large amount of heat generated inside the Earth, and this contributes to the fluidity of magma and the resulting convection currents that help to move tectonic plates and continents.

Nuclear reactions can involve the release or absorption of large amounts of energy. On the other hand, chemical reactions do not involve atomic nuclei directly, rather they occur as a result of interactions between electrons in the outermost shells of atoms and molecules. Relatively small amounts of energy are involved in chemical reactions compared to those in nuclear reactions. A nuclear explosion is a familiar, but thankfully rare, example of the conversion of matter (mass) into other forms of energy. By its very existence, any type of matter possesses energy, bound up in the nuclei of its atoms. Usually we are not aware of this, as the energy is not obviously available from the nuclei of the most common atoms, which tend to be quite stable. The famous equation derived by Albert Einstein is probably the most widely known in science:

$$E = mc^2 \tag{3.1}$$

where E is the energy, m the mass, and c the velocity of light (about 300,000,000 m s^{-1}). The nuclei of certain atoms, called isotopes, of some heavy elements such as uranium-235, undergo the process of nuclear fission, which involves splitting of the atomic nucleus into two or more fragments, with the emission of two or three nuclear particles. The point to note is that the mass of the end products of nuclear fission is ever so slightly less than the mass of the starting materials. Where does this mass go? In accordance with the energy conservation law, some of the mass in the decaying nuclei is converted to other forms of energy. Just how much can be calculated quantitatively using eqn (3.1). Fission is the type of nuclear reaction involved in the Hiroshima bomb. This atomic bomb was constructed in such a way that a small amount of nuclear mass was converted very rapidly to an enormous amount of energy, which is why the Hiroshima explosion was so destructive. We can calculate the amount of energy released by the conversion of mass to other energy forms by using Einstein's equation.

Thus, 1 g (0.001 kg) of matter converts (using $E = mc^2$) to $0.001 \times 300,000,000 \times 300,000,000 = 90,000,000,000,000$ kJ (9×10^{13}) kJ. We saw above that the specific heat capacity of water was 4.184 J K^{-1} g^{-1}, that is the amount of energy required to raise 1 kg of water by 1 degree Kelvin (or Celsius) is 4.184 kJ. Thus, 418.4 kJ is needed to raise 1 kg of liquid water at 0°C to liquid water at 100°C (4.184 × 100). If all the energy released by the conversion of our 1 g of matter could (it usually can't) be used to heat water, it would raise $9 \times 10^{13}/418.4 = 2.15 \times 10^{11}$ kg of liquid water from zero to 100°C. This is an enormous quantity of water, approximately 0.4 Sydarbs or 40% of the volume in Sydney Harbour (a Sydarb, to use a facetiously named measure of water in Australia, is 562 gigalitres or 5.62×10^{11} L) or some 1300 times the volume (1.68×10^8 L) passing over Niagara Falls in a minute. The lesson: a small amount of mass may be converted into a large amount of energy, under certain (restricted) conditions. The $E = mc^2$ equation rearranges to $m = E/c^2$, so that an amount of energy E can be transformed into matter with mass m. This illustrates the point that matter and energy, individually, are not conserved. What is really conserved is called mass-energy. Mass and energy are interconvertible, but total interconversion takes place in special cases only, for example in the early moments of the Big Bang. In a sense, matter can be considered as condensed energy—radiant energy that condensed, eventually to form atoms, as the Universe expanded and cooled after the Big Bang (Ferris 1997).

On the other hand, every time a chemical process results in a release of energy E, there is some reduction in mass, m, equal to E/c^2. This reduction in mass is exceedingly small on the human scale, as c^2 is so large. The effect can essentially be ignored for ordinary chemistry, as it cannot be measured with conventional chemical balances. For example, for the reaction

$$H_2 + \tfrac{1}{2}O_2 \longrightarrow H_2O$$

The heat evolved (E) is 241.8 kJ mol^{-1}. Using $m = E/c^2$ this corresponds to a mass decrease of 2.7×10^{-12} kg. This is essentially undetectable by conventional balances (Laidler 1993). Nuclear energy is also involved when uranium-235 is used in commercial energy production. Controlled nuclear fission is used in nuclear power stations. Neutrons, of which three, on average, are released during nuclear fission of uranium-235, are particles that cause an accelerating chain reaction when they hit nearby uranium atoms. If uncontrolled, this chain reaction could in some circumstances lead to an explosion or certainly a meltdown. The numbers of neutrons used in a gas-cooled nuclear power reactor are carefully controlled in the reactor by boron rods that can be precisely inserted and withdrawn. The boron in the rods absorbs neutrons, thus controlling the number of neutrons that penetrate the uranium fuel of the reactor. Instead of an accelerating chain reaction, a self-sustaining sequence of reactions occurs, a controlled nuclear chain reaction. The critical level is not reached and no meltdown (or explosion) can take place. The heat energy produced is used to drive steam turbines connected to conventional electric generators. The gamma rays and X-rays are unwanted side products, which need to be absorbed by special materials to prevent their release into the environment. Unfortunately, the nuclear fuel spent as well as some of the absorbing materials remain dangerously radioactive, emitting their penetrating life-threatening radiation for up to thousands of years. Their safe disposal continues to cause problems around the world.

Another type of nuclear reaction is nuclear fusion. In this type of reaction atomic nuclei are forced together to form nuclei of a different element. Probably the most familiar nuclear fusion reaction is the one involved in a so-called hydrogen bomb. This is also what takes place in the Sun, where the temperature is such that in a series of reactions four hydrogen nuclei are forced together to form a helium nucleus. One helium nucleus has a mass slightly less than the sum of four hydrogen nuclei (by about 0.7%). The energy conservation law dictates that energy must be conserved in the process, so the extra mass ($m = E/c^2$) is converted to various forms of radiant energy: heat, visible light, radio waves, infrared, X-rays, and gamma rays. Again, the fusion bomb generates products that have long-lasting, deleterious effects on living tissue, although the products in this case are of lesser concern than those from nuclear fission of uranium-235. As with controlled nuclear fission, there is the corresponding possibility of controlled nuclear fusion. This has been attempted by scientists, but as yet has not been achieved in a manner that would allow commercial development of nuclear fusion power stations. The attraction of nuclear fusion is in the greater energy yields and the possibility of lower amounts of less dangerous by-products. Hydrogen is cheap and safe. Work, as the saying goes, is still in progress.

There is a prime example of controlled nuclear fusion that is essential to the existence of life on Earth—the Sun. The fusion of hydrogen to helium in the Sun is what generates the radiation we receive as sunlight. The protesters against the nuclear energy industry who produced signs saying 'Solar not Nuclear' clearly did not understand the origins of sunlight. A more appropriate sign would have read 'Solar *is* Nuclear' but one does appreciate what the protesters were trying to achieve.

Finally, a single object may possess a number of the above energy forms simultaneously. A block of wood thrown into the air will possess kinetic energy because of its motion, gravitational potential energy because of its height above sea level, chemical energy in the wood (which can be burned), heat energy depending on its temperature, and nuclear energy in its atoms (this latter is not readily available from our block of wood, although the other forms of energy may be). The above discussion is concerned with forms and sources of energy in general. What about the energy forms involved in living organisms? Many of the above energy forms are involved in one way or another, but later we will need to consider particularly the conversion of solar (radiant) energy to chemical energy in green plants, and the further conversion of that chemical energy to other forms in animals and fungi that use the plants as foods. In particular, many land animals and plants need to stand upright and have developed specialized structures to resist the force of gravity. Animals use various types of skeletons to do this, and to attach muscles for movement of themselves and other objects. Land plants tend to use cellulose, a large structural molecule made up of thousands of glucose units linked together and formed into specialized, living cell walls. Cellulose, the most abundant biological chemical on Earth, is produced in countless millions of tonnes by plants each year. It is also consumed in similar amounts by animals as food, by humans for a huge number of purposes (a cause of concern to conservationists), and by the natural biological processes involving decay of dead plants and animals (a normal and necessary part of the great recycling of natural resources). All these activities are possible only because of the (mainly) solar energy harnessed and focused by the unique processes of life.

After all this discussion, some people will probably ask, 'What *is* energy?'. Energy is not a substance, an object, or a thing, rather it is a property of all substances, objects, things, and people—a property of all matter. For example, a property of each of us as individual objects is our weight. Usually we are not greatly concerned with the absolute value of our weight. Having decided on some reasonably objective grounds that a certain weight is desirable, we are mainly concerned with changes in our actual weight that take us closer to or further from our ideal. The relationships between the available energy in a meal (in kilojoules) and the amount of weight we are likely to gain from eating that meal (in kilograms) are well known, thanks to quantitative nutritional studies. Increasingly, nutritional information, including energy values, is being included on food packaging. This information allows us to answer questions such as:

How much energy do we burn during an hour of exercise?

How much energy do we take in when consuming a serving of potato chips?

How far and how fast must we run to burn off the kilojoules added after the consumption of a large meal? (A typical fat will yield about 2.5 times the energy as the same weight of a typical carbohydrate. Reasons for the different energy values will be discussed in a later chapter.)

How much electricity can be produced from burning a tonne of coal considering all the energy conversions and heat losses in the process?

How much of a certain explosive will be needed to demolish (safely) a large derelict building?

The answers to all these questions are calculable with considerable accuracy, thanks to many studies in the physical and biological sciences carried out over the last 250 years.

REFERENCES

ARPANSA (2008) *Radiation Protection. Ultraviolet Radiation.* Australian Radiation Protective and Nuclear Safety Agency. http://www.arpansa.gov.au/radiationprotection/Basics/uvr.cfm.

Chang, R. (1994) *Chemistry,* 5th edn. McGraw-Hill Inc., New York, pp. 913–4.

Ferris, T. (1997) *The Whole Shebang.* Weidenfeld & Nicolson, London, Ch. 1, p. 32.

Gribbin, J. (2002) *Science – A History.* Allen Lane/Penguin, London, Ch. 3, p. 95.

Health Thematic Guide (1996) Human Health Overview. *CIESIN Thematic Guide.* http://www.ciesin.org/TG/HH/hh-home.html.

Holick, M.F. (2004) Vitamin D: importance in the prevention of cancers, type 1 diabetes, heart disease, and osteoporosis. *Am. J. Clin. Nutr.* **79**(3), 362–71.

Laidler, K.J. (1993) *The World of Physical Chemistry.* Oxford University Press, New York, Ch. 3, pp. 98–9.

Plimer, I. (2001) *A Short History of Planet Earth.* ABC Books, Sydney, pp. 48–54.

Sagan, C. (1997) *Billions and Billions.* Headline Publishing, London, Ch. 2, p. 23.

4
Which Way? An Introduction to Thermodynamics

Thermodynamics can provide us with broad information of literally universal application. At first sight it may seem too general to be of interest to biologists, dealing with concepts that appear to be a long way from the interactions of molecules in living matter. It can appear abstract, hard to understand, and mathematical. On the other hand, to dismiss it completely is to ignore the processes that underlie the very origins and maintenance of life in the universe. Isn't that enough to justify the study of a little thermodynamics?

Thermodynamics helps us to understand how it is possible to have localized regions of order, for example life, while simultaneously energy in the universe as a whole is becoming more dispersed, more 'disordered'. We are told that the universe is 'running down' in energetic terms, and that eventually its energy will be spread out, dissipated, degraded in intensity, and all that will be left will be cold, dark space and matter. The material I treat here will be helpful when we try to explain later how all these properties of energy come together (or spread out) to keep us alive.

Chemists towards the end of the nineteenth century wanted to know whether or not it was possible to predict the direction of a particular chemical reaction. They knew from the First Law that whichever way a reaction went, energy would be conserved. What was not known was how to predict whether or not it would proceed spontaneously in a particular direction. Such knowledge would be of practical importance, but also would add to the growing theoretical understanding of chemistry. As we will see, it turned out to be possible not only to predict spontaneity, but also to allow quantitative measurement of a number of other very useful functions.

The term thermodynamics was introduced by William Thomson, later Lord Kelvin (1824–1907) in 1849. Thermodynamics as a science was able to begin at that time only because the confusion between heat and temperature had been clarified, and because heat was finally recognized as a form of energy, rather than as an actual substance. As thermodynamics developed it was applied to the study of heat engines, mostly to steam engines, and later to internal combustion engines. Many of the terms used in thermodynamics reflect these origins.

Thermodynamics as it applies to chemistry and thence to biology was developed later, and will be discussed further in Chapter 6.

Thermodynamics is the part of science that deals with all aspects of the interconversion of heat and work. Expressed more generally, it deals with the energy changes (fluxes) involved in processes, and includes all the forms of energy mentioned in earlier chapters. In nearly all cases we will discuss processes involving changes in rather than absolute amounts of energy. In the case of heat, for example, chemists, engineers, and biologists are usually interested in the quantity of heat transferred during a process, rather than in the absolute amounts of heat in the system. When heat is transferred, some of it, but not all as we will see, can be converted to work. For biologists, knowledge of heat production (e.g. in metabolism by biochemical processes) and transfer (e.g. heat gain from and loss to the environment) can be of great use. The other forms of energy can also be studied by the application of thermodynamics.

The relationship between work and energy was introduced in Chapters 1 and 3. As energy is defined as the capacity to do work, both work and energy are expressed in the same units, joules. Next, it is essential to understand the scientific distinction between work and heat.

Heat involves the transfer of energy via thermal motion, which can be described as disorderly molecular motion. Temperature difference determines the direction of heat flow—always from the higher temperature to the lower. To look at it in the opposite sense, the direction of heat flow can be used to define which of two bodies has the higher temperature. When we talk of heat transfer, differences in temperature are *always* involved, although this might not be mentioned explicitly. Heat will flow between two bodies, or between our system and the surroundings, until they reach the same temperature. At this point the bodies or the system/surroundings are said to have reached thermal equilibrium. At the molecular level, thermal motion, a form of kinetic energy, is passed from the molecules in the hot body to those in the colder body until the average thermal motions of their molecules are equal, that is their temperatures are equal. Temperature is a measure of the average kinetic energy of the molecules in the body being measured.

There are three types of thermal motion:

- *translational*, or movement of individual molecules from place to place, as in the movement of molecules of a gas or a liquid, by processes such as diffusion or convection currents.
- *vibrational*, or movement of atoms in molecules joined by chemical bonds, rather like weights joined by a spring, and vibrations of molecules in a solid about their average positions, for example water molecules in solid ice or iron atoms in an iron rod.
- *rotational*, or the spinning motion of molecules around one or more axes in three-dimensional space.

Within certain limits, these motions (equivalent to thermal energy) can be transferred between neighbouring molecules until their average thermal motions are equal and thermal equilibrium is achieved.

Work is the transfer of energy via organized motion (Figure 4.1). When a piston moves as the result of the expansion of a gas in the cylinder, work is done by the system. The effect of the expansion is coordinated and 'purposeful',

50 Introducing Biological Energetics

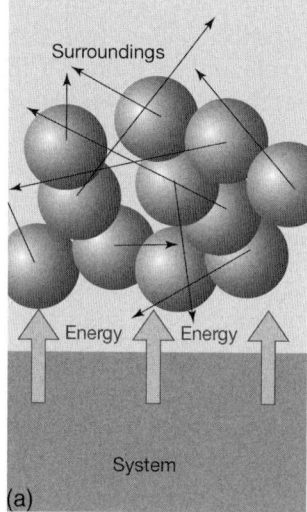 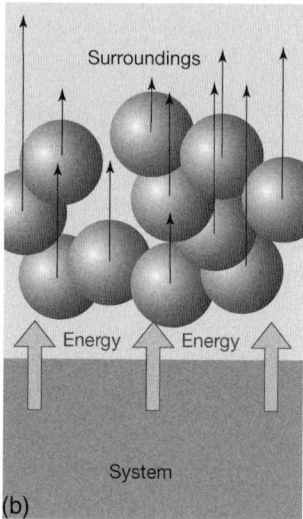

Figure 4.1. (a) Heat involves energy transfer via disorderly, uncoordinated motion and collisions of molecules. Heat transfer does not result in any useful work on the surroundings. (b) Coordinated motion of molecules that results in the performance of work on the surroundings. (From Atkins, P.W. and de Paula, J. *Atkins' Physical Chemistry*, 7th edn, p. 32. © 2002 Oxford University Press.)

acting in a specific direction. Work is done by the system on the surroundings. A weight can be lifted, or a crankshaft turned, in an organized manner. On the other hand, a weight added to the piston would compress the gas, and work is done on the system. Work is also done when the electrons in an electric current are passed in an orderly way around a circuit to drive a motor, and when gravitational potential energy in a weight is used to drive a post into the ground. Certain chemical reactions do work when they rapidly generate gas to propel a shell or bullet. In this case, the source of energy that is tapped is the energy stored in chemical bonds. Chemical reactions can also do work when transferring electrons in what chemists call redox reactions, or when breaking/rearranging chemical bonds in living cells.

Several times in this chapter I use the term reversible. Reversible in thermodynamic terms has a precise and special meaning, but first we need to clarify the meaning of the term equilibrium. A system is said to be in equilibrium with its surroundings if an infinitesimal change in conditions in one direction causes an opposite change in the state of the system.

Here are some examples. Two systems in contact at the same temperature are said to be in thermal equilibrium. The transfer of energy between the two systems is said to be reversible because if the temperature of either is lowered infinitesimally, energy (heat) will flow into the lower-temperature system. If the temperature of either system is increased, heat will flow from the hotter to the cooler system. Similarly, if a gas in a cylinder fitted with a piston has a pressure exactly equal to the external pressure, it can be said to be in mechanical equilibrium with its surroundings. An infinitesimal change in the external pressure in either direction will cause a change in the volume of the gas in the opposite direction. An increase in pressure will cause a decrease in volume and vice versa (Atkins and de Paula 2002). Both the above are examples of thermodynamically reversible processes.

'Reversible processes are finely balanced changes, with the system being in equilibrium with its surroundings at every stage. Each infinitesimal step along a reversible path is reversible, and occurs without degrading the quality of the energy, without dispersing energy chaotically'. (Atkins 1986)

As will be seen, reversible processes are important in the theory of cyclic thermodynamic processes. An example to be discussed is the Carnot cycle, which was important not only in developing the theory of efficiency of heat engines, but which also led to the important concept of entropy. Although some of the following discussion about heat engines may seem a long way from biochemical reactions, the two seemingly unrelated processes are subject to the same laws of thermodynamics. These laws, as I have said previously, apply to all processes and systems that involve energy changes. Let's get a little more formal, with a few definitions. Definitions can be tedious, but are necessary so that there can be no confusion about what is being discussed. The same can be said about defining the conditions under which processes are being studied.

It is important to note that if we are studying thermodynamics, both the system and the surroundings must be considered. A system is the portion of the universe that we are concerned with in the example being studied, say a reaction flask or an experimental rat. The surroundings are the region outside the system where experimental measurements are made. Strictly speaking, the surroundings comprise the rest of the universe. Fortunately, for most purposes consideration of the nearby surroundings, such as the laboratory, are sufficient. We may measure temperature changes or the work being done by our system on the surroundings or by the surroundings on our system. There is often an identifiable boundary between the system and the surroundings, such as a cell membrane, skin, or a glass wall.

An open system is one where both energy exchange and matter exchange between the system and the surroundings can take place. Chemical reactions are carried out in the laboratory in an open flask, and all living organisms from elephants to bacterial cells are examples of open systems. This being the case I will try to use open systems in examples.

A closed system is one that can exchange energy with the surroundings, but not matter. An example of a closed system is a sealed glass flask in a laboratory containing the ingredients of a reaction. Energy, for example heat and light, can pass through the glass in both directions, but there can be no exchange of matter through the boundary, in this case the sealed flask walls. The Earth itself is very close to being a closed system, as the exchange of matter between it and space is minute on the global scale. On the other hand, there are vitally important exchanges of energy between Earth and space and vice versa.

Finally, an isolated system exchanges neither matter nor energy with the surroundings. Isolated systems are of some theoretical interest, as we shall

52 Introducing Biological Energetics

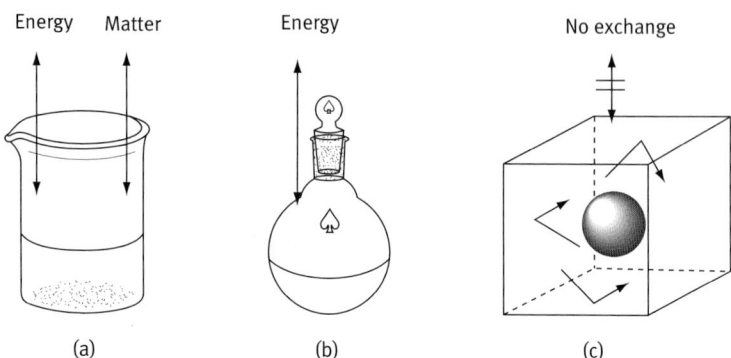

Figure 4.2. Examples of (a) open, (b) closed, and (c) isolated systems. The water plus the beaker is an open system. There are two interfaces, water/glass and water/air, forming the boundaries between the system and the surroundings. Both energy, e.g. heat, and matter, e.g. water vapour, (evaporating out) or air (dissolving in) can be exchanged between (a) and the surroundings. The sealed flask (b) can exchange energy, e.g heat and light with the surroundings, but not matter. The isolated system (c) can exchange neither energy nor matter with the surroundings.

see, but in practice isolated systems do not occur. The only truly isolated system is the whole universe, but that does not help us in any practical way (Figure 4.2).

The need to include the surroundings involved can make some energy studies difficult, but to be rigorous we must do so. I don't wish to pursue thermodynamic rigour too far, but it would be remiss, sometimes misleading, of me not to acknowledge it from time to time. In both open and closed systems there is the possibility of energy exchange either way through the boundary. When such an energy exchange occurs between the system and the surroundings as a result of a temperature difference between them, the energy has been transferred as heat.

This is one definition of heat. When heat flows from the system to the surroundings, the process is called exothermic. An endothermic process is one where heat flows from the surroundings to the system. The terms are probably familiar to anyone who has studied chemistry. Chemists speak of endothermic reactions, that is ones absorbing heat, and exothermic reactions, which give out or liberate heat. The alternative terms exergonic and endergonic are derived from an older unit of energy, the erg. Here, we use the terms exothermic and endothermic in the general sense, applied to all systems and processes, chemical or otherwise. Having defined systems and boundaries, we are in a position to point out that heat transfer is a boundary process, which can be measured as a change in temperature across a boundary between two systems at different temperatures. Work is also a boundary process, for example organized transfer of energy via a movable boundary such as a piston in a cylinder. With the above definitions comfortably out of the way, we are ready to look at ways in which the laws of thermodynamics apply to life.

The First Law of thermodynamics is a statement of the energy conservation law that we discussed in Chapter 1. Energy can change from one form to another during a process, without overall loss. The total energy of the universe remains constant, but can be distributed in many ways between the system and the surroundings. Both heat and work are nearly always involved in energy changes.

To state the First Law in terms of the system and surroundings, where E stands for energy:

$$E_{\text{system}} + E_{\text{surroundings}} = E_{\text{universe}} = \text{a constant}$$

To be able to apply the laws of thermodynamics quantitatively, the concepts and definitions of energy and work that we have carefully discussed are necessary. There is another type of energy that now needs to be introduced: internal energy.

Internal energy is defined as the energy within the system. It has been given the symbol U: in chemistry its unit is kilojoules per mole (kJ mol^{-1}). What forms of energy contribute to U?

Here, U is restricted to those forms of energy involved in chemical processes: bonding, non-bonding, translational, vibrational, and rotational. Bonding and non-bonding energy have their basis in the energy associated with electrons, that is electronic energy. The internal energy U is an extensive property, in that it depends on the quantity of matter (as mass or moles) in the sample under consideration. It is also classified as a state function. This means that a measured value of U depends only on the current state of the system, and not on how the state was reached. See Box 4.1 for more on state functions.

Box 4.1 State functions

State functions (state variables) are represented by capital letters, for example the use of U for internal energy. Several state functions/state variables are required to describe the state of a system. Those listed in the table will be introduced progressively through this chapter.

State function	SI unit	symbol
Volume	litre/L	V
Pressure	pascal/Pa	p
Temperature	Kelvin/K	T
Internal energy	kJ mol^{-1}	U
Enthalpy	kJ mol^{-1}	H
Entropy	J K^{-1} mol^{-1}	S
Gibbs energy	kJ mol^{-1}	G

The state functions of a system do not depend on how the system reached its current state. The route to the current state, however indirect, consisting of many steps or a few, is irrelevant for state functions. This has been confirmed experimentally many times for many systems. Another way of saying this is: all state functions are path independent. An important extension of this is that if our system is taken from a certain initial state (i.e. we know the initial values of the state functions) through a cycle of changes (e.g. pressure changes,

Box 4.1 (continued)

volume changes, temperature changes) and returned to the initial state, the sum of each state variable over the entire cyclic process is zero. Expressed mathematically for volume $\Delta V = 0$. Thus for a four-step process, $dV_1 + dV_2 + dV_3 + dV_4 = 0$. The same applies for the other state variables. Thus, for a cyclic process $\Delta U = 0$, $\Delta T = 0$, $\Delta H = 0$, $\Delta G = 0$, and $\Delta S = 0$.

Importantly, on the other hand, work (w) and heat (q) are not state variables. They *are* dependent on the pathway taken, so Δw and Δq do not equal zero after the system has undergone a cyclic process.

Here is an analogy. A climber may reach the top of a mountain by any of a number of routes. If he is inexperienced, he might take the longer, less steep route. Later he might try to go straight up a vertical face. In each case, the altitude reached, and hence the gravitational potential energy gained, will be the same. Its quantitative value is path independent. State functions are of the same nature—path independent.

One cannot say the same for the amount of work done or the heat expended by the climber. The longer, slower easier route will almost certainly take more work overall, and will probably generate a little more heat over the longer time. The shorter, vertical route will take less work and probably generate less heat overall. However, as the work was (probably) done in a shorter time, the power used (power = work done/time taken) would be greater, and the climber would heat up more *quickly*.

Work done and heat produced in any process depend on the route taken. They are not properties of the system, so they are not state functions.

By convention work and heat are represented by the small letters w and q, respectively. I confidently assert that the path-independence property of state functions is very useful, but rather than detract from the current argument, I will leave the justification until later.

Having whetted your appetite by introducing the new concept internal energy, I must now admit that U cannot be measured directly. It must be calculated from other measurable properties of our system. We actually measure changes in U, or ΔU. This should come as no great surprise, as I have said several times already that we are mainly concerned with changes or fluxes in energy during a process, rather than with absolute values.

The internal energy of a system can be increased in two ways:

1) by transferring heat to it, or
2) by doing work on it.

Mathematically this can be written as:

$$\Delta U = q + w \tag{4.1}$$

where q is the heat transferred and w is the work done on the system.

Now we have achieved something useful, a quantitative statement of the First Law in terms of two variables. This is useful as it is possible to measure both q and w. The internal energy can also be decreased, so to make it clear about whether q and w are being transferred to or from our system, the following conventions have been adopted:

- If heat is transferred to the system, q is positive (e.g. an endothermic chemical reaction).
- If heat is transferred to the surroundings, q is negative (e.g. an exothermic chemical reaction).
- If the system does work on the surroundings, w is negative (e.g. the system undergoes expansion against an external pressure).
- If the surroundings do work on the system, w is positive (e.g. the system is compressed by an external pressure).

Let's look at an example. We have seen in Chapter 3 that work = force×displacement, or $w = F\Delta x$, where force is F and Δx is the displacement. If we consider our system as doing work on the surroundings, by the above convention w is negative, so speaking generally,

$$w = -F\Delta x \qquad (4.2)$$

If we take as our specific example the expansion of a gas against an external pressure, pressure is analogous to force and volume is analogous to displacement, so

$$w = -p\Delta V \qquad (4.3)$$

where p is the external pressure and V is the volume of gas.

At this point I need to introduce yet another thermodynamic state function, enthalpy. Enthalpy is given the symbol H.

Enthalpy is defined as the heat absorbed by a system undergoing a process at constant pressure (strictly speaking, so long as the system does no additional work (Atkins and de Paula 2002)).

Enthalpy measurements can be used to quantify the heat flow into and out of a system in a constant-pressure process. Considering the discussion above, especially the expression for the First Law as $\Delta U = q + w$, you can see why I introduce enthalpy at this point. The need for me to introduce it at all will soon become apparent.

Using eqn (4.1), and q_p to remind us of the constant external pressure, the First Law can be written as

$$q_p = \Delta U - w \qquad (4.4)$$

The external pressure being constant, when our system expands and the only work (w) done is pV work, then using eqn (4.3),

$$q_p = \Delta U - (-p\Delta V) = \Delta U + p\Delta V \qquad (4.5)$$

Thus, the quantity $(\Delta U + p\Delta V)$ is the heat absorbed at constant pressure, q_p.

The heat absorbed at constant pressure (q_p) is defined above as the enthalpy, H, so

$$\Delta H = \Delta U + p\Delta V \qquad (4.6)$$

or

$$H = U + pV$$

As enthalpy is a state function the above development of eqn (4.6) leads to the important conclusion:

At constant external pressure, the amount of heat exchanged during a process is independent of whether or not the overall process occurs directly (in one step) or indirectly (in a number of steps).

This statement is also known as Hess's Law of heat summation, after G.H. Hess (1802–1850), a Swiss chemist who worked mainly in Russia. Hess proposed the law on the basis of a number of chemical experiments; it has since been confirmed by many other workers. The derivation of eqn (4.6) using thermodynamics adds credibility to Hess's law, which was derived previously by experiment, rather than theoretically. This theoretical derivation gives us confidence that Hess's proposal is really a law, in the scientific sense.

If we apply Hess's law to chemical reactions, a useful way to state it is:

'When reactants are converted to products, the change in enthalpy is the same whether the reaction takes place in one step, or in a series of steps.' (Chang 1994)

Expressing this mathematically:

$$\Delta H_{reaction} = H_{products} - H_{reactants} \qquad (4.7)$$

If we can measure $H_{products}$ and $H_{reactants}$ for a chemical reaction, eqn (4.7) can be used to calculate the ΔH of the reaction. The implications may not be immediately obvious, so let's look at this in more detail.

As most biochemical reactions occur at constant pressure, it would be useful to be able to measure the enthalpy change, ΔH, of chemical reactions that take place at constant pressure. This can be done using relatively simple apparatus, consisting of a well-insulated calorimeter with stirrer 'open' to the atmosphere and of known heat capacity, and fitted with a thermometer. A chemical reaction is carried out in, say, water, and the temperature rise is measured. From the temperature rise and heat capacities of water and the calorimeter, it is a simple matter to calculate the heat of reaction. As the reaction was carried out at constant pressure, the heat of reaction will be equal to the enthalpy change, ΔH, for example:

- heat of neutralization: $HCl + NaOH \longrightarrow NaCl + H_2O$ $\Delta H = -56.2$ kJ mol^{-1} (exothermic reaction).
- heat of vaporization: $H_2O_{liquid} \longrightarrow H_2O_{gas}$ $\Delta H = 40.8$ kJ mol^{-1} (endothermic process).
- combustion reaction: $C_6H_{12}O_6 + 6O_2 \longrightarrow 6CO_2 + 6H_2O$ $\Delta H = -2801$ kJ mol^{-1} (exothermic reaction).

We can use the last example to show the importance of Hess's law of heat summation. Suppose the value of $\Delta H_{combustion}$ had been determined by a special calorimeter. The value so obtained would be valid to use in a table showing the calorific (energy) values of foods. This is despite the fact that $\Delta H_{combustion}$ was determined in a single-step combustion experiment, whereas the complete metabolism of glucose to $6CO_2 + 6H_2O$ by a living cell takes place via a large number of biochemical reactions. As H is a state function, the value of ΔH is the same by both routes: -2801 kJ mol^{-1}, as seen in the glucose example. It would be impossible to determine calorific food values if we had to measure ΔH directly using experiments on humans. As we will see later, the ΔH value of -2801 kJ mol^{-1} is close to being the maximum amount of work that combustion or metabolism of glucose can produce. For various reasons, the actual amount of work done during typical animal metabolism is far less than that (about half).

Another useful aspect of Hess's law is that state functions and state function differences (ΔH and ΔU) are additive. This means that from a table of ΔH values, it is possible to calculate ΔH values for an actual reaction without having to do the reaction. This is the theoretical chemists' dream. To do this one obviously needs to get the data to form a ΔH table. A requirement for setting up data tables is that there must be some standard form in which to do the measurements and express the data. Just as there is the SI unit system, which defines the fundamental constants in agreed units, so there is a standard approach to defining enthalpies and other thermodynamic functions. Thus we have the standard enthalpy of formation, which is the reference point for all enthalpy expressions.

The standard enthalpy of formation, ΔH_f^0, is defined as 'the heat change when 1 mole of a compound is formed from its elements at a pressure of one atmosphere'.

The ΔH_f^0 values that appear in tables are usually those measured at 25°C. Even though temperature is not mentioned in the definition, enthalpy values vary with temperature. The use of the superscript ^{0} shows that the measurement was done under standard conditions, and f stands for formation. The units for ΔH_f^0 are kJ mol^{-1}.

Once we know the values for the standard enthalpies of formation it is possible to calculate the enthalpies of reactions using eqn (4.7):

$$\Delta H_{reaction} = H_{products} - H_{reactants}$$

We cannot do this directly, as it is not possible to measure absolute values of enthalpies such as $H_{products}$ and $H_{reactants}$, so we need a set of standard chemical reactions whose $\Delta H_{reaction}$ can be used as standards against which all other $\Delta H_{reaction}$ values can be measured. These are the ΔH_f^0 values that appear in the tables.

Why choose heats of formation as the standard? Any chemical reaction can be written as an algebraic combination of several formation reactions. Let's look at a specific example. Determine ΔH^0 for the reaction:

$$C_6H_{12}O_6 + 6O_2 \longrightarrow 6CO_2 + 6H_2O$$

It is possible to separate this reaction into a series of formation reactions and add them algebraically:

$$C(s) + O_2(g) \longrightarrow CO_2(g) \text{ eq. (a). } \Delta H_f^0 = -393.5 \text{ kJ mol}^{-1}$$

$$H_2(g) + \tfrac{1}{2}O_2(g) \longrightarrow H_2O(l) \text{ eq. (b). } \Delta H_f^0 = -285.8 \text{ kJ mol}^{-1}$$

$$6C(s) + 6H_2(g) + 3O_2(g) \longrightarrow C_6H_{12}O_6(s) \text{ eq. (c). } \Delta H_f^0 = -1277 \text{ kJ mol}^{-1}$$

We need to multiply equations (a) and (b) by 6 to have enough H, O, and C for 1 mole of glucose. The net equation for glucose oxidation is the sum of the equations (a) and (b) and the reverse of (c), so we must write:

$$6C(s) + 6O_2(g) \longrightarrow 6CO_2(g) \quad (6 \times \text{eq. (a). } \Delta H_f^0 = 6(-393.5) = -2361 \text{ kJ})$$

$$6H_2(g) + 3O_2(g) \longrightarrow 6H_2O(l) \quad (6 \times \text{eq. (b). } \Delta H_f^0 = 6(-285.8) = -1715 \text{ kJ})$$

$$C_6H_{12}O_6(s) \longrightarrow 6C(s) + 6H_2(g) + 3O_2(g) \quad (-1 \times \text{eq. (c). } -\Delta H_f^0 = -(-1277)$$
$$= 1277 \text{ kJ})$$

Net: $C_6H_{12}O_6(s) + 6O_2(g) \longrightarrow 6CO_2(g) + 6H_2O(l) \quad \Delta H^0_{(reaction)} = -2799 \text{ kJ}$

The value of $-2{,}799$ kJ compares very well with that of 2801 kJ derived directly from the combustion of glucose, illustrating the value of Hess's law. Looking at the reverse reaction, that is the formation of glucose from carbon dioxide and water, we have a summary of what happens in photosynthesis. Hess's law predicts that the amount of energy needed to produce 1 mole of glucose from its elements (C, H, and O) is $+2801$ kJ, whatever the mechanism might be. The reaction is highly endothermic, and the energy for photosynthesis comes from the Sun in the form of electromagnetic radiation in the first instance. The synthesis of glucose is certainly not direct, not a simple reversal of the net equation above, as we will see in Chapter 10.

The above discussion is all about enthalpy and heats of reaction, a section of chemistry known as thermochemistry. As shown by the above examples, changes in enthalpy are easier to measure than changes in internal energy U. The whole calculation process works in practice because the data in thermodynamics tables were derived by strict use of standard conditions and conventions, as agreed internationally. The somewhat tedious precision of definitions, and insistence on standard conditions, is wholly justified in this and other parts of science.

Where are we now overall in our discussion of thermodynamics? Our objective in this section is to derive a means to predict the spontaneity of a process, particularly that of a chemical reaction. The above discussion has shown us how to do some useful calculations about the energy of chemical reactions, but we still cannot answer two questions:

1) Will a given process or a particular chemical reaction occur spontaneously?
2) Will a decrease in energy between reactants and products alone guarantee that a chemical reaction will be spontaneous?

It would be very convenient if the answer to the second question was 'yes'. We would just need to calculate or experimentally determine the change in energy for a process, for example the enthalpy change. If there was a decrease, we could then say the process was spontaneous.

Unfortunately, this is not always the case. Certainly, many chemical reactions occur spontaneously and are also exothermic, giving out energy to the surroundings in the form of heat. The energy of the system will decrease as a result of the reaction. An example is the hot packs often used by athletes or arthritis sufferers. Anhydrous calcium chloride or magnesium sulphate is used in hot packs. Thus, when the calcium chloride makes contact with water:

$$CaCl_2(s) \xrightarrow{H_2O} Ca^{2+}(aq) + 2Cl^-(aq) \quad \Delta H_{solution} = -83 \text{ kJ}$$

Unfortunately for simplicity, some reactions that occur spontaneously are endothermic, absorbing heat from the surroundings, as in the case of a cold pack that uses ammonium nitrate:

$$NH_4NO_3(s) \xrightarrow{H_2O} NH_4^+(aq) + NO_3^-(aq) \quad \Delta H_{solution} = 26 \text{ kJ}$$

Thus, the First Law on its own cannot predict spontaneity. Another predictor of spontaneity is needed, preferably one that covers all possible cases.

We can't discuss the historical development of thermodynamics here. It was long and tortuous, and somewhat confusing even for many of the major contributors. It is of necessity quite mathematical. Physicists and physical chemists are applied mathematicians. They prefer to express their work using the rigorous methods of mathematics. This might be daunting to the rest of us, but the results can be profoundly insightful. In the mid-nineteenth century, the search for a predictor of spontaneity was a great challenge. At some stage, the idea of a new state function entered the debate. Why a state function? As outlined above, state functions are useful in that they are path independent. If they can be measured or calculated, state functions may be applied generally to provide information about the state of a system.

Another feature of state functions, such as U, is that if a process is carried out such that the system is returned to its original state, $\Delta U = 0$. This is always the case for state functions when we consider the whole process as a cycle returning to the original conditions. The overall change in the state function U, internal energy, is zero for such a cyclic process. This is despite the fact that during the process heat changes (Δq) and work changes (Δw) are involved in the overall cyclic process. The process may be carried out in several ways, each involving different values of q and w because q and w are not state functions, being path dependent. The overall result for the cyclic process will always be the same for state functions, so $\Delta U = 0$.

The idea of a cyclic process was introduced for the case of heat engines by the French engineer Sadi Carnot (1796–1832). Carnot was studying the efficiency of heat engines. He realized that heat engines could only work if two temperatures were involved, say a higher T_h and a lower T_c. Heat is extracted from a reservoir at T_h, used to do work, and delivered to a heat reservoir at T_c. Then the engine has to be returned to the original conditions, ready to carry out another cycle of work. Not only that, but the process must be reversible. This is analogous to our description of a state function that goes through a cycle and returns to its original conditions, making $\Delta U = 0$.

Carnot wanted to know theoretically the maximum work (expressed as the highest achievable efficiency) that could be extracted from a heat engine working between the two temperatures. In a book published in 1824, he showed that

$$\text{efficiency} = \frac{1 - T_c}{T_h} \text{ or efficiency} = \frac{T_h - T_c}{T_h}$$

where the temperatures are in Kelvin. Immediately, this tells us that no heat engine can be 100% efficient. This would require $T_c = 0$ K, in which case, efficiency $= T_h/T_h = 1 = 100\%$.

Absolute zero or 0 K is unachievable in practice. In reality, it can be seen that the higher the T_h and the lower T_c, the closer efficiency comes to 1, or to 100%. Let's put some numbers into the efficiency equation, say 400 K for T_h and 300 K for T_c:

$$\text{efficiency} = \frac{400 - 300}{400} = 0.25 \text{ or } 25\%$$

This is the maximum theoretical efficiency for an ideal reversible heat engine, not allowing for real losses of heat or friction losses. The efficiency of heat engines is largely determined by thermodynamics, and clever design plays a minor role. Heat is quite an inefficient source of work unless the temperature between the engine and its surroundings ($T_h - T_c$) is large; the larger the difference, the more efficient the engine. You can't drive a heat engine with warm water, even an ocean of it. For a heat engine, the lesson is to use high-pressure steam at the maximum practicable temperature as the high-temperature reservoir, and the coolest temperature practicable for the cold reservoir.

In the case of a diesel engine high temperature and high compression make for high efficiency. Rudolph Diesel (1858–1913) knew his thermodynamics! His engines achieved approximately twice the efficiency of contemporary steam engines. Even today, diesel-type engines, which work at high temperatures and pressures, are the most efficient heat engines for motor vehicles. Their fuel-saving virtues are promoted as helping, at least to some extent, the economy and the environment, and reducing the rate of consumption of our fossil fuels.

How does all this help in the search for a new thermodynamic function to test for spontaneous processes?

Carnot's eponymous cycle involved temperature and a reversible process involving heat transfer, where the starting conditions are left and finally returned to at the end of the cycle. We have said that if a process starts and stops in the same conditions, the overall changes in state functions for the process total zero. Carnot's ideas sparked activity in the area from researchers such as Kelvin himself and physicist Rudolf Clausius (1822–1888).

In 1854, Clausius showed, by detailed analysis of the Carnot cycle, that for an engine operating reversibly,

$$\int \frac{dq_{rev}}{T} = 0$$

where dq_{rev} specifies that the heat q must be for a reversible process. The temperature, T, is the on the absolute (Kelvin) scale. Clausius realized that the quantity dq_{rev}/T had special significance. It had the form of a state function, in that its integral over a full cycle = 0, that is $\int \frac{dq_{rev}}{T} = \frac{\Delta q_{rev}}{T} = 0$.

Heat alone, q, is not a state function, and over a full cycle Δq does *not* equal 0.

Clausius worked and published in the area over the following years, and in 1865 produced a paper of great importance in which he introduced the term entropy, to which he gave the symbol S. The term is derived from *en*, in, and *trope*, transforming. Clausius wanted the new word to be as close as possible to the word energy, as they are so closely related. Entropy was a new state function (Laidler 1993).

Clausius formally defined the infinitesimal change in entropy dS by the relationship:

$$dS = \frac{dq_{rev}}{T}$$

For a measurable change, ΔS, between two states, the above expression integrates to

$$\Delta S = \int \frac{dq_{rev}}{T} = \frac{q_{rev}}{T}$$

When the heat is in joules and the temperature in Kelvin, the units of entropy are joules per Kelvin ($J\,K^{-1}$). The molar entropy has units of joules per kelvin per mole ($J\,K^{-1}\,mol^{-1}$). Entropy is therefore *not* a form of energy, which is measured in joules.

For a complete Carnot-type cycle, $\int dS = 0$, that is $\Delta S = 0$.

This work may not seem particularly interesting, but thermodynamic theory, fully developed and understood, enabled the chemists and biologists of the future to understand chemical and biological energetics at the most fundamental level. Let us press on. What is the significance of all this theory?

It turns out that for the case of an isolated system, that is a system that exchanges neither matter nor energy with the surroundings, entropy can tell us something about the spontaneity of a process.

For an isolated system, a process will occur spontaneously if it results in an increase in entropy.

Stated another way:

The entropy of an isolated system increases in the course of a spontaneous change. Thus $\Delta S_{\text{total}} > 0$.

The statements are alternative ways of expressing the Second Law of thermodynamics. There are others, some of which I will mention at the end of this chapter. We have a new state function, entropy, stated quantitatively in terms of temperature T and heat q. It is not a form of energy. This still leaves us with the questions: What *is* this new state function called entropy? What is its physical meaning? What is its nature?

The concept of entropy arose from the mathematical treatment of the developing science of thermodynamics. It was essentially forced on the early researchers, and as such can be described as 'a useful mathematical object that provides insight into the nature of change in the material world' (Haynie 2001). That might satisfy the more mathematically minded, but probably doesn't help the rest of us understand much about entropy as a concept.

For the moment, consider entropy as a measure of the degree of disorder, or the breadth of distribution, of energy. Consider the Clausius definition of entropy:

$$\Delta S = \frac{q_{\text{rev}}}{T}$$

By convention, if heat is added to the system, $q > 0$. Thus if the heat transferred to a system (q) is positive, the entropy change involved in the process will be inversely proportional to the absolute temperature. This in turn implies that if a given amount of heat, q, is transferred to a system at high temperature, it has less effect on the entropy change than the same amount of heat, q, transferred to the system at a lower temperature. Does this make logical sense, considering entropy as a measure of the degree of disorder of energy? Thermal (heat) energy is proportional to the absolute temperature, so the molecules of a cold object are not moving all that much. Addition of heat to a cold object will increase degree of energy disorder (increase the entropy) to a relatively large extent. Addition of the same amount of heat to the same object when it is hot (already quite thermally disordered and higher in entropy) will have proportionately less effect on the entropy. It is therefore reasonable to say that ΔS is proportional to the heat added (q) and inversely proportional to the absolute temperature, T, that is ΔS is proportional to $\frac{q}{T}$.

And for the special case of a reversible process: $\Delta S = \frac{q_{\text{rev}}}{T}$. I think we can accept this as reasonable.

All our reasoning about entropy and the Second Law so far has used the principles of classical mechanics. The nature of entropy remains somewhat of an enigma.

It was not until the German physicist Ludwig Boltzmann (1844–1906) proposed a statistical approach to the Second Law that a physical description of the nature of entropy became possible (Box 4.2).

> **Box 4.2 The reality of entropy?**
>
> Boltzmann was a strong believer in the existence of atoms, roughly in the form that we understand them today. This was not universally the case in Boltzmann's time, and he was in conflict in this matter with some of the influential contemporary scientists, notably Ernst Mach and Wilhelm Ostwald (Laidler 1993, p.160). Even Clausius did not accept Boltzmann's statistical interpretation of the Second Law (Laidler 1993, p. 158). The great Kelvin was apparently never really convinced that the idea of entropy was necessary (Laidler 1993, p. 106). I include these remarks to show that even the powerful minds of early thermodynamicists struggled with concepts that are often glibly presented in textbooks with little or no comment on their origins.

After early attempts to explain the Second Law on the basis of classical mechanics, Boltzmann came to realize that for a large assembly of molecules, entropy was related to probability. In 1877 he published a paper in which he proposed his now-famous relationship between entropy and probability:

$$S = k_b \ln W$$

where k_b is Boltzmann's constant and W (not to be confused with the lower case w for work) is the number of possible molecular configurations corresponding to a given state of the system (Laidler 1993, p. 161) This is the statistical definition of entropy. The molecular configurations are called microstates. Boltzmann assumed that all microstates having the same total energy have the same probability of being occupied by the molecules of the system, hence the statistical relationship. For a given set of conditions, W is usually very large for a large number of molecules, so that the probability that the molecules of a particular system will occupy a large number of equi-energy microstates is usually very high. This means that the molecules, and their respective energies, will tend to be distributed over a large number of available microstates. Boltzmann's equation thus implies that the entropy S of such a spread-out assembly of molecules will be large. This leads to the idea that an increase in the entropy of a system is accompanied by a spreading out or dissipation, or increase in the disorder of the energy of the molecules of the system. All these terms have been used to describe the increase in entropy that occurs as a result of the availability of a large number of microstates. As the temperature is lowered, the total energy of the system is lowered, fewer microstates are available, so the value of W becomes lower. From the relationship $S = k_b \ln W$ we see that the entropy must decrease also.

This is consistent with the statement above that ΔS is proportional to $\frac{q}{T}$. We can now make statements about the nature of entropy that make conceptual sense.

Entropy can also be visualized as:

A measurement of the degree of dispersal of the energy in a system, as a function of temperature.

Compare the molar entropies of water at 0°C and 1 atmosphere pressure:

$$S_{[H_2O(s)]} = 41 \text{ J K}^{-1}$$

$$S_{[H_2O(l)]} = 70 \text{ J K}^{-1}$$

$$S_{[H_2O(g)]} = 188 \text{ J K}^{-1}$$

All three states of water are at the same temperature (why this condition?). The number of molecules in each case is the same, 1 mole. The molecules that contain the energy under consideration have different degrees of order, depending on which of the three states of matter they occupy.

Water molecules in ice are highly ordered in a crystalline lattice, in water vapour the molecules are free to move about, and the liquid state is intermediate, so the above entropy values are consistent with our idea of entropy relating to the degree of energy disorder.

Some different ways to visualize entropy include, the more positive the entropy:

- the wider the distribution of energy in the system among its possible microstates
- the lower the intensive factor of the energy is
- the more disordered the energy is
- the more randomly the energy is distributed
- the less 'concentrated' the energy is
- the more 'spread out' the energy is.

I was careful to mention the word 'energy' in each statement about entropy. I believe to relate entropy to 'disorder', without stressing that it is primarily energy that is being described, is misleading. The distribution of physical entities, such as molecules in a gas or liquid, or even clothes in a teenager's room, may become more disordered but that distribution is a result of the random distribution of energy.

Physicist Richard Feynman (1918–1988) defines energy disorder as 'the number of ways the insides can be arranged so that from outside, the system looks the same'. I like that one.

Other statements about energy and entropy can be made:

- Energy of all types spontaneously disperses if it is not hindered from doing so.
- The dispersal of energy among a large number of microstates does not mean that the energy is 'smeared' over a number of these microstates. Each microstate contains all of the energy in the system. All the vast number of microstates are equally accessible to the molecules of the system, and this is why the molecules become so widely distributed—because they can, and the large degree of probability engendered by the availability of so many states ensures that they will do so (given time and a suitable pathway).
- Irreversible processes are spontaneous processes that generate entropy. They increase the entropy of the universe and cause degradation in the quality/intensity of the energy. Examples of irreversible processes include cooling of an object, free expansion of a gas, and melting of an iceblock.

- 'Reversible processes are finely balanced changes, with the system being in equilibrium with its surroundings at every stage. Each infinitesimal step along a reversible path is reversible, and occurs without degrading the quality of the energy, without dispersing energy chaotically, and without increasing the entropy of the universe. Reversible changes do not generate entropy, but they may transfer it from one part of the universe to another. Living organisms are highly ordered, low-entropy systems, but they grow and are sustained because their metabolism generated excess entropy in the surroundings.' (Atkins 1986 pp. 98–9)

I hope you now have some concept of entropy. It will feature from time to time, but for now let us continue with the search for a spontaneity test.

In the case of an isolated system, if $\Delta S_{total} > 0$, the process will be spontaneous. This statement is all very well for isolated systems, but most systems are not isolated. What about real systems, open systems such as laboratory reactions and living organisms, and closed systems that can only exchange energy with the surroundings? What are the indicators of spontaneity in such processes? Just as the First Law could not help us with the question of spontaneity, entropy alone can't help for the open systems of interest to chemists and biologists.

Entropy *must* be included when we are dealing with the thermodynamics of real processes, but as yet we have not shown how. The intuitive reader might guess what is coming next. Surely, not another state function! I'm afraid so.

To obtain a really useful, universally applicable test for spontaneity, we must use both enthalpy H and entropy S combined, to define a new state function, the Gibbs energy, G. This is sometimes called the Gibbs free energy, and was previously widely known just as the free energy. It is named after American physicist J.W. Gibbs (1839–1903), who did a great deal of important work on the application of thermodynamics to chemical reactions.

I can promise that G will be the last state function I will discuss (although there is at least one other). The Gibbs energy is defined as:

$$G = H - TS$$

where H, T, and S have already been defined. Using calculus notation, the infinitesimal dG is

$$dG = dH - TdS - SdT$$

For an isothermal process, that is one at constant temperature, usually the case in biological reactions,

$$dT = 0$$

so

$$dG = dH - TdS$$

For a measurable change between states, we can integrate to get

$$\Delta G = \Delta H - T\Delta S$$

66 Introducing Biological Energetics

At last, this is the spontaneity condition we have been seeking!

The term $T\Delta S$ has the units of energy, joules, because $\Delta S = \frac{q_{rev}}{T}$ so $\frac{Tq_{rev}}{T} = q_{rev}$, which as heat has the unit joules.

We can now generalize to say that for conditions of constant pressure and temperature:

- If $\Delta G < 0$ (is negative) the process will be spontaneous.

  For a chemical reaction, if $\Delta G < 0$ the reaction will be spontaneous in the forward direction and can do work in that direction.

- If $\Delta G > 0$ (is positive) the process will not be spontaneous.

  For a chemical reaction, if $\Delta G > 0$ the reaction will be spontaneous in the reverse direction and can do work in the reverse direction.

- If $\Delta G = 0$ the system is at equilibrium.

  For a chemical reaction, if $\Delta G = 0$ the reaction is at equilibrium and can do no work in either direction.

Thus, the spontaneity of a process or a chemical reaction is determined by a balance between the enthalpy change, the entropy change, and the temperature at which the process or reaction takes place.

Just as we could use a Hess's law approach to determine ΔH and ΔU for chemical reactions, we can use ΔH and ΔS to find ΔG values for chemical reactions, and thus predict whether or not the reaction will be spontaneous.

Just as we defined ΔH_f^0 values for chemical reactions, and used them in Hess's law to calculate $\Delta H_{reaction}$ the enthalpy of reaction, we can define ΔG_f^0 values and use them to find the ΔG of a reaction, $\Delta G_{reaction}$. For example, determine $\Delta G_{reaction}$ for the following, at 25°C/298 K:

$$2H_2(g) + O_2(g) \longrightarrow 2H_2O(l)$$

	$H_2(g)$	$O_2(g)$	$H_2O(l)$	
ΔH_f^0	0	0	−286	kJ mol^{-1}
S^0	131	205	70	J K^{-1} mol^{-1}
ΔG_f^0	0	0	−237	kJ mol^{-1}

From these data and the balanced equation we can calculate:

$$\Delta H_{reaction} = 2(-286) - (2\times 0 + 1\times 0) = -572 \text{ kJ}$$
$$\Delta S_{reaction} = 2(70) - (2\times 131) + 205 = -327 \text{ J K}^{-1} = -0.327 \text{ kJ K}^{-1}$$
$$\text{(to kJ from J)}$$
$$\text{As } \Delta G = \Delta H - T\Delta S$$
$$\text{Then } \Delta G_{reaction} = -572 - (298 \text{ K})(-0.327 \text{ kJ K}^{-1}) = -474.6 \text{ kJ}$$

(Note that ΔH_f^0 and ΔG_f^0 are zero for all elements, in this case H_2 and O_2. Data from Aylward and Findlay (1994)).

As $\Delta G_{reaction}$ is negative, the reaction is spontaneous and exothermic, yielding a maximum of 474.6 kJ of work. This is useful information. It tells us that in principle the 474.6 kJ could be used to 'drive' another reaction that was *not* spontaneous. It is possible, given certain conditions, to use one reaction to make another proceed in the 'reverse' direction. Such a process is called 'coupling' and is essential in the metabolism of all living organisms. Coupling is especially important in anabolic pathways, where the organism needs to build large molecules such as proteins or polysaccharides. The building of these molecules is endothermic, requiring a net input of energy, which is delivered by means of the coupling mechanism. Much more of this will be discussed in Chapter 6.

Let's next consider the purely scientific significance of the Second Law in a little more detail. In chemistry, the Second Law has a rather obvious statistical basis, that is it applies because of the large number of molecules usually involved. A mole of water, 18 g, consists of about 6.02×10^{23} molecules. They don't all have exactly the same amount of energy, but the 18 g as a whole behave in a predictable and reproducible way.

Although Boltzmann derived the relationship $S = k_b \ln W$, it was James Clerk Maxwell who first realized the statistical nature of the Second Law. In a letter of 1870, Maxwell commented 'The second law of thermodynamics has the same degree of truth as the statement that if you throw a tumblerful of water into the sea, you cannot get the same tumblerful out again.' Think about that for a while.

A good example of the statistical nature of the Second Law is the spontaneous mixing of two gases at constant temperature. Consider two gases, such as oxygen and nitrogen, in separate glass vessels, kept separated by a glass panel (even a litre of each would contain billions of molecules). If the panel were to be removed, what would happen over time? Note that the two gases chosen will not react chemically. Most people would soon arrive intuitively at the correct answer—that the oxygen and nitrogen molecules would eventually become evenly distributed throughout the whole combined volume of the two vessels. Why should this be so? There is no temperature gradient to cause the molecules to move from one part of our system to the other. The final, mixed state has a much higher probability of existing than for each of the two gases to remain in its own vessel and this provides sufficient 'driving force'. The entropy of the system is greater after the spontaneous process of mixing has occurred. Each gas has a huge number of extra microstates available to it once the glass panel is removed. The number of microstates or molecular arrangements available, say at standard temperature and pressure, is astronomical. Similarly, we know that odours, which consist of volatile molecules to which our noses are sensitive, will spread through a room and a crystal of purple potassium permanganate placed at the bottom of a glass of water will eventually colour the whole volume evenly, although this may take some time if the water is not stirred. All these are examples of spontaneous processes.

At this point it is useful to reiterate that all exchanges of energy are characterized by both an intensive factor (a measure of the 'driving force' or 'quality' of the energy) and an extensive (or capacity) factor (a measure of the quantity of energy) involved in the process under study. For heat energy, the intensive factor is temperature. The extensive factor is the total quantity

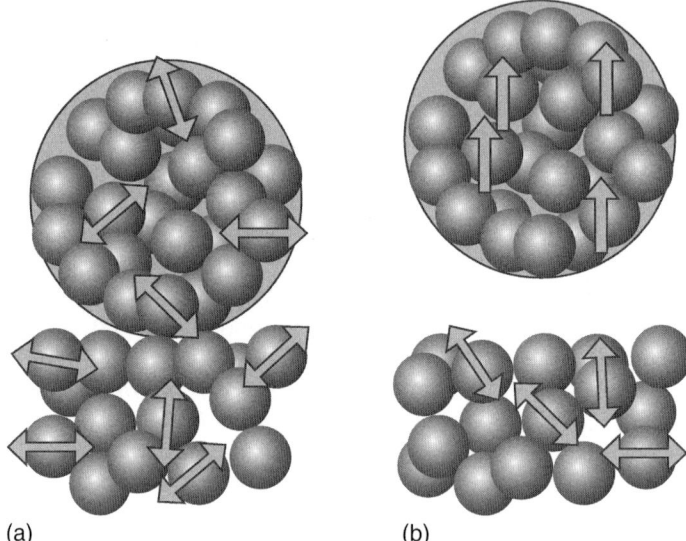

Figure 4.3. Irreversibility and the Second Law of thermodynamics. (a) A ball resting motionless on a surface. Its molecules are undergoing chaotic thermal motion, so the ball doesn't move in any direction. (b) The ball flies spontaneously upwards, its molecules having moved in a coordinated, directed manner. This latter occurrence is highly improbable and would contravene the Second Law. (Illustration from Atkins, P.W. and de Paula, J. *Atkins' Physical Chemistry*, 7th edn, p. 92. © 2002 Oxford University Press.)

of heat involved, which depends on the mass and specific heat capacity (see below) of the substances involved. To illustrate the difference between intensive/quality (temperature) and extensive (heat) properties, consider mixing a 200 mL cup of boiling water with 100 L of water at 37°C. Although the intensity of heat in the cup (measured by its temperature) is much greater than that in the bath (try dipping a finger in each!) the total (extensive) heat capacity of the bath is by far the greater. Pouring the cup of boiling water into the bath will drastically reduce the temperature, a measure of the heat intensity, of the 200 mL water added, but will hardly affect the temperature of the much more extensive 100 L of bath water.

During heat transfer the direction of spontaneous energy flow is always from objects of higher temperature to objects of lower temperature, never in the opposite direction. If the latter were the case, perpetual motion machines would be possible. Heat could be readily transferred from a cooler body to a hotter one, making it hotter still, and the cooler one correspondingly cooler. Overall, energy would be conserved, satisfying the First Law of thermodynamics, but the 'extra' heat gained by the hotter body could be made to do work, for example in a steam engine. This would be weird. It would be similar to the random movement of molecules in a ball on the floor spontaneously becoming organized in a small area and doing work on the floor, causing the ball to leap spontaneously into the air (Figure 4. 3).

The above examples are against all experience and would violate the Second Law of thermodynamics. The Second Law concerns the dispersal of energy. For any type of process, the direction of spontaneous change is from regions of high-energy intensity to accessible regions of lower intensity. Energy spreads out, and in the process its intensity is degraded and its capacity to carry out work is correspondingly reduced. We have seen that this is accompanied by an increase in entropy. That is the way energy behaves. The direction of spontaneous change is an arrow pointing us to the future, and the future is reached by spontaneous changes from the present. An unavoidable price we pay for reaching the future is the expenditure of some of our currency of high-intensity energy in favour of increasing the entropy of the universe. Fortunately, the Sun will continue to favour the Earth with a reasonably steady supply of energy of usable intensity for another 5 billion years, although the universe as a whole is becoming cooler and its energy distribution more chaotic. Eventually, entropy will rule.

The idea of intensity invokes feelings of something being concentrated, intensified. Recall people who are intense or excited by something. They give off a sense of concentrated, pent-up 'energy', of being fully focused on something, of almost bursting to release some built-up pressure. How do they reach such a state, and what happens when they release the flood of emotion? Firstly, something has to pump them up to the high-intensity levels so that their performance is powered by pent-up passion. When the passion is released, they will return to their personal 'ground state', their intensity gone, their energy dispersed, the work done, nothing 'in the tank'.

Consider any process, say one of the six examples below. How has the intensity factor been raised to the level where it causes the process to occur with a large negative ΔG? The intensity, the 'concentration of energy', has to come from somewhere. It won't occur spontaneously. By now we realize high-intensity energy doesn't emerge out of nowhere. Let's look at the waterfall example. What gives the water its gravitational potential energy, which converts to kinetic energy to drive a turbine at the bottom? The Sun provides enormous amounts of thermal energy to evaporate water from oceans or lakes, which subsequently falls as rain or snow on higher ground. These sources feed a number of small streams, ultimately flowing into the much larger river, still well above sea level. Radiant energy from the Sun is the 'ultimate' source for our concentrated, more intense potential energy source at the waterfall rim.

The intensity derives from the nature of the collecting system. The shape and height of the river and its tributaries focus or intensify the energy source. If the river system had not concentrated the water into a relatively intense flow, not as much work could have been done to drive a turbine. The same amount of water falling on a large area of flat land at sea level could not be harnessed directly to do useful work, as the intensity of its energy would be too low. Ultimately, the same amount of energy would be involved, but the flat land would dissipate it directly, as heat and by evaporative processes. The waterfall gives us the option of using the intensity of the collected water to do work. The waterfall will eventually flow to sea level and its work will end up as heat, but at least via a route that could do some work via a turbine.

70 Introducing Biological Energetics

The total amount of available heat in an Olympic swimming pool of warm (35°C) water is considerable, but it could not drive a heat engine when the ambient temperature is 20°C. Its maximum theoretical efficiency would be $(308-293)/308 = 15/308 = 4.87\%$. Allow for heat losses and friction, and the piston wouldn't move. The intensity of the heat, its quality, is not sufficient for the purpose.

Photosynthesis, where the energy in solar radiation ends up as energy concentrated in glucose molecules, is another obvious example, but one that needs much more background than we have covered so far. We should be in a position to tackle photosynthesis in Chapter 10.

To summarize some familiar examples involving spontaneous energy flow:

1) For heat energy in general, the intensive factor is temperature. The extensive factor is the total quantity of heat involved, which depends on the mass and specific heat capacity of the substances involved.

Direction of spontaneous energy flow—from high temperature to lower temperature.

2) For a hot gas moving a piston against a resistance in the surroundings, the intensive factor is the pressure and the extensive factor is the change in volume as the piston moves.

Direction of spontaneous energy flow—in the direction of the piston movement (increasing volume).

3) For water in a waterfall or a dam, the intensive factor is the gravitational potential energy, which is proportional to the height through which the water falls or to the depth of the dam where the turbine is located. The extensive factor is the mass of water flowing per unit time.

Direction of spontaneous energy flow—from higher to lower elevations (higher to lower gravitational potential).

4) For electricity, the intensive factor is the voltage, and the extensive factor is the number of electrons flowing, that is the current in amperes.

Direction of spontaneous energy flow—from higher to lower voltages.

5) For a wound-up spring, the intensity factor is related to degree of tension in the spring, and the extensive factor depends on the size of the spring.

Direction of spontaneous energy flow—in the direction of unwinding the spring.

6) For a chemical reaction (at constant temperature and pressure) the intensity factor is the Gibbs energy and the extensive factor is the number of moles involved in the reaction.

Direction of spontaneous energy flow—from higher to lower levels of the Gibbs energy. ΔG will be negative.

To generalize for all processes, the intensive factor in an energetic process is a quantitative measure of the tendency for energy to flow spontaneously in a particular direction. That direction is from high intensities to lower intensities.

This is a general phenomenon, if there is a suitable pathway available. This is not always the case, at least in the short term.

Spontaneous does not mean fast. Thermodynamics cannot tell us how fast a process will occur, only whether or not it *can* occur. Thus, thermodynamics will tell us that a cube of sugar sitting on the kitchen table can spontaneously be oxidized to water and carbon dioxide, but don't wait around for it to happen! Similarly, the example used above in a calculation of mixing hydrogen and oxygen has a $\Delta G = -474.6$ kJ, but the mixture of the two gases could sit on a bench for months without reacting substantially. Diamond spontaneously rearranges its structure to form graphite, but the time under standard conditions would be a million years or so. Factors other than pure thermodynamics are at work here, but more of that later.

For the sake of completeness, I need to discuss a property of substances referred to above. The specific heat capacity at constant pressure, c_p (previously called the specific heat) of a substance is the quantity of heat required to raise the temperature of 1 g of the substance at room temperature by 1 K.

The degree superscript (°) isn't used for degrees Kelvin, e.g. one writes 273 K not 273°K. The units are joules per degree Kelvin per gram ($J\,K^{-1}\,g^{-1}$). For water the value is 4.184. The specific heat capacity is the most convenient unit to use when we want to compare heat capacities of a number of substances because it is measured per gram rather than per mole. Other substances have specific heat capacities different from that of water. For pure iron it is 0.45 and for aluminium, 0.9. The important point here is that two bodies of the same mass, at the same temperature, but made of different materials, do not require the same amount of heat energy to increase their temperature by 1 K by virtue of the fact that they have different specific heat capacities. To heat a gram of water at 25 to 26°C will take $4.18/0.45 = 9.3$ times the amount of heat energy needed to raise a gram of iron through the same temperature range, and so on for substances of differing heat capacity.

Looking at the situation from the other direction, a gram of water cooling from 26 to 25°C, will transfer 9.3 times the amount of heat to the surroundings than would a gram of iron. What would be more effective as a bed warmer, a kilogram iron pan or a kilogram of water, each starting at 100°C?

Some other specific heat capacities

Ethanol	2.42
Copper	0.385
Nacl	0.864
O_2(gas)	0.918

The specific heat capacity of water is larger than that of most other substances. This and other properties of water are of the utmost importance in biology. The special properties of water relating to biology will be discussed later. Water is the most abundant substance in most living organisms, and is essential to all.

Finally, specific heat capacity is related to the final law of thermodynamics. If the magnitude of entropy is inversely related to absolute temperature, entropy must approach 0 as the temperature approaches 0 K. What happens to the value of entropy at a temperature of absolute zero, or 0 K? The most ordered entity we can imagine is a perfect crystal at 0 K. A perfect crystal at 0 K has an entropy of 0. This is a statement of the Third Law of Thermodynamics.

Such a situation is used to define an absolute entropy scale for any substance. If the heat capacity c_p can be evaluated for a substance over a sufficient range of temperature, its absolute entropy can be calculated. In biological studies we are mainly interested in changes in entropy rather than absolute entropies, as discussed above.

What does the above introduction to thermodynamics tell us about biological energy? The thermodynamic principles discussed apply to all types of energy exchange. It is easier to explain the principles using familiar, everyday examples. Consider the main points in this chapter:

- The direction of spontaneous energy flow in a system is always from higher-energy intensities to lower ones; always from higher values of G to lower values, such that ΔG is negative.
- The more negative the value of ΔG for a process or reaction, the more energy it can provide and thus the more work it can carry out.
- The ideas of intensity and entropy have given us a sense of direction of energy flow.
- High-energy intensity/low-entropy systems tend to move spontaneously to a low-energy intensity/high-entropy state.
- Thermodynamics tells us nothing about the rate (speed) of a spontaneous process.
- 'Entropy is time's arrow.' (Lowenstein 2000).

GENERAL REFERENCES

Tinoco, I., Sauer, K., Wang, J.C., and Puglisi, J.D. (2002) *Physical Chemistry–Principles and Applications in Biological Sciences*. Prentice-Hall, Upper Saddle River, NJ.

REFERENCES

Atkins, P.W. (1986) *Physical Chemistry*, 3rd edn. Oxford University Press, Oxford, pp. 98–9.
Atkins, P.W. and de Paula, J. (2002) *Atkins' Physical Chemistry*, 7th edn. Oxford University Press, Oxford, pp. 40–1, 45.
Aylward, G. and Findlay, T. (1994) *SI Chemical Data*, 3rd edn. Wiley & Sons, Milton, Queensland, Chs 4–6.
Chang, R. (1994) *Chemistry*, 5th edn. McGraw-Hill Inc., New York, p. 222.
Haynie, D.T. (2001) *Biological Thermodynamics*. Cambridge University Press, Cambridge, p. 52.
Laidler, K.J. (1993) *The World of Physical Chemistry*. Oxford University Press, New York, pp. 83, 102–107.
Lowenstein, W.R. (2000) *The Touchstone of Life*. Penguin Books, London, p. 9.

5
The Building Blocks

So far we have discussed energy and some of its applications mainly in terms of relatively large objects, such as whole animals or plants, waterfalls, steam engines, and movements of the Earth. It soon becomes clear that we can't explain the detailed working of living organisms without looking at their underlying structures—their organs, tissues, and cells. A deeper look at the substances that make up these structures, to determine their individual roles, will reveal a whole new dimension, full of unique concepts and challenging problems. By providing an introduction to the world of atoms and chemical compounds, this chapter lays the foundation for later discussions about biomolecular structure and function. Of the enormous number of possible molecules, relatively few have been selected to provide the needs of living organisms. By the end of this chapter we will have gone some way towards explaining why.

Humans consist mainly of water, plus other compounds containing the elements listed below, and that's about all. The elements C, H, O, and N contribute over 99% of atoms in the human body (Table 5.1; Garrett and Grisham 1999).

Certain trace elements that make up less than 0.01% are essential for proper function, for example iron and iodine in humans. Some 15 elements are present in all living things, and another eight to ten are found in particular organisms (Campbell *et al.* 2008).

What properties unite H, O, C, and N and render them so appropriate to the chemistry of life?

1) Their ability to form covalent bonds by electron sharing.
2) Their ability to form an enormous range of molecular types.
3) The ability of some of the molecules to transfer energy to others.

I will concentrate initially on atomic structure, chemical bonding, and molecular structure to rationalize the subsequent treatment of chemical reactivity. It will be a highly selective treatment that assumes some prior knowledge of chemistry. Much of it will probably be familiar; the aim is to provide some useful reminders.

Chemistry is the branch of science that deals with reactions between substances to form different substances, with the various properties of these substances and with the energy generated or absorbed in chemical processes. Synthetic chemists need to understand chemical principles and apply them to make new or improved materials such as plastics, drugs, tastier and more nutritious foods, more efficient fuels, or better paints and adhesives. Other

Table 5.1. Percentage of component elements within the human body.

Element	Percentage*
Hydrogen	63
Oxygen	25.5
Carbon	9.5
Nitrogen	1.4
Calcium	0.31
Phosphorus	0.22
Chlorine	0.08
K, Na, S, Mg	0.15

*Percentage of the total number of atoms in the body.

chemists are interested in why certain reactions proceed spontaneously, that is why they react without outside influences, while other reactions do not, instead needing to be heated or to be subjected to high pressures, or need a catalyst to speed them up. Still others study the factors that influence the rate or the energy changes involved in chemical reactions. Together with physics, chemistry is considered a basic science. The other sciences, astronomy, geology, biology, etc., are based on the laws of chemistry and physics, which are the fundamental scientific laws of the universe—at least from our perspective as humans. Mathematics is the language of vital importance for a full understanding of science. Many mathematicians would argue that mathematics is a science in itself, and I have no problem with that. Fortunately for children and beginners in science, a great deal of fascinating introductory material can be explained without recourse to any mathematics at all.

Biological chemistry is the study of any aspect of the chemistry of biological molecules. Much of this is done *in vitro*, that is in the test-tube. In many cases it is still necessary for scientists to study pure, well-characterized substances in relatively simple reactions and experiments to be able to follow what is happening and to develop hypotheses. Subsequently, the knowledge gathered may be applied to study reactions and systems of greater complexity. The ultimate aim of many such studies is to understand the reactions occurring in actual, living systems. This is the field of biochemistry.

The building blocks of the substances studied by chemists are atoms and groups of atoms. The idea that matter was composed of infinitesimally small particles, eventually to be called atoms, can be traced back to the early Greeks, notably Democritus in the fifth century BC. For centuries the idea of the atom was mere speculation, another of the key scientific concepts that had to wait many years for acceptance. The concept of the atom had its first real impact on chemistry early in the nineteenth century, largely due to the atomic theory proposed in 1803 by John Dalton (1766–1844). Atoms were not actually proven to exist until much later than that, about the end of the nineteenth century. We now know that there are some 92 different atom types that occur naturally. These are called the natural elements. Each element has been given a symbol, consisting of one or two letters, to represent a single atom of that element.

Thus, the symbol for oxygen is O, while sulphur is S, nitrogen N, helium He, and so on. An atom is the smallest part of an element that still has all the properties of that element. How small are we looking when we speak of atoms? Scientists would answer somewhat less than one nanometre (a billionth of a metre, or 10^{-9} m) but this is impossible to visualize. A water molecule consists of three relatively small atoms combined, still very small. To get another idea of the scale of atoms and molecules, consider the number of molecules in 18 g of water—about two dessertspoonfuls (at least in my kitchen). The population of Earth is about 6.8 billion, that is 6.8×10^9 or 6,800,000,000. The number of molecules in 18 g of water is about 6.02×10^{23}. Dividing 6.02×10^{23} by 6.8×10^9 gives about 8.85×10^{13}, so the number of molecules in two spoonfuls of water is a staggering 8.85×10^{13} or 88,500,000,000,000 times the human population on Earth.

Suppose that the whole Earth, all the vast continents, mountain ranges, deserts, seas, and oceans that make up our planet, were to be shrunk to the size of an orange. Once you have come to terms with that, imagine the orange shrunk by the same proportion, and you will have reached the size of a hydrogen atom. I'm not sure whether or not that helps, but there it is. Considering the vast difference in scale between our world of the centimetre and kilometre, and that of the atomic world it is not surprising that some of the rules are different, and some of the concepts seem strange.

Atoms themselves consist of even smaller units known as elementary particles or subatomic particles. The main elementary particles are the negatively charged electrons, which speed around a central nucleus consisting of positively charged protons and neutral, uncharged neutrons. The negative charge on the electron is of the same magnitude as the positive charge on the proton, although the proton is actually much heavier (about 1840 times) than the electron. In the atoms of a particular element, the number of electrons equals the number of protons, so the overall charge on the atom is normally zero. The number of protons is characteristic of a particular element and is called its atomic number. A neutron is about the same weight as a proton. The sum of protons and neutrons in an atomic nucleus is called its mass number.

The simplest atom is that of the element hydrogen, symbol H. Extensive studies of hydrogen atoms reveal that they are mostly 'space', with a tiny one-proton nucleus surrounded by a single, even smaller, electron, the movement of which is restricted to a three-dimensional 'probability cloud'. The probability terminology derives from quantum mechanics, as electrons and other entities of atomic and subatomic dimensions can only be located with a limited degree of probability. At this level of treatment, the cloud description is apt, as one way we can visualize the electron in a hydrogen atom in its lowest energy state is to imagine it moving around the nucleus so that it occupies a kind of cloudy, blurred, spherically symmetrical volume of space called an atomic orbital. The density of such a cloud at any given distance from the nucleus is proportional to the probability of finding an electron there (Figure 5.1). The highest probability of finding an electron in the hydrogen atom is in a spherically symmetrical region of the orbital, 0.0529 nm from the nucleus. This distance is known as the first Bohr radius of the hydrogen atom. It was calculated using both quantum mechanics and the Bohr model of the hydrogen atom (Ball 2003). Niels

76 Introducing Biological Energetics

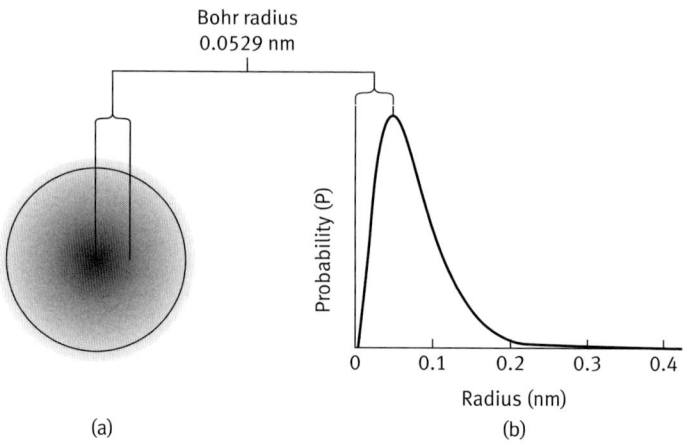

Figure 5.1. (a) A visual representation of the electron density distribution in a hydrogen 1s atomic orbital. The circle represents the boundary surface, which encloses about 90% of the total electron density. (b) A plot of the probability (P) of finding a electron on a spherically symmetrical surface of radius r, vs r (distance from the nucleus). The probability of finding the electron at the nucleus is zero, increases to a maximum at the first Bohr radius, and decreases toward zero as r approaches infinity. Both (a) and (b) provide the same information, (a) having been derived from the data in (b).

Bohr (1885–1962), a Danish physicist, was one of the founders of modern physics and a pioneer of quantum mechanics.

Hydrogen is by far the most abundant element in the universe (about 73%) and can be considered as its fundamental atomic building block. As we saw in Chapter 3, the major reaction in the Sun is nuclear fusion involving the formation of helium from hydrogen. Helium is the second most abundant element in the universe at about 25% (Ferris 1997). For this fusion reaction to proceed, the high temperatures and pressures found within the Sun are required. As a result of nuclear fusion reactions, a large amount of energy is being continuously released in the form of electromagnetic radiation of various wavelengths—microwaves, visible, infrared, ultraviolet, and gamma rays—and is emitted in all directions into space. We on Earth receive our share of this solar radiation. Nuclear fusion of hydrogen to helium occurs in the many stars similar to our Sun that are scattered throughout the universe. Most of the heavier atoms having more complex nuclei were formed within violently exploding stars called supernovae during the various stages of stellar evolution. The supernovae scattered their debris, enriched in heavy elements, throughout the universe around them. This became components of younger stars, such as our Sun, to be followed eventually by the planets and us. The elements found on Earth and the other planets mainly had their origins long before the birth of the solar system. As astronomer Carl Sagan said, we are literally made of starstuff.

If we could see it, a single hydrogen atom would probably look something like the cloud in Figure 5.1(a), without the circle. In principle, the electron could be found anywhere in its spherically symmetrical 1s orbital from close to the nucleus out to infinity, but it is useful in the case of hydrogen to draw a circle enclosing about 90% of the total electron density in the orbital, called a boundary surface diagram. This allows us to represent the hydrogen atom as a circle of definite size and to visualize two hydrogen atoms being brought together to form a hydrogen molecule, H_2. As we will see, the exercise of visualization becomes more complex for atoms larger than helium. One way to gain an understanding of atomic structures is to describe a number of them in order of complexity, then come back and rationalize what has been taking place. In terms of increasing atomic number, the order of the elements is shown in Table 5.2.

I have arranged the first 20 elements in this way for reasons that will become apparent shortly. The atomic numbers tell us the number of protons and also the number of electrons in the atom, but no detail about how the electrons are arranged around the nucleus. One way to describe the arrangement of electrons is to assign to each atom type an electron configuration. An electron configuration is a summary of quite a lot of information about the arrangement of electrons in an atom. As already described, hydrogen has its single electron in a 1s orbital. Its electron configuration is $1s^1$. Helium is next, has two electrons in the 1s orbital and an electron configuration of $1s^2$, pronounced 'one-s-two'. The 1s orbital is now 'full'. Quantum theory tells us that each orbital can contain a maximum of two electrons only. That's a rule for all orbitals. Lithium is next, so what happens here? The next orbital available is another s-type, also spherically symmetrical, located another 'layer' out from the nucleus and designated 2s. The electron configuration of lithium is $1s^2\ 2s^1$. Beryllium has one more electron, so it becomes $1s^2\ 2s^2$. The 2s orbital now has its maximum complement of two electrons, so what happens to boron? Quantum theory now requires the introduction of a new type, a p atomic orbital. There are in fact four atomic orbital types, designated s, p, d, and f, but the only ones that really concern us are the s- and p-types because most (but not all) of the atom types of interest in biology have their bond-forming electrons in s- or p-type orbitals. For the p and other atomic orbital types a similar approach to that used for hydrogen is adopted to generate a boundary surface diagram that helps us to visualize the shape and orientation of the orbitals. The shapes and numbers of the atomic orbitals have been established by the application of quantum theory. There is one shape for s orbitals (spherically symmetrical), three for p orbitals (dumbell

Table 5.2. Atomic numbers and symbols of the first 20 elements, arranged in groups 1A –VIIIA.

1A	IIA	IIIA	IVA	VA	VIA	VIIA	VIIIA
1 H							2 He
3 Li	4 Be	5 B	6 C	7 N	8 O	9 F	10 Ne
11 Na	12 Mg	13 Al	14 Si	15 P	16 S	17 Cl	18 Ar
19 K	20 Ca						

78 Introducing Biological Energetics

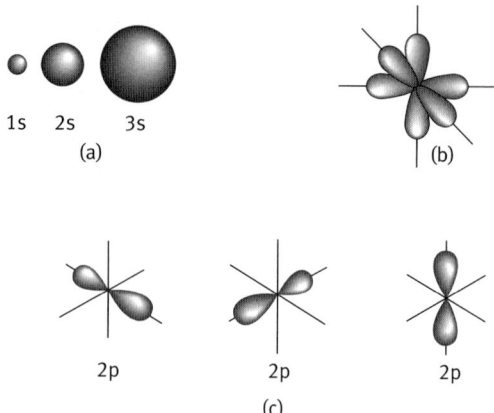

1s   2s   3s
(a)

(b)

2p   2p   2p
(c)

Figure 5.2. Shapes of s and p orbitals, shown as boundary surface diagrams. (a) Three s-type orbitals showing spherical symmetry and increasing size from shell 1 to shell 3. (b) The three mutually perpendicular 2p orbitals. (c) The individual 2p orbitals, one for each axis in three-dimensional space.

shaped, one for each axis in three-dimensional space), five shapes for d orbitals, and seven for f orbitals (not shown) (Figure 5.2).

As each orbital can accommodate a maximum of two electrons, the three p orbitals can together hold up to 6 electrons, the five d orbitals a total of 10, and the seven f orbitals a total of 14 electrons.

Table 5.3 shows the complete electron configurations of elements up to atomic number 20. Each element has one electron and one proton more than the preceding one. The orbitals fill in the order shown for reasons of energy. Because they are lower in energy, s orbitals fill before p orbitals in the same shell.

Atoms have their electrons arranged around the nucleus in orbitals that are grouped into shells. Each shell is identified by the principal quantum number, n. $n = 1, 2, 3, 4$, etc., numbering from the nucleus outwards. Shell 1 has the lowest energy level, and shells 2, 3, etc. are of increasing overall energy level. Each shell is 'allowed' to contain only a certain maximum number of electrons, which is twice the number of orbitals in the shell. Each shell has its own combination of orbitals. Table 5.3 shows that for $n = 1$ there is a 1s orbital, for $n = 2$ there are 2s and 2p orbitals, and for $n = 3$ there may be 3s, 3p, and 3d orbitals present. In any particular shell, a group of equivalent orbital types is called a subshell. Thus, the 2s orbital is one subshell and the three 2p orbitals comprise the other subshells of shell 2.

Another important property of atomic orbitals is their relative energy levels. As always in our examination of systems, of whatever nature, their behaviour is determined by energy considerations. It is no different for atomic structure. The energy levels of electrons in atomic orbitals increase as the distance of the orbital from the nucleus increases, that is, as n increases. The relative energies of s, p, and d atomic orbitals in a multi-electron atom are shown in Figure 5.3. One cause for the splitting of electron shells into subshells in multi-electron

The Building Blocks 79

Table 5.3. Electron configurations of elements 1 to 20. Electrons are arranged in shells designated by the principal quantum number, n. (n = 1, 2, 3, etc.). For n = 2 and above, s and p subshells occur.

n	IA	IIA	IIIA	IVA	VA	VIA	VIIA	VIIIA
	1 H							2 He
1	$1s^1$							$1s^2$
	3 Li	4 Be	5 B	6 C	7 N	8 O	9 F	10 Ne
1	$1s^2$	$1s^2$	$1s^2$	$1s^2$	$1s^2$	$1s^2$	$1s^2$	$1s^2$
2	$2s^1$	$2s^2$	$2s^2 2p^1$	$2s^2 2p^2$	$2s^2 2p^3$	$2s^2 2p^4$	$2s^2 2p^5$	$2s^2 2p^6$
	11 Na	12 Mg	13 Al	14 Si	15 P	16 S	17 Cl	18 Ar
1	$1s^2$	$1s^2$	$1s^2$	$1s^2$	$1s^2$	$1s^2$	$1s^2$	$1s^2$
2	$2s^2 2p^6$	$2s^2 2p^6$	$2s^2 2p^6$	$2s^2 2p^6$	$2s^2 2p^6$	$2s^2 2p^6$	$2s^2 2p^6$	$2s^2 2p^6$
3	$3s^1$	$3s^2$	$3s^2 3p^1$	$3s^2 3p^2$	$3s^2 3p^3$	$3s^2 3p^4$	$3s^2 3p^5$	$3s^2 3p^6$
	19 K	20 Ca						
1	$1s^2$	$1s^2$						
2	$2s^2 2p^6$	$2s^2 2p^6$						
3	$3s^2 3p^6$	$3s^2 3p^6$						
4	$4s^1$	$4s^2$						

atoms is the repulsive interactions between electrons in the different orbitals, resulting in different energy levels.

Take time to look at Table 5.3 and Figure 5.3 and consider their implications. I have taken Figure 5.3 only to shell 4, sufficient to cover the elements of greatest biological interest. Figure 5.3 explains the order of filling electron orbitals shown in Table 5.3. The lower-energy orbitals fill first. As we will see, the same principle applies when electrons are involved in forming bonds in chemical compounds. Formally, each element can be considered as being 'built' from its predecessor by addition of one electron, one proton, and one or more neutrons. Table 5.3

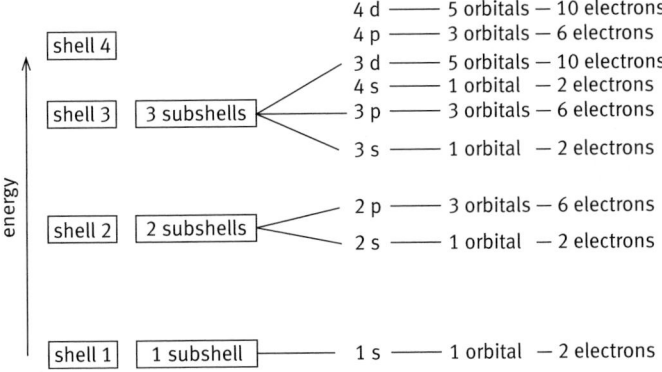

Figure 5.3. Relative energy levels of atomic orbitals in multi-electron atoms.

shows this build-up process explicitly for electrons. A look at the full atomic structure will show that the nuclei of successive elements increase by one proton and one or more neutrons. Thus:

Helium is hydrogen + 1 electron + 1 proton + two neutrons.

Lithium is helium + one electron + one proton + two neutrons.

Oxygen is nitrogen + one electron + one proton + one neutron.

Chlorine is sulphur + one proton + two neutrons, and so on, all the way up to uranium, atomic number 92, the 'heaviest' natural element.

The eight vertical groups 1A to VIIIA in Tables 5.2 and 5.3 emerge when the elements are grouped according to similarities in their chemical properties.

Very briefly, some typical periodic trends are:

Group 1A elements (H excepted) are highly reactive metals with a combining power, or valence, of 1 (see below).

Group IIA elements are metals with a valence of two.

Group VIIA elements are non-metallic, have a valence of 1, and dissolve in water to form acids.

Group VIIIA elements are chemically inert gases.

All these chemical properties and others can be explained in terms of the electron configuration of the atoms involved. The periodicity in properties reflects the periodicity in electronic structure. I will now introduce the rest of the elements in a manner that will help us to understand much more about atomic and molecular structure.

When scientists in the nineteenth century began systematic studies of the elements, it was discovered that these could be formed into groups having similar chemical characteristics. Eventually, all the elements were classified into what is called the periodic table of the elements. Tables 5.2 and 5.3 are abbreviated forms of the periodic table. A reasonably complete version is shown in Figure 5.4(a). The table is arranged into horizontal rows, called periods, and vertical columns, called groups. We could spend a considerable amount of time discussing this theoretically important and practically useful classification and its implications, but instead I refer those who are interested to comprehensive periodic tables that appear in chemistry texts (e.g. Chang 1997, Chapter 8; see also Box 5.1).

Note the dual numbering of the groups, Roman and Arabic. I will use the former. The s-, p-, d-, and f-block notations refer to the electrons in the respective atomic orbitals. Elements in the same vertical group have the same number of valence electrons. Groups 1A and IIA are metals, tend to donate electrons to elements in group VIIA to form ionic compounds, and so achieve the stable eight-electron configuration of the group VIIIA noble gases. Group VIIA elements tend to accept electrons to form anions and achieve the Group VIIIA configuration. Elements toward the middle of the table tend to form covalent bonds. As we ascend the periodic table, the nuclei increase in size, and the ratio of neutrons to protons increases. Oxygen has 8 neutrons and 8 protons, iodine has

Period	1/IA	2/IIA	3	4	5	6	7	8	9	10	11	12	13/IIIA	14/IVA	15/VA	16/VIA	17/VIIA	18/VIIIA
	1 H 1.0079																	2 He 4.003
2	3 Li 6.941	4 Be 9.012											5 B 10.81	6 C 12.01	7 N 14.01	8 O 16.00	9 F 19.00	10 Ne 20.18
3	11 Na 22.99	12 Mg 24.31											13 Al 26.98	14 Si 28.09	15 P 30.97	16 S 32.07	17 Cl 35.45	18 Ar 39.95
4	19 K 39.10	20 Ca 40.08	21 Sc 44.96	22 Ti 47.87	23 V 50.94	24 Cr 52.00	25 Mn 54.94	26 Fe 55.85	27 Co 58.93	28 Ni 58.68	29 Cu 63.55	30 Zn 65.39	31 Ga 69.72	32 Ge 72.61	33 As 74.92	34 Se 78.96	35 Br 79.90	36 Kr 83.80
5	37 Rb 85.47	38 Sr 87.62	39 Y 88.91	40 Zr 91.22	41 Nb 92.91	42 Mo 95.94	43 Tc (98)	44 Ru 101.07	45 Rh 102.91	46 Pd 106.42	47 Ag 107.87	48 Cd 112.41	49 In 114.82	50 Sn 118.71	51 Sb 121.76	52 Te 127.60	53 I 126.90	54 Xe 131.29
6	55 Cs 132.91	56 Ba 137.33	La-Lu	72 Hf 178.49	73 Ta 180.95	74 W 183.84	75 Re 186.21	76 Os 190.23	77 Ir 192.22	78 Pt 195.08	79 Au 196.97	80 Hg 200.59	81 Tl 204.38	82 Pb 207.2	83 Bi 208.98	84 Po (209)	85 At (210)	86 Rn (222)
7	87 Fr (223)	88 Ra (226)	Ac-Lr	104 Rf (261)	105 Db (262)	106 Sg (263)	107 Bh (262)	108 Hs (265)	109 Mt (266)	110 Uun	111 Uuu	112 Uub	113 Uut	...				
	S block		d block										p block					

Lanthanides	57 La 138.91	58 Ce 140.12	59 Pr 140.91	60 Nd 144.24	61 Pm (145)	62 Sm 150.36	63 Eu 151.96	64 Gd 157.25	65 Tb 158.93	66 Dy 162.50	67 Ho 164.93	68 Er 167.26	69 Tm 168.93	70 Yb 173.04	71 Lu 174.97
Actinides	89 Ac (227)	90 Th 232.04	91 Pa 231.04	92 U 238.03	93 Np (237)	94 Pu (244)	95 Am (243)	96 Cm (247)	97 Bk (247)	98 Cf (251)	99 Es (252)	100 Fm (257)	101 Md (258)	102 No (259)	103 Lr (262)
	f block														

Figure 5.4. (a) A periodic table showing the symbol, atomic number, and atomic weight/atomic mass of each element. (Illustration from Atkins, P. and de Paula, J. *Atkins' Physical Chemistry*, 7th edn. © 2002 Oxford University Press).

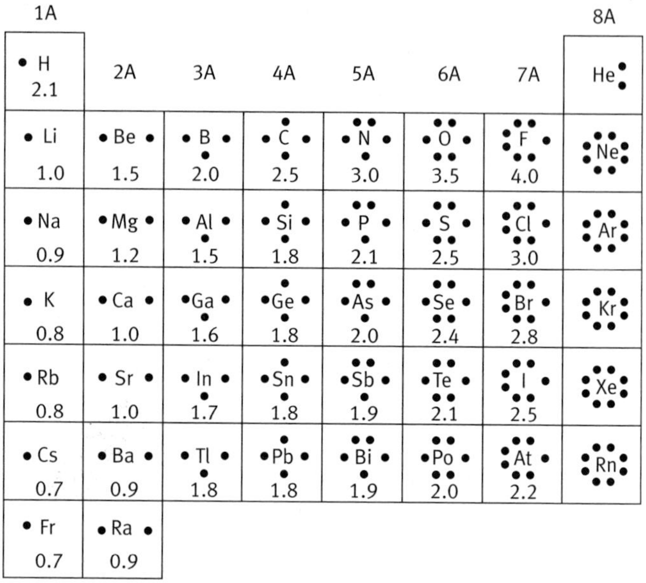

1A	2A	3A	4A	5A	6A	7A	8A
•H 2.1							He:
•Li 1.0	•Be• 1.5	•B• 2.0	•C• 2.5	•N• 3.0	•O• 3.5	:F• 4.0	:Ne:
•Na 0.9	•Mg• 1.2	•Al• 1.5	•Si• 1.8	•P• 2.1	•S• 2.5	:Cl• 3.0	:Ar:
•K 0.8	•Ca• 1.0	•Ga• 1.6	•Ge• 1.8	•As• 2.0	•Se• 2.4	:Br• 2.8	:Kr:
•Rb 0.8	•Sr• 1.0	•In• 1.7	•Sn• 1.8	•Sb• 1.9	•Te• 2.1	:I• 2.5	:Xe:
•Cs 0.7	•Ba• 0.9	•Tl• 1.8	•Pb• 1.8	•Bi• 1.9	•Po• 2.0	:At• 2.2	:Rn:
•Fr 0.7	•Ra• 0.9						

(b)

Figure 5.4. (b) A partial periodic table of the elements. The d-block and f-block elements are not shown. The symbol, and electron configuration (dots) of the outer electron shell (the valence shell) of each element are shown. Numbers under each symbol are the Pauling electronegativity values.

74 neutrons and 53 protons, while radium-226 has 138 neutrons and only 88 protons. Why this trend?

Imagine the repulsive energy between 88 positively charged protons crowded closely together in the radium nucleus. It is not surprising that the nucleus tends to fly apart, which as we know is exactly what natural radium-226 nuclei do. They undergo radioactive decay, emitting an α-particle (a helium nucleus) to form radon-86.

To stabilize proton/proton repulsion, neutron numbers are increased in large nuclei. There are short-range attractions between proton/proton, proton/neutron, and neutron/neutron that stabilize the nucleus, and extra neutrons assist in these processes, to stabilize larger nuclei. The main points I want to emphasize about the elements and their periodic grouping concerns their chemical reactivities, which are related, in a periodic or repeating manner, to their electron configuration.

Let's examine this in some detail. Helium is the second simplest element. Its atoms consist of a nucleus containing two protons and two neutrons, surrounded by two electrons. Its electronic configuration is $1s^2$ so the s orbital, which also comprises the first shell ($n = 1$), is full. The combination of two 1s electrons in helium is energetically low, and therefore is stable. In general, atoms with filled subshells of s and p orbitals are the most stable, and the most chemically inactive. Helium is inactive chemically. Let us follow this trend. Addition or removal of electrons would move helium away from having the ideal stability number of two electrons, so helium is a gas consisting of individual He atoms. Helium is the first of a series of unreactive, or inert, gases—helium, neon, argon, krypton, xenon, radon—found to the right of the periodic table in Group VIIIA (Figure 5.4). The unreactive nature of these elements intrigued the early chemists, who called them the noble gases. All the noble gases have filled s and p subshells. All except helium ($1s^2$) have an ns^2np^6 outer electron configuration, where n is the shell number. This gives them all eight electrons in their outer shells. In atoms, it is usually these outer shell electrons, the valence electrons, which are involved in the formation of chemical bonds. Helium and the other noble gases are quite unreactive, and

---

Box 5.1 Elements, compounds, atomic weights, and isotopes

What distinguishes a chemical compound from an element? Here are a few reminders.

An element consists of one type of atom only. It has an individual number of protons and electrons that defines its chemical properties. All carbon atoms have six protons and six electrons, even though some have more than six neutrons (isotopes). The smallest possible amount of an element is a single atom. For a compound, a major characteristic is its constant chemical composition, reflecting the type of atoms and kinds of bonding in the compound. Thus, water is a compound consisting of two atoms of hydrogen combined with one atom of oxygen. Water always has that composition, and the formula H_2O. Other compounds have their own unique composition and structure. Acetic acid is always CH_3COOH and sulphur dioxide SO_2. The smallest possible amount of a compound is called a molecule. The combination of symbols representing a molecule is called a molecular formula (usually abbreviated to formula). Two atoms of oxygen combine to form a molecule with the formula O_2. Proteins or DNA molecules are very large compared with the simple compounds above, but each has its own characteristic composition. There has been drawn up a scale of relative atomic masses, commonly called the atomic weights, of the elements. The atomic mass is one of the prime characteristics of each element in the periodic table. The actual values of atomic masses in grams are extremely small and need not concern us. What scientists use in their calculations is usually the average relative atomic mass of the element found naturally on Earth. The standard to which other elements are compared is carbon-12, an isotope taken to have an atomic mass of 12.0000 atomic mass units (amu). An amu is also called a dalton (Da). The atomic weight of the hydrogen atom on this scale is 1.008 Da, helium is 4.003 Da, and lead 207.2 Da. The dalton is named after John Dalton, to whom is attributed the first enunciation of the modern atomic theory. A check of the periodic table will show the atomic weight of carbon to be 12.01, not 12.0000. This is because carbon that is found naturally on Earth, and many other elements, consists of a mixture of isotopes. An isotope of an element, say carbon, has a different number of neutrons in the nucleus from the most abundant one. Carbon on Earth has three isotopes: carbon-12

> **Box 5.1** (continued)
>
> (98.9%) has six protons and six neutrons; carbon-13 (1.1%) has six protons and seven neutrons; carbon-14 (a trace) has six protons and eight neutrons. All have six protons and electrons, so they are all carbon atoms. The average percentage of each isotope found on Earth is called its natural abundance. The average weight of the three carbon isotopes, taking into account their natural abundance, comes out to be 12.01. Many elements have isotopes, some of which are radioactive, for example carbon-14. Radioactive isotopes can be useful in medicine to follow, for example, iodine and other atoms used to label molecules administered to patients. They are also useful for dating purposes in geology, palaeontology, and archaeology, and for following particular atoms in chemical and biochemical reactions.
>
> When we need to know the relative weight of a molecule of a chemical compound (its molecular weight) we simply look at the elements in the chemical formula, check their atomic weight from the periodic table, and add these together in the correct proportions. Thus water is H_2O, consisting of two H atoms (2×1 Da) plus one oxygen atom (16 Da) = 18 Da. For carbon dioxide we have $CO_2 = 12 + (2 \times 16) = 44$ Da. For the sugar glucose, $C_6H_{12}O_6$ we have $(6 \times 12) + (12 \times 1) + (6 \times 16) = 180$ Da. The molecule of glucose is about 180 times the mass of an atom of hydrogen. All pure chemical compounds are characterized by a molecular formula and an exact molecular weight.

don't (with a few exceptions) form chemical bonds to other atoms. From now on I will call the numbers two and eight stability numbers for convenience. When they react to form chemical bonds, elements in each horizontal period tend to achieve the same number of electrons as the nearest noble gas in group VIIIA.

A careful look at Figure 5.4 will show this trend:

Row 1: contains H and He (stability number 2)

Row 2: Li, Be, B, C, N, O, F, Ne.

Neon, the noble gas in Group VIIIA of the row, has a stability number of 8. The elements early in the row therefore seek to achieve the two-electron configuration of He and the later ones will adopt the eight-electron configuration of Ne in their outer shell. Similarly:

Row 3 elements (Na, etc.) seek to adopt the stability number of neon or argon (8).

Row 4 elements (K, etc.) seek to adopt the stability number of argon or krypton (8)

Row 5 elements (Rb, etc.) seek to adopt the stability number of krypton or xenon (8).

Note the pattern. The stability numbers refer to the outermost electrons in the atom mentioned. The innermost electrons are there, but are not explicitly identified. The prevalence of the number eight has led to a rule of thumb for stability numbers called the octet rule:

Many elements tend to react in such a way as to achieve an octet of electrons in their outer shell.

The number of outer shell electrons determines the combining power or valence of an element. Carbon has four valence shell electrons, and needs four more to make the stable eight configuration of neon. Carbon therefore has a valence of four, so methane has the formula CH_4. Nitrogen (five electrons) needs three more, making its valence three, and ammonia has the formula NH_3. Oxygen has six outer electrons, its valence is two, and the formula for water is H_2O.

It happens that the majority of the elements on Earth are not found as free atoms, that is as the pure, uncombined elements as shown in the periodic table. Most are so chemically reactive that they form compounds with other elements. Many compounds on Earth are found as various components of soils, sand, rocks, and minerals—metal oxides, nitrates, sulphates, phosphates, sulphides, chlorides, etc. that were formed during early history of the Earth.

The rules of formation of chemical compounds are well known. They follow the trends in reactivity that we have just touched on, so let's look at what happens when atoms come in contact with other atoms.

Hydrogen, we know, is a gas. It normally does not occur as single H atoms floating about, rather as two H atoms combined to form a molecule, H_2. Why? The overall energy of the molecule H_2 is less than that of the separate H atoms. How do we know this? Let's look at the situation in reverse. To separate the combined atoms in the H_2 molecule into two individual hydrogen atoms requires an input of a considerable amount of energy, about 436 kJ for 2 g of hydrogen, which corresponds to heating it to about 1575°C. This means that hydrogen can and does exist as individual H atoms in the Sun and other stars.

Why should two hydrogen atoms combine to form H_2? Why should they 'need' to form a bond between them? The answer lies in one of the properties of the electron orbitals around atoms. We have seen that atoms have lowest energies and are most stable when they possess the stability number of electrons in their valence shells. For hydrogen, the stability number for electrons is two. Two electrons will fill its 1s orbital and achieve the stable $1s^2$ electron configuration of helium. One hydrogen atom alone has only one such electron. If another hydrogen atom is available, it will also have a one-electron cloud and an energy-favourable compromise is struck. The two hydrogen atoms share the two electrons. Electron sharing with a nearby hydrogen atom isn't difficult, provided certain quantum requirements are met. The two electrons weave a different kind of cloud that embraces both hydrogen nuclei (Figure 5.5).

Applying the molecular bonding approach called molecular orbital (MO) theory, this new electron arrangement is called a molecular orbital. This theory is an approach that treats the MO as covering the whole molecule. The MO in the H_2 molecule predicts that the two negatively charged electrons will spend slightly more of their time close to the positively charged nuclei. This explains the shape of the MO and electron contour map shown in Figure 5.5(b).

The stability number of two for hydrogen is satisfied and the new hydrogen molecule is lower in energy than the two separate hydrogen atoms by about 436 kJ mol^{-1}. The electron density between the positively charged nuclei

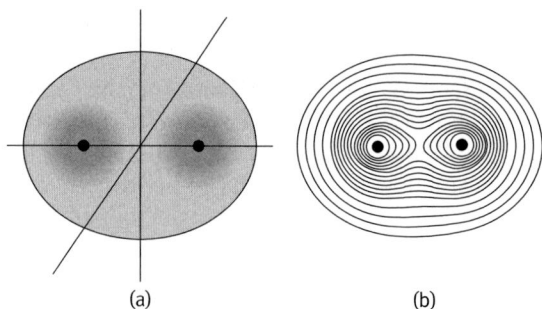

Figure 5.5. (a) The atomic orbitals of two H atoms combine to form an overall bonding molecular orbital (MO) in H_2. (b) A contour map showing the electron density in an alternative way to (a). The black dots indicate the positions of the H nuclei.

serves to reduce their mutual repulsion, and the resulting net attraction is manifested as what we call a chemical bond. This bond has an associated energy known as the bond energy. The distance between the two H nuclei is called the bond length and is such that the overall energy of the molecule is at a minimum under the prevailing conditions. For H_2 the average bond length is 7.4×10^{-11} m or 0.074 nm. The two individual 1s H orbitals actually combine to form two MOs that have different energy levels. This is a general rule—the number of MOs formed is the same as the number of atomic orbitals that combine to form them. The MO with the lowest energy 'accepts' the two electrons and forms the bond. This is called the bonding orbital. The second MO is called the antibonding orbital, and it holds no electrons in the case of a hydrogen molecule under ground state conditions. Representations of the formation and the energetics of the formation of bonding and antibonding MOs in H_2 are shown in Figure 5.6.

MOs must possess both bonding and antibonding orbitals, as proposed by the MO approach to quantum theory, which is described briefly here.

Electrons possess wavelike properties. When two waves of the same type come in contact, they experience interference. The waves may interact such that the resulting wave has either an enhanced amplitude (constructive interference) or diminished amplitude (destructive interference) (see Appendix A). Bonding orbitals correspond to constructive interference, while antibonding orbitals correspond to destructive interference. The result in bonding orbitals is significant electron density between the bonding nuclei, leading to net attraction and bond formation. In antibonding orbitals the electron density decreases to zero between the nuclei (Figure 5.6(a)). The hydrogen molecule is conventionally written as H–H, where the dash represents a chemical bond involving the sharing of two electrons. Bonds formed by sharing electrons between atoms are called covalent bonds. The prefix *co-* in this instance means mutual, indicating the electron-sharing nature of the bonds. The type of covalent bond formed in hydrogen is called a sigma (σ) bond because of the type of symmetry in the bonding MO. Many bonds in carbon-based molecules are of the σ type.

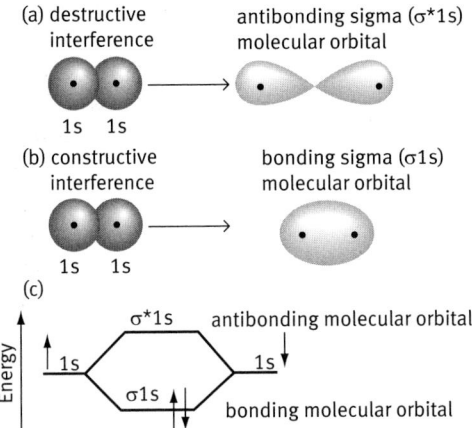

Figure 5.6. Covalent bond formation in hydrogen. The 1s orbitals of two H atoms may form (a) an antibonding orbital (σ^*_{1s}) by destructive interference or (b) a bonding molecular orbital (σ_{1s}) by constructive interference. (c) A molecular orbital energy diagram showing the relative energy levels of the two molecular orbitals in hydrogen H_2. In the ground state electronic configuration of H_2 shown, both electrons are accommodated in the lowest energy orbital, the bonding orbital, which is lower in energy than the separate 1s atomic orbitals present before bond formation. To satisfy quantum mechanical theory, the pair of bonding orbital electrons must be of opposite spin, as shown by the arrowheads.

Oxygen atoms combine to form O_2 molecules. They do this for the same reason that hydrogen does—the energy of O_2 is lower than that of unbonded O+O atoms. In this case, however, the required stability number of eight dictates that two pairs of electrons need to be shared, leading to the representation O=O and what is called a double bond.

Once again one dash represents one shared electron pair. Alone, a single oxygen atom has only six electrons in its outer electron shell. It needs two more from elsewhere: in the case of O_2 these come from the second oxygen atom. The original oxygen atom shares two electrons of its own to satisfy the stability number of the second one, so two pairs of electrons are in fact shared, giving rise to the double bond. The bond length for O_2 is 12.1×10^{-11} m or 0.121 nm. The bonding in oxygen and in other multiple bonds again involves the formation of MOs from the atomic orbitals.

For nitrogen, a similar stability number requirement of eight leads to three electron pairs being shared and a triple bond is formed:

$$N{\equiv}N$$

or N_2. A nitrogen atom alone has only five electrons in its outer shell and needs three more. Triple bonds are about as far as two atoms go in electron-pair sharing. Single, double, and triple bonds of the electron-sharing type are commonly found in biomolecules and are all classified as covalent bonds. Covalent bonds have bond energies in the hundreds of kilojoules per mole range, which are formally reported as average bond enthalpies (Table 5.4).

Table 5.4. Average covalent bond enthalpies at 25°C (ΔH, kJ mol^{-1}).

Single bonds	C–C	346	C–O	358	C–H	414	O–H	463
Double bonds	C=C	614	C=O	804	C=N	615	N=N	470
Triple bonds	C≡C	839	C≡N	890	N≡N	945		

Source: Aylward and Findlay (1994), p. 115.

Compounds consisting exclusively of covalent bonds are known as molecular compounds because they occur as discrete molecular units or molecules (cf. ionic compounds, see below).

Carbon tends to form four covalent bonds in many organic molecules. It quite readily forms stable covalent sigma (σ) bonds to hydrogen atoms or to other atoms, including other carbon atoms. This property is the basis of the almost limitless structures we find in organic chemistry. Broadly, organic chemistry is the chemistry of carbon-containing compounds. When a carbon atom forms four covalent single σ bonds to other atoms, as in methane CH_4, the atoms are arranged in space so that each of the four attached hydrogen atoms is as far from its neighbour as possible. It has been shown experimentally that the four bonds in methane are identical, with a bond angle of 109.5°, and point towards the corners of a regular tetrahedron, a solid comprising four faces, each of which is an equilateral triangle. The bonds assume this energetically favoured geometry because carbon is able to form four 'combination orbitals' from one 2s orbital and three 2p orbitals, called sp^3 hybrid orbitals (pronounced s-p-three). The four hybrid orbitals are all of the same shape and energy level. Let's see how hybrid orbitals form.

The electron configuration of carbon is $1s^2\ 2s^2\ 2p^2$. The valence electrons are those in the 2s and 2p orbitals, so these are involved in the discussion below. The electron configuration can be written in a slightly different way, to include information on electron spin. The quantum rule is that electrons in equivalent orbitals (e.g. the three 2p orbitals) tend to have parallel ($\uparrow\uparrow$) rather than anti-parallel ($\uparrow\downarrow$) spins as far as possible. For uncombined carbon atoms in the ground state, the configuration is

$\uparrow\downarrow$		$\uparrow$	$\uparrow$	
2s		2p$_x$	2p$_y$	2p$_z$

The 2p$_x$, 2p$_y$, and 2p$_z$ designations refer to the orientation of the p orbitals along the three axes of three-dimensional space. Because the carbon atom has two unpaired electrons, in this configuration it could only form two σ bonds with electrons having opposite spins with, for example, hydrogen, to form CH_2. This is a very unstable species, as it would have only six electrons in the valence shell instead of the ideal eight. Although there is an energy cost involved, carbon is able to 'promote' one of the 2s electrons into the vacant 2p$_z$ orbital, so the configuration becomes:

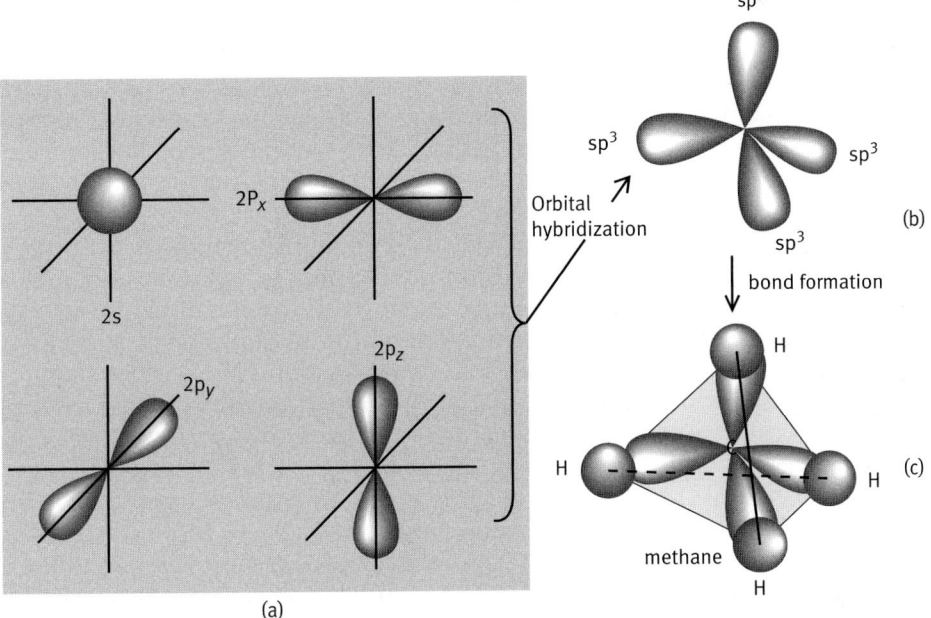

Figure 5.7. Formation of sp³ hybrid orbitals in carbon atoms. (a) The one 2s and three 2p orbitals of carbon before hybridization. (b) Hybridization of carbon to form an sp³ orbital. (c) Formation of methane from one tetrahedrally oriented sp³-hybridized orbital and the s orbitals of four hydrogen atoms.

2s		$2p_x$	$2p_y$	$2p_z$
↑		↑	↑	↑

There are now four unpaired electrons in the carbon atom, each of which is capable of forming a covalent bond to, say, H. To do this most effectively, the 2s and the three p orbitals hybridize, to form four equivalent sp³ orbitals. The hybridization process is depicted in Figure 5.7.

The resulting geometry is tetrahedral, the ideal arrangement energetically for four equivalent bonds around a single central atom. Methane therefore adopts that shape:

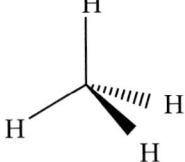

Repulsion between electrons in neighbouring orbitals also has an effect on molecular shape, so repulsion between the electrons in the four σ bonds in methane also favours a tetrahedral arrangement. Energetically, this makes sense, as such an arrangement puts the attached atoms as far apart in space as possible, and minimizes what is called steric strain. This is especially important for attached groups larger than hydrogen atoms, where such energetically unfavourable groups tend to push the bonds around the central carbon atom away from the ideal tetrahedral shape.

Hybridization is used only to explain the bonding known to occur in molecules. The assumption is that hybridization occurs in an atom only if it is involved in bond formation to another atom. Hybridization could be considered as being induced by the necessity of forming covalent bonds. It does not apply to isolated, unbonded atoms. As seen in the example of carbon, hybridization results in orbitals that have markedly different geometries from those of 'natural' atomic orbitals. At least five types of hybridization have been proposed to account for the observed geometries of molecules (Chang 1994). The reason that hybrid orbitals form at all is the familiar, all-pervasive drive for lower overall energy. Hybridization can only occur between atomic orbitals of similar energy level. Even though the formation of hybrid orbitals initially requires an input of energy for electron promotion, this energy and more is recovered during the subsequent bond formation and favourable geometry. I cite the example of carbon sp^3 hybridization in methane leading to the formation of four covalent bonds in a manner that satisfies the octet stability rule and has the bonding electrons with their attached groups as far apart in space as possible.

Our rather brief introduction to chemical bonding so far has been limited to single bonds, with only passing mention of multiple bonds. A check of the functional groups in Table 5.5 will show that several contain double bonds between adjacent atoms, such as C=C and C=O, and P=O. There are also other multiple bond types present in biomolecules, such as C=S and C=N. The properties and influence of multiple bonds in biomolecules are important and warrant a closer look.

The hybridization concept is also useful for molecules having double and triple bonds.

In the case of ethylene, $H_2C=CH_2$, we have the promotion of a 2s valence electron in carbon, as before, to give:

↑		↑	↑	↑
2s		$2p_x$	$2p_y$	$2p_z$

However in the case of ethylene, the subsequent hybridization is different.

On the way to making covalent bonds with another carbon atom and four hydrogens to form ethylene, three equivalent hybrid sp^2 orbitals (pronounced s-p-two) and an unhybridized 2p orbital emerge transiently. The electron configuration at this stage can be represented as

Table 5.5. Functional groups and linkages important in biomolecules.

Functional group/ Linkage	Structural formula	Examples
Alkane (saturated hydrocarbon)	–C–C–	linear, branched, cyclic
Alkene (unsaturated hydrocarbon)	C=C	trans, cis, cyclic; trans and cis fatty acids, cholesterol
Alcohol	–CH(H)–OH	Carbohydrates, lipids, amino acids
Aldehyde	–CH=O	Reducing sugars, e.g. glucose
Ketone	>C=O	Pyruvate; citric acid cycle metabolites
Carboxylic acid	HO–C(=O)–R	Organic, amino and fatty acids; proteins; lipids
Ester	R_1O–C(=O)–R	Lipids of eukaryotes and bacteria; acetyl groups
Amine (primary)	–NH_2	Free amino acids; peptides; proteins (often as –NH_3^+)
Thiol	–SH	Proteins often in oxidized form [–S–S–]
Thioester	R_1S–C(=O)–R	Acetyl-coenzymeA; biosynthesis of fatty acids; energy metabolism.
Phosphate ester	⁻O–P(=O)(O⁻)–O–C–	DNA, RNA

92 Introducing Biological Energetics

Table 5.5. (*continued*).

Functional group/ Linkage	Structural formula	Examples
Phosphoanhydride	$^-\text{O}-\overset{\overset{\text{O}^-}{\|}}{\underset{\underset{\text{O}}{\|\|}}{\text{P}}}-\text{O}-\overset{\overset{\text{O}^-}{\|}}{\underset{\underset{\text{O}}{\|\|}}{\text{P}}}-\text{O}^-$	ATP, ADP, and other nucleotides
Mixed acid anhydride	$-\overset{}{\underset{\underset{\text{O}}{\|\|}}{\text{C}}}-\text{O}-\overset{\overset{\text{O}^-}{\|}}{\underset{\underset{\text{O}}{\|\|}}{\text{P}}}-\text{O}^-$	Energy metabolism, e.g. acetyl phosphate

R refers to the rest of the molecule. Thus RCOOH refers to any carboxylic acid and RNH_2 to any primary amine.

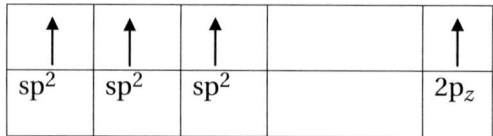

i.e. three sp^2 hybrid orbitals.

The boundary surface diagram of these orbitals is shown in Figure 5.8.

Figure 5.8 also shows the formation of a C to C sigma bond by 'end on' overlap of two sp^2 orbitals on neighbouring atoms. The p_z orbitals overlap 'sideways' to form what is called a π bond. The π orbital has two characteristic electron distributions, each roughly banana shaped, one above and one below the plane of the three sp^2 bonds. The two carbon and the four hydrogen atoms lie in the same plane (Figure 5.8(c)). Molecules with double bonds are 'flat' (coplanar) in the region of the double bond. No rotation is possible around the C=C bond. This rigidity is common to multiple bonds, and has an effect on the geometry around such bonds. In molecules such as unsaturated fatty acids, two possible arrangements across the bond are possible: the *cis* and *trans* configurations (Figure 5.8(d) and (e)). The physical and chemical properties of molecules having *cis* vs *trans* double bonds are different. A *cis/trans* interconversion is involved in the chemistry of the visual process in mammals.

Most single bonds, for example the C–C in ethane, H_3C-CH_3, or in general R_3C-CR_3, allow for rotation of attached groups, as shown below, adding to the flexibility of molecules:

The final type of carbon atom hybridization is called sp.

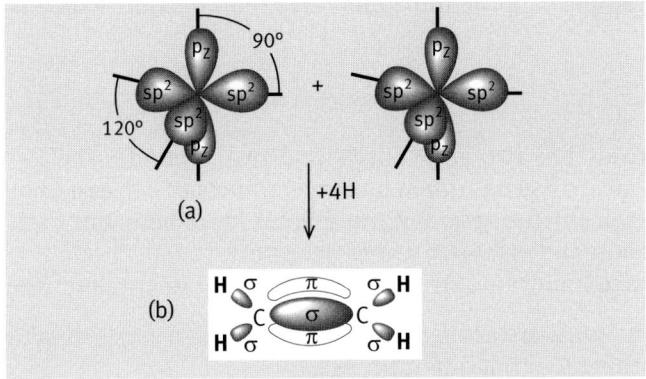

Figure 5.8. (a) Boundary surface diagrams of three hybrid sp² and one unhybridized p_z orbital. The sp² orbitals have planar geometry, with 120° angles; the p_z orbital is at 90° to the sp² plane. (b) Formation of ethylene, $HC_2=CH_2$, by the formation of one C–C sigma (σ) bond and four C–H σ bonds by 'end on' sp² orbital overlap, plus one C–C π bond by 'sideways' $2p_z$ orbital overlap. (c) A representation of the π bond showing the planar arrangement of the atoms involved. (d) Arrangement of hydrocarbon chains (⋀⋀⋀) on the same side (*cis*) of the double bond and (e) on opposite sides (*trans*) of the double bond. Antibonding orbitals are not shown.

The sp hybridized configuration for carbon is

↑ ↑		↑	↑
sp hybrid		$2p_y$	$2p_z$

In acetylene, the two carbons are joined by the overlap of two sp orbitals to form a C–C σ bond and each hydrogen forms a C–H σ bond. The $2p_y$ orbitals of each carbon overlap sideways to form one π bond and the $2p_z$ orbitals form a second π bond at right angles to the first. The result is a linear molecule with a triple bond. Only the sp σ bond orbitals are shown in the lower structure:

$$H-C\equiv C-H$$

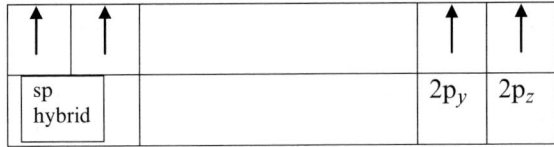

spσ spσ spσ

Another example of sp hybridization is carbon dioxide, which is also a linear molecule:

$$\overset{\delta^-}{O}=\overset{\delta^+}{C}=\overset{\delta^-}{O}$$

Carbon dioxide has two sp σ bonds and two π bonds. The C=O bonds are polarized because oxygen is much more electronegative than carbon (see below).

I won't discuss the formation of more bond types here, but I will draw on the above discussion and extend it where necessary.

To summarize for atomic structure and bonding in covalent molecules:

- Electrons in atoms are arranged systematically in a series of allowable energy levels according to strict quantum rules.
- Energy ultimately determines the type of bonding needed to form stable molecules from a combination of atoms.
- Atomic orbital hybridization allows for an increased number of stable molecular geometries.
- Bonding determines structure, and structure determines the three-dimensional geometry of molecules.
- Bonding determines chemical properties, for example via functional groups, and thereby chemical reactivity.
- Complex molecules have a range of structural features that determine their overall functionalities.

There is another major way in which atoms may obtain the stability number of electrons. Some achieve this is by fully donating one, two, three, or even four electrons to other atoms, of a kind that are capable of fully accepting electrons. Strong donors of electrons are found in Groups 1A and IIA of the periodic table. These are metals, and metals in general tend to donate electrons. Good acceptors of electrons are some of the non-metals in Groups V–VII. As an example of ionic bonding, consider sodium chloride.

Solid sodium chloride consists of the elements sodium, Na, and chlorine, Cl. The sodium atom donates one electron to a chlorine atom, so in this case $Na^+ Cl^-$ rather than Na–Cl is formed. The Na^+ and the Cl^- are called ions. In general, ions are atoms or groups of atoms that possess one or more electronic charges. A crystal of salt consists of billions of these sodium ions Na^+ and an equal number of chloride ions Cl^-, held together by their opposite electrostatic (coulombic) charges (Figure 5.9).

Once again the reason ions are formed is to do with lower energy. The sodium atom has one electron more than the nearest stability number and the chlorine atom has one fewer, so the formation of Na^+ and Cl^- satisfies the stability numbers of each. Similarly, other atoms may gain up to three or lose up to four electrons for the same purpose of minimizing energy. These tendencies of the various atoms to gain or lose, or to share, electrons are the reason for the specific formulae we see in chemical compounds. Sodium chloride is $Na^+ Cl^-$. Aluminium needs to lose three electrons to achieve its stability number, chlorine needs to gain only one, and the resulting compound has the formula $AlCl_3$ in the form Al^{3+} and $3Cl^-$.

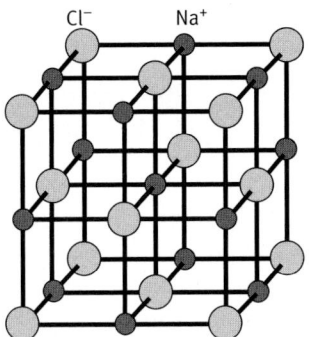

Figure 5.9. Sodium chloride crystal lattice structure. (Illustration from Crowe, J., Bradshaw, T., and Monk, P. *Chemistry for the Biosciences*, p. 43. © 2006 Oxford University Press.)

Ionic compounds in the solid state consist of the various ions packed in one of a number of three-dimensional arrays, with positive and negative ions alternating so that electrical neutrality is maintained locally. They tend to form structures with highly regular arrangements of their ions, known as crystal lattices. Ionic compounds tend to be harder than molecular ones and melt at much higher temperatures. Sodium chloride, molecular weight 58.5, melts at 801°C. The extended crystal lattice structure is held together by electrostatic forces and energy is required to break the lattice apart. Rather than use the term 'bond strength' when discussing the force that holds ionic compounds together, the term 'lattice enthalpy' is used. Sodium chloride and magnesium chloride, $MgCl_2$, have lattice enthalpies of 788 and 2523 kJ mol^{-1}, respectively.

Many covalent compounds also form crystals, but these are held together by much weaker intermolecular bonds—secondary bonds between the individual molecules. Molecular compounds tend to be softer and (unless they decompose first) melt at lower temperatures than ionic ones.

Covalent and ionic bonds are the 'strongest' types found in chemical compounds, and are called primary bonds. Primary bonds hold molecules together and maintain their integrity as compounds. Ionic compounds, if the compounds are soluble in water, tend to dissociate into individual ions, each of which is surrounded and stabilized in the energetic sense by a number of water molecules. Such solutions conduct electricity and a number of ions have properties essential to living systems. Examples include the involvement of sodium (Na^+) and potassium (K^+) ions in nerve transmission, and calcium ions (Ca^{2+}) in muscle contraction and in a number of enzyme systems. Magnesium ions (Mg^{2+}) activate many enzymes and magnesium is essential in chlorophyll. Bicarbonate (HCO_3^-) is an important buffer and phosphate in ATP/ADP processes is involved in cellular energy transformations.

Chemists have found it useful to classify elements by the power of their atoms to attract electrons to themselves. This property is called the electronegativity of the element and its value is often included in comprehensive versions of the periodic table. The idea and method of calculation of electronegativity were devised by the great chemist and Nobel Laureate Linus Pauling (1901–1994). A number calculated from measured bond energies to represent electronegativity has been assigned to each element. These values range from 4 for fluorine to 0.7 for the metal francium (Figure 5.4(b)). Electronegativity values are related to orbital energies in the atoms involved. They may be used to predict the relative ionic/covalent components of chemical bonds. A useful rule of thumb is: the greater the difference in electronegativity between two elements, the higher the percentage ionic character of the bond formed between them.

Conversely, the closer two elements are in electronegativity, the closer the bond between them is to being fully covalent (Aylward and Findlay 1994). Sodium (electronegativity 0.9) and chlorine (3.0) form sodium chloride, classified as completely ionic. Carbon (2.5) and hydrogen (2.1) in hydrocarbon molecules such as the gas methane, CH_4, have C−H bonds that are essentially fully covalent, or non-polar. Compounds with predominantly non-polar bonds include such hydrocarbons as petrols and oils, fats, and lipids. Compounds with C−O (2.5/3.5), O−H (3.5/2.1), and other bonds having electronegativity differences between their atoms of around 1 unit are about 20% ionic and are classified as being polar. Compounds with predominantly polar bonds include water, carbohydrates, amino acids, nucleic acids, and parts of many proteins. Polar bonds are present in most of the functional groups in biomolecules, the peptide linkage in proteins and the phosphodiester linkages in DNA and ATP to name just a few. Polar compounds tend to be soluble in water, whereas non-polar compounds are not.

Considering the full range of bonding discussed above, we can conclude that the degree of electron sharing in chemical bonds constitutes a continuum (Freeman 2008). At one extreme are the fully covalent, non-polar compounds that have bonds involving equally shared electrons. Next are bonds with a range of polarities. At the other extreme are fully ionic compounds, which don't share electrons but donate/accept them. The range of properties conferred by such a bonding continuum has been fully utilized by living organisms, as demonstrated by the corresponding range of functionalities exhibited by biomolecules:

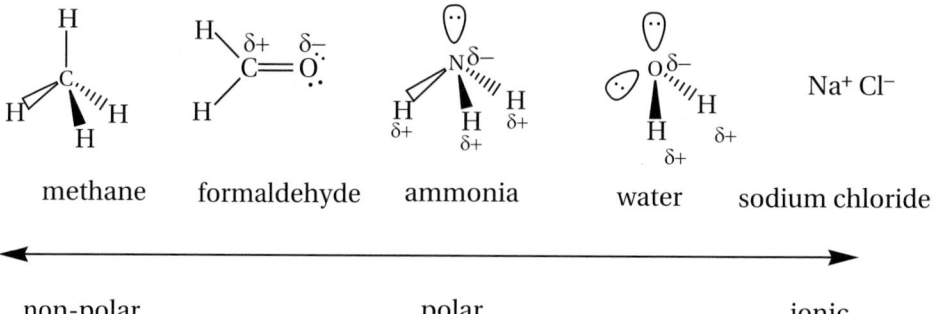

Polar bonds are also the basis of the one of most important types of secondary bond found in chemistry, the hydrogen bond. As an example of the formation of hydrogen bonds consider the water molecule. Although we represent it as H–O–H, this is not the whole story, and to explain the properties of water molecules,

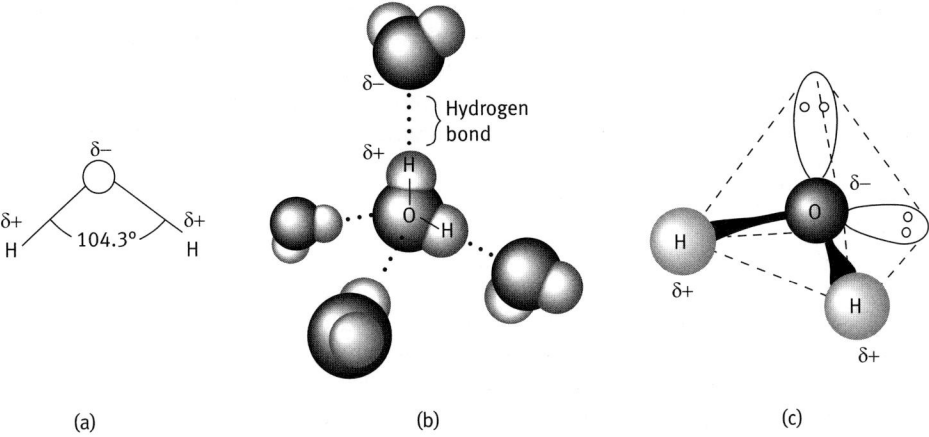

Figure 5.10. Some effects of differences in electronegativity. (a) The O–H bonds in water are permanently polarized, the bonding electrons spending greater time around the more electronegative oxygen. (b) Partial positive charges on hydrogen atoms are attracted to the partial negative charges on oxygen to form hydrogen bonds. (c) The two non-bonding molecular orbitals containing lone pairs of electrons make the water molecule effectively tetrahedral (dotted lines) in shape.

we need to go a little deeper into the stability number of oxygen. The oxygen atom has the electron configuration $1s^2\ 2s^2\ 2p^4$, and eight protons to balance these in the nucleus. Each hydrogen atom has one electron and one proton in the nucleus. The electrons shared to form each of the O–H σ bonds in H–O–H spend some of their time around the hydrogen nuclei, to satisfy the hydrogen stability number, and some of their time around the oxygen nucleus, to satisfy the oxygen stability number. However, when they are around the oxygen nucleus, the negatively charged electrons experience the net attractive influence of the eight positively charged protons. The net influence of these eight protons is not eight full units. The 1s and 2s electrons 'shield' the bonding electrons in the 2p orbitals from the full effects of eight protons. Nevertheless, as a result of the extra attraction, compared with that of only one proton when they are around the hydrogen nucleus, the shared electrons spend slightly more time near the oxygen nucleus, and slightly less time near the hydrogen nuclei. This confers a permanent net negative charge around the oxygen nucleus and a slight but permanent net positive charge around each hydrogen nucleus (Figure 5.10(a)) The Greek lower case symbol delta (δ) is conventionally used to indicate a small part of something, in this case a partial positive (δ+) or partial negative (δ−) charge, instead of a full electronic (−) or proton (+) charge.

Hydrogen bond donor atoms (δ+) of nearby water molecules will align to be as close as possible to the lone pairs of electrons (δ−) in the non-bonding oxygen orbitals. The actual bond strength of each hydrogen bond is quite low, normally being about 10–15% of covalent or ionic bond strengths, hence the term 'secondary bond'. Despite the relative weakness of hydrogen bonds (and other secondary bonds) they are important in helping to determine the overall shapes of many biomolecules. One reason for this is that some large biomolecules have a number of hydrogen bonds acting in a cooperative manner and holding them in the correct shape. This is a situation similar to that in velcro—a large number of weak linkages acting in cooperation to form a strong overall join. In large biomolecules such as proteins, polysaccharides, and DNA, linear chains are often coiled and/or multiply-folded into single and double helices or into quite compact, globular shapes.

Electrons in non-bonding orbitals, called lone pairs, play important roles in many molecules. Water, alcohols, and ammonia are good examples. In ammonia, NH_3, the nitrogen atom is sp^3 hybridized and forms three sigma bonds with the three hydrogen atoms. The remaining sp^3 orbital is occupied by a lone pair of electrons. The overall shape, if one includes the lone pair, is close to tetrahedral. This geometry is revealed when the lone pair accepts a proton H^+, to form the ammonium ion NH_4^+:

Ammonia and amines, for example RNH_2, are thus classified chemically as bases, as they can accept protons via the lone pairs of electrons. There is a lot more that could be said about bonding and structure in biomolecules, and some of it will be introduced in succeeding chapters as necessary.

Having dealt rather briefly with thermodynamics in Chapter 4 and with bonding and structure in molecules in this chapter, we are now in a position to discuss chemical reactions in some detail. Why should compound A plus compound B yield compound C and not compounds X and Y? In most cases reactions are planned with a particular product in mind. Because of the extensive knowledge built up by chemists over many years, the expected result is often achieved. In many cases a feasible mechanism can be proposed in terms of bond-breaking and bond-making and the properties of functional groups. The proposed mechanism can be tested with further experiments.

I will now concentrate on a few chemical principles that will be helpful when we come to deal with the enzyme-catalysed reactions of metabolism. The same principles apply.

Some of the reaction types that are catalyzed by enzymes will be taken as examples.

It is useful to discuss chemical reactivity and mechanisms of carbon compounds in terms of various functional groups. Functional groups have properties that depend on the type of bonding involved. A particular functional group can be expected to behave, in terms of its reactivity, in essentially the same way wherever it may occur in a molecule. There are exceptions to this, especially when two or more functional groups are close together in the molecule. This is called the neighbouring group effect, and can be important in fine tuning the properties of biomolecules. Functional groups important in biological molecules are listed in Table 5.5, and will be encountered extensively in later chapters.

Many of the functional groups in Table 5.5 contain the carbonyl group, >C=O, which has a type of double bond we have not encountered. The bond occurs between two different types of atom and this has important consequences. As the nature of the bonds largely determines the reactivity typical of the carbonyl group, it is useful to consider this bonding in some detail. We will look at formaldehyde, HCHO, the simplest carbonyl compound, as an example:

$$2p_z(C) + 2p_z(O) \longrightarrow \pi$$

(a)   (b)   (c)   (d)

The C–O and C–H sigma bond core structure of the carbonyl group is shown in (a).

The molecule can be considered as built from an sp^2 hybridized carbon atom forming three sigma bonds. In (b) and (c) the remaining $2p_z$ orbitals of C and O overlap sideways to form a π bond. Because the more electronegative oxygen dominates in the π MO the C=O bond is polarized, giving rise to the charge distribution, as shown in (d). This makes the C=O bond susceptible to attack by nucleophilic reagents. Nucleophilic regents include OH^-, CN^-, H^- (hydride), and other 'electron-rich' species. Nucleophilic addition to a carbonyl group is one of the most common and useful reactions in organic chemistry. Table 5.5 lists five biologically relevant functional groups that incorporate the >C=O functionality that I will use in examples of reaction mechanisms. Reaction mechanisms are reasonably detailed descriptions used by chemists to explain how reactions form the products observed.

1) The reduction of ketones to form alcohols.

The overall reaction can be generalized to:

$$\begin{array}{c} R \\ \diagdown \\ C=\ddot{O} \\ \diagup \\ R_1 \end{array} + H_2 \longrightarrow \begin{array}{c} R \\ \diagdown \\ HC-OH \\ \diagup \\ R_1 \end{array}$$

The chemical method involves use of a metal hydride such as $NaBH_4$, a powerful reducing agent that is best used in the absence of water. The primary reductant is the hydride ion, H^-, which attacks the electron-deficient carbon atom:

$$R\underset{R_1}{\overset{}{C}}=\ddot{\overset{..}{O}} \xrightarrow{\text{H from NaBH}_4} R\underset{R_1}{\overset{H}{C}}\overset{}{O^-Na^+} \xrightarrow{H_2O} R\underset{R_1}{\overset{H}{C}}\overset{}{OH}$$

The curly arrows show the movement of a pair of electrons; hydride transfer is a two-electron process. The final alcohol product is formed by the addition of water to the reaction mixture.

The biochemical analogue is the reduction of pyruvate to lactate during anaerobic metabolism of carbohydrate. The enzyme involved is lactate dehydrogenase (LDH):

$$\underset{\text{pyruvate}}{H_3C\underset{^-OOC}{\overset{}{C}}=\ddot{O}} \underset{NADH + H^+ \rightarrow NAD^+}{\overset{\text{lactate dehydrogenase}}{\rightleftharpoons}} \underset{\text{lactate}}{H_3C\underset{^-OOC}{\overset{H}{C}}OH}$$

The NADH is a coenzyme utilized here by LDH to carry out the net transfer of two hydrogen atoms to pyruvate. The vital role of this coenzyme in metabolism will be discussed in Chapter 11.

In the biochemical reaction, the overall result, reduction of a carbonyl group to an alcohol, is the same, but there are important differences from the chemical mechanism.

- The reaction is faster. Although NaBH$_4$ reductions can be quite fast by organic chemistry standards, they are not as rapid as the enzyme-catalyzed reaction.
- The conditions are greatly different. The organic reaction occurs best in the absence of water, usually in an organic solvent. The reagent NaBH$_4$ is destructive to biological tissues. The enzyme-catalyzed reaction occurs in an aqueous but protected environment, involving the enzyme active site and the coenzyme NADH.
- The enzyme-catalyzed reaction is much more selective. If one were to try the reduction of lactic acid using NaBH$_4$, the –COO$^-$ group would be at least partly reduced as well. Of necessity, selectivity is a feature of enzyme-catalyzed reactions. The process is again a two-electron reduction involving hydride (H$^-$) transfer, an essential step being hydride transfer from NADH.

2) The base-catalyzed hydrolysis of an ester.

The overall reaction can be generalized to:

$$\underset{R_1O}{\overset{R}{>}}C=O \xrightarrow{NaOH/H_2O} \underset{Na^+ \; ^-O}{\overset{R}{>}}C=O \; + \; R_1-OH$$

The chemical method typically uses sodium hydroxide in aqueous ethanol, for example to hydrolyze fatty acid esters. The curly arrow mechanism is:

$$\text{(a)} \quad \rightleftharpoons \quad \text{(b)}$$

$$\text{(b)} \quad \rightleftharpoons \quad \text{(c)} \quad + \quad R_1O^- \quad \longrightarrow \quad \text{(e)} \quad + \quad R_1OH \quad \text{(f)}$$

The final step involves the transfer of a proton from acid (c) to the strong base RO^- (b) to form an alcohol R_1OH (f).

A specific example is the hydrolysis of triglycerides (triacyl glycerols), the main fat storage molecules in many cells. In vivo, the reaction is catalyzed by lipases.

triglyceride $\xrightarrow{OH^- \text{ or lipase}}$ glycerol + fatty acid salt [a soap]

This chemical reaction is commonly called a saponification, as it is used to make soap from animal fats. The fatty acid shown is palmitic ($CH_3(CH_2)_{16}COOH$) and the counterion in the soap is Na^+.

The examples above provide some idea of the way electrons in compounds can move, becoming involved in breaking and forming bonds to produce new compounds. In addition to a reasonable mechanism to form the observed products, a comprehensive study of a chemical reaction should provide information on aspects such as the kinetics, energetics, and extent of the reaction, the influence of conditions (solvents, pH, salt concentration, temperature, pressure), and the effects of catalysts and inhibitors. These will be dealt with in Chapter 6.

Biomolecules perform a large number of roles and constitute a structurally diverse group. The overall structure of each molecule is such that it is suited to whatever role it plays in the living organism. Chemists and biochemists are very interested in these structure–function relationships. Why, for example, should some small molecules, such as ATP (adenosine triphosphate), be highly reactive and capable of providing energy to drive important reactions? What causes other molecules, such as cellulose, to be large and quite unreactive, but to have high tensile strength, making them useful as structural elements in plant cell walls, capable of withstanding great forces? Imagine the stresses experienced by large trees in a hurricane or by seaweed in a violent coastal storm.

What we need to consider are the determinants of chemical reactivity in the ATP example and those of overall three-dimensional structure in cellulose. Both of these properties are largely determined by the types of chemical bonds in the molecule, the way the bonds are arranged, and the elements involved. The type of bonding largely determines the structure, shape, and reactivity of molecules, and this is why I have spent considerable time on these aspects of chemistry.

Glucose has the molecular formula $C_6H_{12}O_6$. Unfortunately, this doesn't tell us much about its three-dimensional structure, which is important. In fact, there are at least 24 sugars with the molecular formula $C_6H_{12}O_6$. If we consider the alternative ring forms that these sugars may adopt, the number rises to 48. How are the atoms arranged? What kind of bonds are involved? Obviously the simple molecular formula is inadequate because each of these sugars has it own unique properties. This means that the three-dimensional structure of each sugar is unique to that sugar alone. The various sugars possess some similar properties; being classified as sugars, they naturally share some structural and reactivity features. There are several ways to represent the three-dimensional structures of molecules on a two-dimensional page. Three ways are shown below, using β-D-glucose as an example. The first gives some idea of its shape. Thus, we see the formation of a six-membered ring, containing five carbon atoms (not shown explicitly here) and one oxygen atom. The second is more a realistic representation that is referred to as the 'chair form' for obvious reasons.

The third figure gives a better idea of the three-dimensional shape, but takes considerable time to draw. All are abbreviated forms in which not all the atoms are explicitly represented—the carbon atoms of the ring and the hydrogen atoms attached to them are not shown in the first two. These forms are often used by chemists for convenience when they want to draw structures quickly in lectures or discussions. One disadvantage of the above structures is that they greatly exaggerate the bond lengths between atoms. This is done for purposes of clarity and ease of recognition. A fourth representation of glucose, known as a space-filling model, most realistically represents the shape of the glucose molecule (Figure 5.11). Note, however, that it is not possible to see all the atoms, and their bonding relationships are not clear, in a space-filling model projected in two dimensions. Also it would be impractical to draw this by hand. The chair form is overall much more useful, and similar diagrams may be used to depict almost any molecular structure. Space-filling models are useful, however, when we wish to see how two molecules fit together and are best constructed by computer, so they can be spun round and looked at from any direction by the use of simple commands. This technique is known as molecular graphics and it is very useful in many areas of chemistry and biochemistry. Quantitative calculations involving energy can be carried out and recent years have seen a remarkable increase in the use of computers for examining the three-dimensional structures of proteins, enzymes, and drugs, their shapes and their interactions. A whole new subdiscipline of molecular modelling has emerged. Some recent drugs have in fact been designed by computer, for example the influenza drug Relenza. Using the known three-dimensional structures of certain 'target' molecules to which the drug attaches itself, a drug can be designed to have the specific shape and size to do so. This approach could lead to much more rational drug design than in the past, when chemists spent huge amounts of effort to carry out structural modifications of existing drugs in the hope that some of the modifications would cause improvements in effectiveness or fewer side effects. It was rather a hit-and-miss approach, not efficient, and therefore costly. Rational drug design using computer modelling of molecules will, it is hoped, reduce the rate of increase in the cost of new drugs.

I have spent considerable space discussing spontaneous processes, energy liberated or absorbed during reactions, bonding, structure, and reactivity in molecules. I have extolled the virtues of having sound theoretical explanations for the experimental results we observe and of having a quantitative approach to science. Does this brief theoretical background allow us to rationalize two reactions that are the very essence of life?

$$\text{Respiration: } C_6H_{12}O_6 + 6O_2 \rightarrow 6CO_2 + 6H_2O \quad (5.1)$$

$$\text{Photosynthesis: } 6CO_2 + 6H_2O \rightarrow C_6H_{12}O_6 + 6O_2 \quad (5.2)$$

We know experimentally that reaction (5.1) is spontaneous as written and vice versa for reaction (5.2). Why?

Reaction (5.1) produces 12 moles of gas from 1 mole of solid and 6 moles of gas. Even qualitative thinking points to a significant increase in entropy, as the number of microstates available to the 12 moles of products is enormously

104 Introducing Biological Energetics

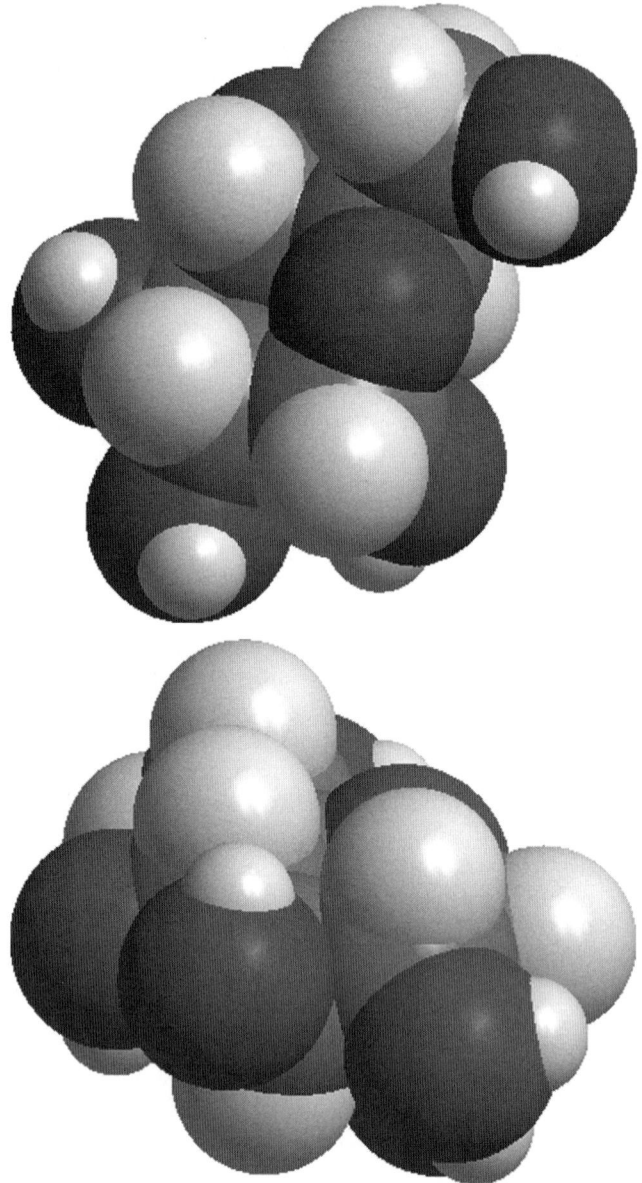

Figure 5.11. Space-filling—models of -β–D- glucopyranose. They are not easy to interpret in two dimensions

more than those available to the 7 moles of reactants. This alone is not enough. Enthalpy changes must be considered. From Hess's law and using average bond enthalpies it is possible to predict the approximate enthalpy of a reaction. For reaction (5.1) average bond enthalpies yields a ΔH^0 value of 2,799 kJ mol^{-1}, compared with an experimental value of 2801 kJ mol^{-1} (Chapter 4, p. 58).

This quantitative result is satisfying, but still leaves the nagging question of why the bonding in the molecules O_2 and H_2O should be of higher bond dissociation energy and higher stability, and of lower potential energy than those in glucose and oxygen.

Some relevant bond dissociation energies (in kJ mol^{-1}) are C=O in CO_2 799, C−H 414, O−H 460, C−O 351, and C−C 347.

In water and carbon dioxide the bonding electrons are closer to the more electronegative oxygen nucleus than they are to carbon in the non-oxidized compounds.

$$\overset{\delta+}{H}\overset{\overset{\delta-}{O}}{\diagup}\overset{\delta+}{H} \quad \overset{\delta-}{H}\overset{\delta+}{O}=\overset{\delta-}{C}=\overset{\delta-}{O}$$

The bonds are polar and have high bond-dissociation energies (O−H 460 and C=O 799). In glucose, non-polar C−C and C−H bonds predominate, with lower bond dissociation energies (C−C 347, C−H 414, C−O 351). To emphasize this, glucose can be drawn as

$$HO-\underset{H}{\overset{H}{C}}-\underset{OH}{\overset{H}{C}}-\underset{OH}{\overset{H}{C}}-\underset{H}{\overset{OH}{C}}-\underset{OH}{\overset{H}{C}}-\overset{H}{C}=O$$

In terms of bond energy levels, the C atom in carbon dioxide is involved in two sp-hybridized σ−bonds and two π−bonds. The σ−bond has a high degree of s-character relative to p-character. In glucose, most C atoms have sp^3 hybridized σ−bonds, which have a high degree of p-character relative to s-character. Figure 5.3 shows that p-orbitals are higher in potential energy than s-orbitals in the same shell. As a result, hybrid sp^3 orbitals are of higher potential energy than sp-orbitals. Thus C−H bonds are of higher potential energy than C=O bonds.

In summary, glucose has more bonding electrons in non-polar, higher potential energy bonds than those in water and carbon dioxide. This higher potential energy is 'tapped' during the process of direct oxidation during combustion, releasing 2801 kJ mol^{-1}, and generating more stable, lower potential energy carbon dioxide and water.

During the metabolism of glucose some, but not all, of this potential energy is tapped to provide biological energy. The rest of this book largely concerns how living organisms achieve this feat and how the photosynthetic organisms drive the reverse reaction (5.2) to generate the potential energy in the first instance.

Glucose and other carbohydrates are already partially oxidized. Let's look at the progression of oxidation from the least oxidized carbon compound, a hydrocarbon, through to the fully oxidized carbon dioxide:

```
     H                    OH                                    OH              O
     |                    |                                     |               ‖
H — C — H   [O]→   H — C — H   [O]→   H — C = O   [O]→   C = O   [O]→   C
     |                    |                    |                    |            ‖
     H                    H                    H                    H            O

    (a)                  (b)                  (c)                  (d)          (e)
hydrocarbon           alcohol             aldehyde         carboxylic acid   carbon dioxide
 (methane)           (methanol)         (formaldehyde)      (formic acid)
    890                  726                  571
```

The sequence from (a) to (e) shows the hypothetical oxidation of a hydrocarbon. All five compounds exist and are stable. The overall equation is:

$$CH_4 + 2O_2 \longrightarrow CO_2 + 2H_2O + \text{heat}$$

One could oxidize compounds (a) to (d), separately, to CO_2. Each oxidation would release less energy than the preceding one. The numbers below (a), (b), and (c) show the standard heats of combustion (kJ mol^{-1}). A typical carbohydrate would fit between (b) and (c). This energy release relative to extent of oxidation is well illustrated in the differences in energy produced by the oxidation of different foods (Table 5.6).

Note that the values are for complete oxidation per gram to water and carbon dioxide. They give an indication of the relative energy contents of the foodstuffs, on a weight-to-weight basis, if fully metabolized. Thus, a typical fat will yield about 2.5 times the energy as the same weight of a typical carbohydrate. Considering their respective degrees of oxidation, the values are in the same sequence as the compounds (a) to (d) above, with the average dietary protein about midway between carbohydrate and alcohol. Such data can be used to estimate the number of kilojoules consumed in a diet of given composition. This example illustrates that a quantitative knowledge of thermodynamics can have a very practical application, one that impacts on our daily lives. Note the high-energy value of alcohol.

Let's look at a summary of the biological oxidation of a hydrocarbon, taking methane as the example to make the comparison simple. (In a typical cell, a hydrocarbon chain would be broken into two-carbon fragments by the process of

Table 5.6. Amount of energy produced by the oxidation of different foods.

Substance	Energy (kJ g^{-1})
Fat	37
Alcohol	29
Protein	23
Carbohydrate	16

Source: Haynie (2001), p. 12.

β-oxidation, then fed as acetyl-coenzyme A into the tricarboxylic (TCA) cycle (see Chapter 11.))

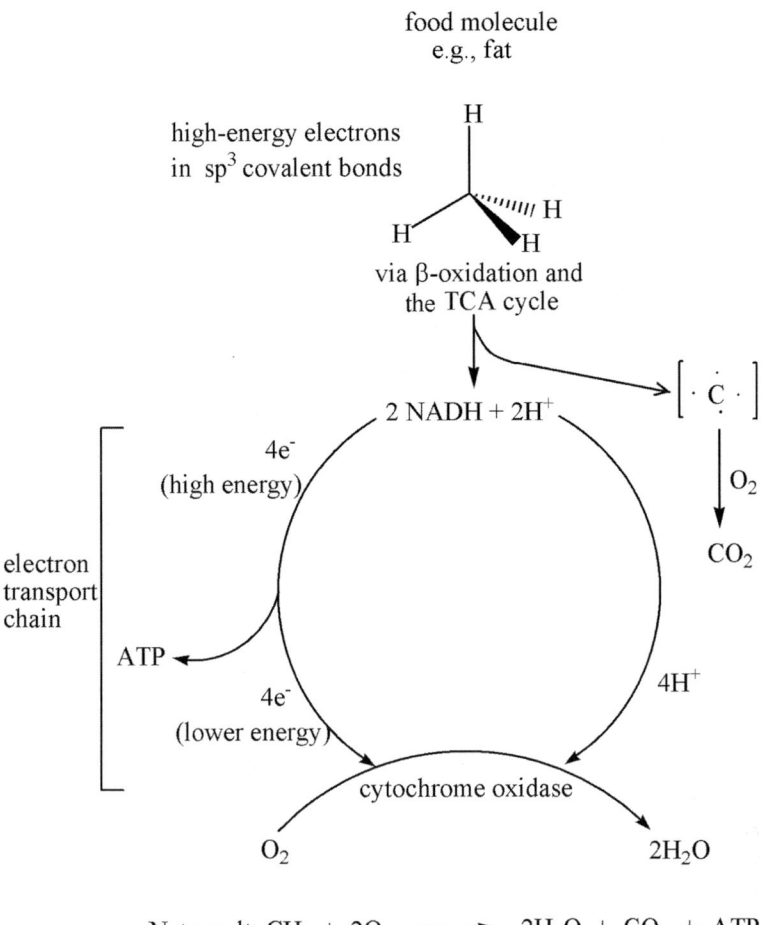

Net result: $CH_4 + 2O_2 \longrightarrow 2H_2O + CO_2 + ATP$

In terms of products, the net result is the same as for direct combustion, but some of the energy is diverted to form ATP, to the advantage of the cell. More of this will be discussed later, but here we should take note of the chemical principles used in such biological oxidations. Electrons in high potential energy orbitals are separated from their atoms and used via the electron transport chain to help drive the synthesis of ATP. The protons formed go to providing the protonmotive force for the same purpose. Finally, the electrons and protons recombine (in lower potential energy orbitals) by bonding to the ultimate electron acceptor, oxygen. The carbon is oxidized to carbon dioxide, which is typically liberated via the TCA cycle. Overall, this is beautifully simple, a solution to an energy supply problem that was solved by natural selection some 2 billion years ago.

Oxidation has been a common reaction on Earth since levels of oxygen became significant as a result of photosynthesis. Free oxygen is a reactive substance. Combustion of organic materials in an excess of oxygen generates H_2O, CO_2, SO_x, and NO_x as end products because they are stable. Similarly, some inorganic compounds and metals, if heated in air, will stop at the oxide stage: $CaCO_3$ yields CaO and CO_2; $Ca(OH)_2$ yields CaO and water; iron forms iron oxides.

We will see in Chapter 10 that the concept of oxidation can be extended from the direct addition of oxygen, as in the examples above, to a more fundamental process. Oxidation as a process involves a net donation of electrons to another species. Oxygen just happens to be the best-known electron acceptor. Thus, the conversion of iron II to iron III is an oxidation:

$$Fe^{2+} \rightleftharpoons Fe^{3+} + e$$

The opposite and linked process is reduction and involves a net acceptance of electrons. Thus, a common oxidizing agent, permanganate, is reduced in acid solution:

$$MnO_4^- + 4H^+ + 3e^- \rightleftharpoons MnO_2 + 2H_2O$$

Electron transfer in these and the many other redox reactions involves energy changes. A number of such reactions are utilized by living cells in important metabolic processes such as respiration.

Life has brought us a long way from the hydrogen atom.

GENERAL REFERENCES

Keeler, J. and Wothers, P. (2003) *Why Chemical Reactions Happen*. Oxford University Press, Oxford.

REFERENCES

Aylward, G and Findlay, T. (1994) *SI Chemical Data*, 3rd edn. John Wiley & Sons, Brisbane, pp. 115–6.
Ball, D.W. (2003) *Physical Chemistry*. Brooks/Cole-Thomson Learning, Pacific Grove, CA.
Campbell, N.A., Reece, J.B., Urry, L.A., et al. (2008) *Biology*, 8th edn. Pearson Benjamin Cummings, San Francisco, p. 32.
Chang, R. (1994) *Chemistry*, 5th edn. McGraw-Hill Inc., New York, pp. 291–332, 394.
Ferris, T. (1997) *The Whole Shebang*. Weidenfeld & Nicolson, London
Freeman, S. (2008) *Biological Science*, 3rd edn. Pearson Benjamin Cummings, San Francisco, p. 22.
Garrett, R.H. and Grisham, C.M. (1999) *Biochemistry*, 2nd edn. Saunders College Publishing, Fort Worth, p. 6.
Haynie, D.T. (2001) *Biological Thermodynamics*. Cambridge University Press, Cambridge, Ch 1, p. 12.

6
How Fast, How Far? Chemical Kinetics and Equilibrium

An aim of this book is to demonstrate how energy is accessed and utilized to sustain life on Earth. Chemical reactions play a major role in this, so the more we understand about chemistry, the easier the task should become. Let's reflect on what needs to be addressed. Ideally, scientists would like to be able to predict the likely result of any proposed reaction. That would provide confidence that chemical theory is soundly based. It would encourage further work to refine the ideas—no one would suggest that everything is known. Experience has shown that deeper understanding leads to better prediction, with the result that knowledge is continually expanding.

Much of what we need to know about chemical reactions can be summarized in a few words:

Which way?
How fast?
How far?
What?
How?

For any given reaction, these aspects are totally integrated and any order of treatment will be somewhat arbitrary. How far have we come so far in the search to understand chemical reactions?

Chapter 4 dealt rather briefly with *which way*, by introducing the Second Law of thermodynamics and a means to determine the direction of a spontaneous change—a negative ΔG is required for any spontaneous process, with contributions from enthalpy and entropy. Entropy might be conceptually challenging, but its effects are very real and in some processes a dominating influence.

Chapter 5 described the way in which atoms combine to form molecules: the importance of relative electron energies and the formation of molecular orbitals. The section on mechanisms provided some insight about the rearrangements of bonding electrons to form new compounds. It partially answers the *how* question.

The *what* question is mostly about identifying the products of a reaction. Predicting the products beforehand is a harder proposition, but ultimately the one that provides the richest rewards. A systematic knowledge of reaction mechanisms is important for full understanding and prediction.

The remaining questions—'*how fast, how far*'—are the topics to be addressed in this chapter.

110 Introducing Biological Energetics

How fast is the topic of chemical kinetics. I won't treat this topic formally, concentrating instead on the basic concepts in the context of biological reactions.

Consider the simple organic reaction between chloromethane and hydroxide ion in water:

$$HO^- + CH_3Cl \longrightarrow CH_3OH + Cl^- \qquad (6.1)$$

This reaction proceeds spontaneously and the products, methanol and chloride ion, can be isolated in high yield. This is far from the whole story, so let's look at the details.

1) We have seen in Chapter 4 that if ΔG is negative, a reaction will be spontaneous. That is the case here, as $\Delta G^0 = -105.6$ kJ mol^{-1}. This answers the question *which way* quantitatively. It is a large value, confirming the high yield of products. As a comparison, it can be shown that any reaction that goes to 99.9% completion at 25°C has a ΔG of -17.1 kJ mol^{-1} (Streitwieser and Heathcock 1981).
2) How fast? This is our interest here. What factors determine the rate of a reaction?

Though ΔG^0 in reaction (6.1) is large in favour of the products, the reaction is actually quite slow. A 0.05 M solution of chloromethane in 0.1 M aqueous NaOH solution reacts to only to the extent of about 10% in 2 days at room temperature (Streitwieser and Heathcock 1981). Obviously, favourable thermodynamics is not enough. There must be a suitable pathway available. Chemical reactants usually encounter an energy barrier before any reaction takes place. This is particularly the case in organic reactions when covalent bonds are broken and formed. This energy barrier is formally called the activation energy for the particular reaction. The values of activation energies can vary greatly from reaction to reaction. A useful way to visualize the energy changes during the course of a reaction is to plot a reaction profile. Potential energy is plotted vs the reaction coordinate to represent the progress of a reaction (Figure 6.1).

High activation energies lead to slow rates, and vice versa. The ΔG^* for reaction (6.1) is about 105 kJ mol^{-1}.

The average kinetic energy of molecules at room temperature is about 2.5 kJ mol^{-1}. However, this is an average value, and molecules are constantly colliding

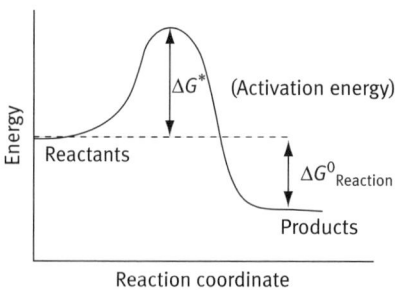

Figure 6.1. A reaction profile for a one step chemical reaction that has a negative $\Delta G_{reaction}$ and activation energy ΔG^*. Reactions with high values of ΔG^* tend to be slower and *vice versa*.

and exchanging energy. There will be a few with energies above 105 kJ mol^{-1}, hence the 10% reaction over 2 days at room temperature. At a higher temperature the average kinetic energy of the molecules will be greater, a larger number will achieve the activation energy in a given time, so the rate will increase (Figure 6.2). Thus, the rate of reaction (6.1) at 50°C is about 25 times that at 25°C.

Under any set of conditions, molecules have a distribution of kinetic energy depending on the absolute temperature. As the temperature is increased, the distribution of energy broadens, and shifts to higher-energy values (Figure 6.2) The curves are skewed slightly in the high-energy direction; a small percentage of molecules has very high kinetic energy. The result is that, even when the activation energy is high, some reactions will occur at room temperature.

Reaction (6.1) provides a useful example. It answers the which way, how fast, and how far questions quantitatively. It is important to note that the magnitude of $\Delta G_{reaction}$ tells us nothing about the rate. The sign of $\Delta G_{reaction}$ tells us the direction of change (which way) and its magnitude informs us of the position of equilibrium (how far). The position of equilibrium will be considered in some depth later in the chapter.

Consider another example. If I were to mix 2 L of hydrogen and 1 L of oxygen in a glass flask and seal it, what would happen? At room temperature, nothing much. One could observe the flask for days and still see no change. However, if I were to arrange to generate a small electric spark inside the flask, you would certainly notice a difference. There would be a flash, a few drops of water would form in the flask, and the temperature would rise slightly, showing that some energy was liberated overall. Temperature is a reflection of the average kinetic energy in a molecular system. If the glass of the flask were too thin it might collapse—3 moles of gas becoming 2 moles of liquid water would result in a partial vacuum inside. The equation for this reaction is:

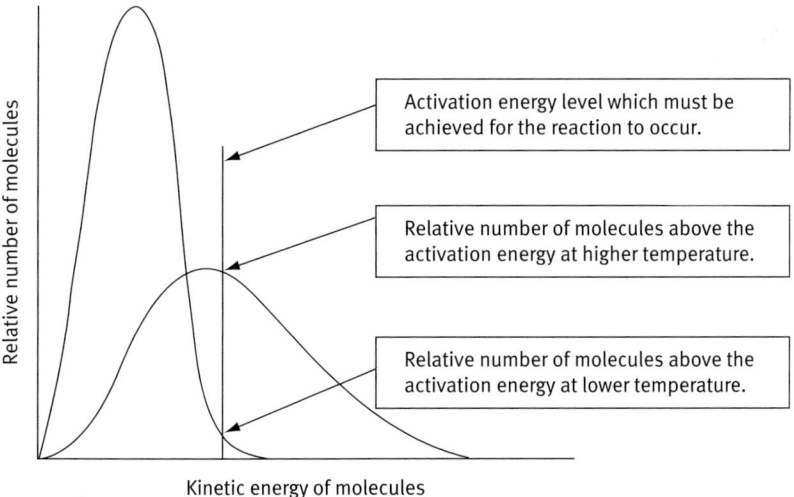

Figure 6.2. Distribution of molecular kinetic energy. At higher temperatures, a greater number of molecules will be above the activation energy level, resulting in a faster reaction rate.

$$2H_2(g) + O_2(g) \longrightarrow 2H_2O(l) \tag{6.2}$$

The spark provides enough energy to initiate the reaction, but why was energy needed to set it off under the prevailing conditions, and once started, why did it continue?

The covalent bonds in hydrogen and oxygen molecules are quite stable at room temperature and pressure. To react, the molecules must be brought into intimate contact and any repulsion between them must be overcome. In the case of hydrogen and oxygen at room temperature, the energy distribution is such that an insignificant number of molecules possess the activation energy. Once the kinetic energy of a few more molecules of hydrogen and oxygen has been boosted above the activation energy level by the energy from the spark, there is enough heat energy liberated by the reaction itself to break a few more H–H and O=O bonds, which rearrange to form more water, liberating more energy to break more bonds, and so on. All this happens very fast so that the hydrogen and oxygen form water literally in a flash.

The involvement of the spark is another example of the statement that a chemical reaction will proceed only if there is a suitable pathway available for it to do so. Many chemical reactions need an energy boost to get them going at a reasonable rate at room temperature. Some, like the formation of water above, once given the initial boost will generate enough energy to keep going without the need for any more external energy. Such reactions liberate energy overall and are therefore exothermic. Other reactions might need to be heated continuously to provide sufficient activation energy for the molecules to be converted to products at a reasonable rate. Most chemical and biochemical reactions need some activation energy; some need much more than others. In the case of industrial processes, slow reaction rates may not be economical. One solution to this problem is to raise the temperature of the entire reaction, increasing the average kinetic energy of the molecules. More molecules will achieve the activation energy, so more will react in a given time. A useful rule of thumb is: for each 10°C increase in temperature, the rate of a reaction will approximately double.

Another approach is to use a catalyst, which alters the rate of a chemical reaction. Catalysts have proved to be extremely valuable. Intense efforts by the chemical and pharmaceutical industries have produced an array of catalysts, many of which are protected by patents. The reaction between hydrogen and oxygen mentioned above may be brought about at room temperature by a catalyst such as finely powdered platinum metal. To merely say that catalysts speed up a chemical reaction by lowering the activation energy is an oversimplification. Usually the detailed mechanism is changed, so that the new, catalysed pathway has lower activation energies (Figure 6.3).

For a reaction to proceed spontaneously under the given conditions, the Gibbs energy change, ΔG, must be negative. Note that the free energy change for the overall reaction is the same with or without the catalyst. Usually, only a small amount of catalyst is required, much less than the stoichiometric amounts of the reactants. This is because the catalyst, once it has done its job, can be released from the products to catalyse more reactant molecules. A true catalyst

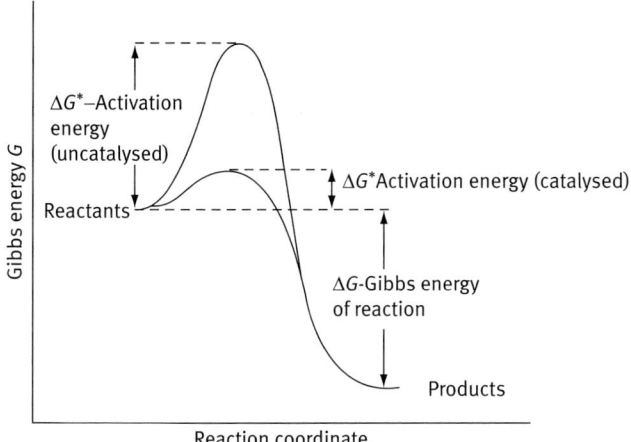

Figure 6.3. Reaction coordinate for uncatalysed and catalysed reactions. The activation energy, ΔG^*, is lower for the catalysed reaction, so the reaction rate is increased. The Gibbs energy change of the overall reaction, $\Delta G_{reaction}$, is not altered by the presence of a catalyst.

is not permanently changed by the reaction (6.1) catalysed. This is important industrially, as catalysts are usually expensive. Catalysis is vitally important in biology, as the enzymes that catalyse and modulate biochemical reactions are normally present in quite small amounts, and are a precious resource for the organism involved. Much more about enzymes, their very special properties, and the energetics of enzyme-catalysed reactions will be discussed in Chapter 8.

We haven't yet addressed the question 'why should activation energy be required at all?'

As we have seen, the rate of a reaction depends on the fraction of molecules above the activation energy. It also depends on the concentration of reacting molecules because concentration directly determines the probability that any two molecules will collide and initiate a reaction. Reaction rates are directly proportional to concentration, so if we let the rate of reaction equal v, then

$$v \propto [\text{reactants}]$$

or

$$v = k\,[\text{reactants}]$$

where k is the rate constant. Square brackets are used from here on to denote the concentration in moles per litre (mol L^{-1}).

The rate of reaction (6.3) depends on two reactant concentrations and is classified as a second-order reaction.

$$\text{HO}^- + \text{CH}_3\text{Cl} \xrightarrow{k} \text{CH}_3\text{OH} + \text{Cl}^- \qquad (6.1)$$

$$v = k\,[\text{HO}^-][\text{CH}_3\text{Cl}]$$

Reaction (6.1) probably involves a single-step mechanism. The short picture is the direct displacement of Cl$^-$ by HO$^-$ (a). The route via a transition state (b) is more realistic and more informative.

114 Introducing Biological Energetics

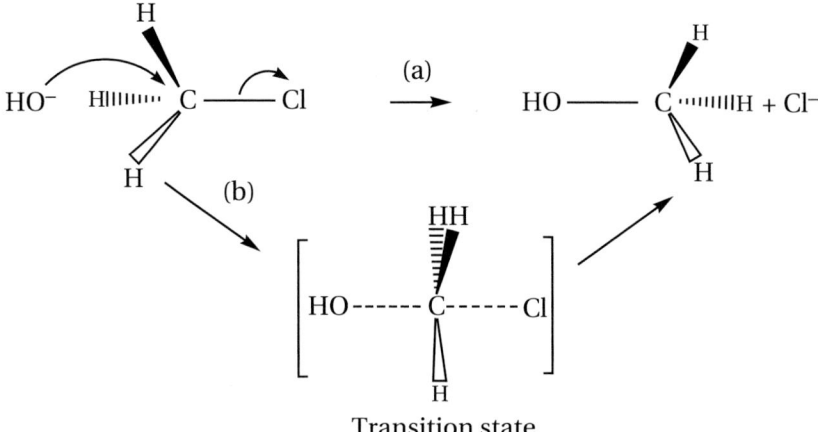

The transition state model explains much of the energetics and kinetics of reactions. In general, as the reaction proceeds along the reaction coordinate, the potential energy increases until it reaches a maximum, which determines the activation energy ΔG^*. In molecular terms, the energy maximum corresponds to the transition state. The transition state is the position of maximum energy partly because to achieve it there often is a crowding of different reacting species into a relatively small volume. In the case of reactions such as (6.1) the transition state involves a fleeting situation where all the groups attached to the central carbon atom are coplanar, as shown in route (b) above, where steric interference is at a maximum. The dotted bonds indicate that as the C–Cl bond is breaking, the HO–C bond is forming in a concerted manner. The molecule is on its way to being flipped inside out like an umbrella, or inverted. This inversion process can have a particularly large effect on ΔG^* when the attached groups are bigger than the hydrogens in reaction (6.1). For a reaction with a single transition state as proposed for reaction (6.1), the formation and breakdown of the transition state will be the rate-determining step of the reaction.

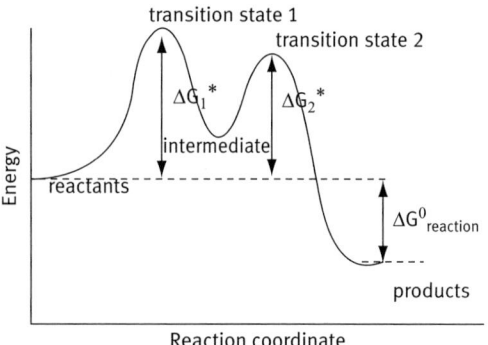

Figure 6.4. Profile of a reaction involving one relatively stable intermediate species. The rate-determining step in such a reaction will be formation of the transition state with the greater activation energy, ΔG^*. In this example formation of transition state 1 is rate determining.

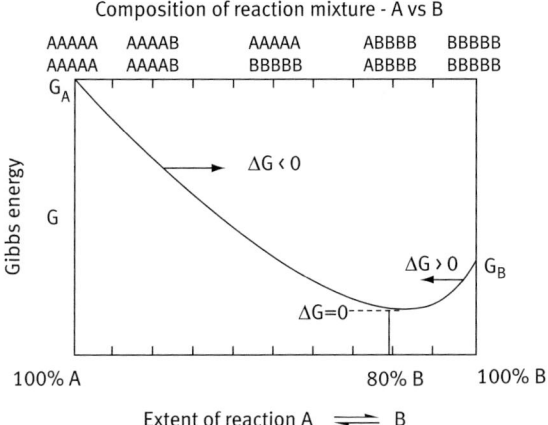

Figure 6.5. A plot of Gibbs energy vs extent of reaction for the conversion of compound A to compound B. The composition at equilibrium is 80% B. For any composition below 80% B, $\Delta G < 0$ and the reaction will be spontaneous in the A–B direction. For any composition above 80% B, it will be spontaneous in the B–A direction. At equilibrium, 80% B, $\Delta G = 0$, and the composition will remain constant.

Some reactions are more complex than (6.1) and may involve a relatively stable intermediate species. Such a reaction profile is shown in Figure 6.4. There would be two activated complexes, and the one with higher activation energy (ΔG_1^*) would be the rate-determining step. There are many examples of such stable intermediates in enzyme-catalysed reactions such as the protease chymotrypsin (see Figure 8.15).

To answer *how far*, we must deal with the state of equilibrium for chemical reactions. This will allow us to discuss the principles of chemical reactions as applied to metabolism, especially what needs to be considered when a number of reactions are involved in metabolic pathways. It is an interesting and useful fact that few chemical reactions proceed only in one direction. To some extent, most reactions are reversible. Most often we write reactions so that they proceed from left to right on the page. This is just a convenient convention. The reactants appear on the left of the equation and the products on the right, again by convention. To indicate a reaction is at least partially reversible, a double arrow is used:

$$A \rightleftharpoons B \tag{6.3}$$

These arrows alone don't tell us much. To understand the full implications of a reversible reaction, including the energetics involved, we need to express what is happening quantitatively, just as we have with ΔG, ΔH, etc.

As an introduction to this, let's look at a graphical way to represent Gibbs energy changes in a reaction.

Figure 6.5 shows a plot of Gibbs energy G vs extent of reaction. Several points are important to note:

- The reaction as written will be exothermic and will proceed spontaneously ($\Delta G < 0$) from a maximum value of G, when the composition is 100% of A molecules, towards the formation of B.
- When the composition reaches 80% B/20% A, the slope of the curve is zero ($\Delta G = 0$) and the reaction will be at equilibrium.

- If we were to start with 100% B, the reaction would take place in reverse (B ⟶ A) until the composition reached 80% B and 20% A.
- The plot is a curve with a minimum point in G rather than a straight line going from 100% A to 100% B. This is in part because there is a contribution to G arising from the mixing of A and B molecules. The A/B ratio changes continuously during the course of the reaction A ⟶ B, so its contribution to G will vary during the course of the reaction (see the composition diagram at the top of Figure 6.5).
- As soon as the reaction starts, there will be some B formed. The concentration of B will increase until the forward reaction A ⟶ B occurs as the same rate as the reverse reaction (B ⟶ A). This occurs at 80% B/20% A in our example.
- At 80% B, the situation is described as a dynamic equilibrium. Despite the constant overall composition, some individual molecules of B are reacting to form new molecules of A, at the same rate that some molecules of A are reacting to form new molecules of B. At the molecular level, plenty is happening, but overall the reaction is going nowhere.
- As the magnitude of G varies with the concentration of A relative to that of B, we can calculate the value of G at any [A]/[B] ratio. Importantly for our understanding of biological reactions, the curve in Figure 6.5 shows that ΔG for such a reaction will be greatest the farther the reaction is from equilibrium. This has two implications for regulation of metabolism: (i) a reaction with a large negative ΔG is often a site for regulation by allosteric control and (ii) reactions close to equilibrium can be reversed in some metabolic pathways. This is useful if the cell needs to change the pathway from a catabolic to an anabolic role (see Chapter 11).

How do we express all this information quantitatively? Firstly, a few more words about rates of reaction.

For the reaction:

$$A \underset{k_2}{\overset{k_1}{\rightleftharpoons}} B \tag{6.4}$$

k_1 and k_2 are the rate constants for the forward and reverse reactions, respectively. Let the molar concentrations of reactant A and product B be indicated by square brackets [A] and [B].

The rate of the forward reaction is given by: $v_{\text{forward}} = k_1[A]$.
The rate of the reverse reaction is given by: $v_{\text{reverse}} = k_2[B]$.

At equilibrium, the forward and reverse rates are equal, so at equilibrium,

$$k_1[A] = k_2[B] \tag{6.5}$$

This rearranges to:

$$\frac{\text{rate constant for the forward reaction}}{\text{rate constant for the reverse reaction}} = \frac{k_1}{k_2} = \frac{[B]}{[A]} = K_{\text{eq}} \tag{6.6}$$

To make the treatment applicable to all chemical reactions, I will change from the simple reaction (6.4) used so far to a general expression for a reaction:

$$aA + bB \rightleftharpoons cC + dD \tag{6.7}$$

where the reactants A and B, plus the products C and D are present in their stoichiometric (balanced equation) ratios a, b, c, and d, respectively. Expressing the equilibrium constant for this reaction leads to:

$$K_{eq} = \frac{[C]^c[D]^d}{[A]^a[B]^b} \quad (6.8)$$

The overall Gibbs energy change can be shown (but not here) to be:

$$\Delta G = \Delta G^0 + RT \ln \frac{[C]^c[D]^d}{[A]^a[B]^b} = \Delta G^0 + RT \ln K_{eq} \quad (6.9)$$

where ΔG^0 is the standard free energy change for the reaction (that is, the ΔG under standard conditions) T is the absolute temperature, R the gas constant (Boltzmann's constant, k_b, per mole) and $[A]^a$, etc. means the molar concentration of reactant A raised to the power 'a'.

Equation (6.10) consists of two parts: a constant term, ΔG^0, which depends only on the particular reaction taking place, and a variable term, $RT \ln = \frac{[C]^c[D]^d}{[A]^a[B]^b}$, which depends on the temperature, concentrations of the reactants and products, and the stoichiometric relationships a, b, c, and d.

At equilibrium, the forward reaction is exactly balanced by the reverse reaction, so $\Delta G = 0$.

It follows that at equilibrium:

$$\Delta G^0 = -RT \ln \frac{[C]^c[D]^d}{[A]^a[B]^b} = -RT \ln K_{eq} \quad (6.10)$$

where the subscript 'eq' signifies equilibrium, that is the concentrations refer to the concentrations at equilibrium. If the equilibrium constant can be determined, ΔG^0 can be calculated, and vice versa. The standard states agreed by convention and used in most thermodynamics tables are:

pressure: 10^5 Nm^{-2} = 10^5 Pa (pascal)

temperature : 273.15 K or 25°C

For pure solids and liquids the standard state is the pure solid or pure liquid. For solutes the standard state is 1.00 M concentration.

Different standard conditions have been adopted by biologists, who use 25°C and pH=7.0, as these are close to physiological conditions for many organisms. When the biological standard states are used, this is indicated by the addition of a prime ('), after the symbol. Thus ΔG^0 becomes $\Delta G^{0\prime}$, ΔH^0 becomes $\Delta H^{0\prime}$, and ΔS^0 becomes $\Delta S^{0\prime}$. Thermodynamic values can vary substantially between the two standard states because the normal thermodynamic standard state concentration for H$^+$ is 1 M, or pH=0. This is obviously unsuitable for most biochemical reactions, which take place in buffered solutions close to pH 7. Major differences will occur between ΔG^0 and $\Delta G^{0\prime}$ when H$_2$O and H$^+$ are involved in the reaction. This is the case in many biological reactions. Rather than incorporate specific terms for [H$^+$]

and [H_2O] in equations such as (6.10), they are usually incorporated into the $\Delta G^{0\prime}$ and K_{eq} values. When water and protons are *not* involved, $\Delta G^0 = \Delta G^{0\prime}$.

We'll see in Chapter 11 that ΔG values under conditions in the cell (physiological conditions) can be markedly different from the ΔG^0 or $\Delta G^{0\prime}$ values under standard conditions. It is useful to bear in mind that the ΔG of a chemical reaction is just another way of expressing the equilibrium constant, and vice versa. Because the relationship between ΔG and K_{eq} is logarithmic, small changes in ΔG result in large changes in K_{eq}.

To gain some idea of the relative magnitudes of ΔG^0 and the corresponding K_{eq}, let us use some specific values.

Applying equation (6.10)

$$\Delta G^0 = -RT \ln K_{eq}$$

Using values of 2.26×10^{-4}, 1, and 2.26×10^4 as examples of K_{eq}, we can draw up a table to show how the values of ΔG^0 vary with changes in K_{eq}.

ΔG^0 (kJ mol^{-1})	K_{eq}*
+20.9	2.26×10^{-4}
0	1
−20.9	2.26×10^4

*K_{eq} has no units. It is the ratio of k_1/k_2 so is a simple number.

A value of $K_{eq} > 10^4$ ($\Delta G^0 < -20.9$ kJ) means that the rate constant for the forward reaction is $>10^4$ times that of the reverse reaction. As a rule of thumb, this means the reaction goes to completion as written.

Conversely, if $K_{eq} < 10^{-4}$ ($\Delta G^0 > +20.9$ kJ) the reverse reaction will be favoured to completion.

The above relationships between ΔG^0 and K_{eq} can be summarized to:

If $\Delta G^0 < 0$ $K_{eq} > 1$

If $\Delta G^0 = 0$ $K_{eq} = 1$

If $\Delta G^0 > 0$ $K_{eq} < 1$

Suppose we have a reaction which has come to equilibrium. Equation (6.10) shows that if some change moves the reaction away from equilibrium, this will stimulate a change in the direction of restoring the equilibrium. The system will change to restore the equilibrium concentrations of the reactants and products. This is known as Le Chatelier's principle (after French chemist H.L. Le Chatelier (1850–1936)).

What factors influence the position of equilibrium? Firstly, let us consider the effect of temperature on the equilibrium constant.

Using eqn (6.10), $\Delta G^0 = -RT \ln K_{eq}$, and $\Delta G^0 = \Delta H^0 - T\Delta S^0$, we can show that

$$\ln K_{eq} = -\Delta G^0/RT = -(\Delta H^0/R)(1/T) + \Delta S^0/R \quad (6.11)$$

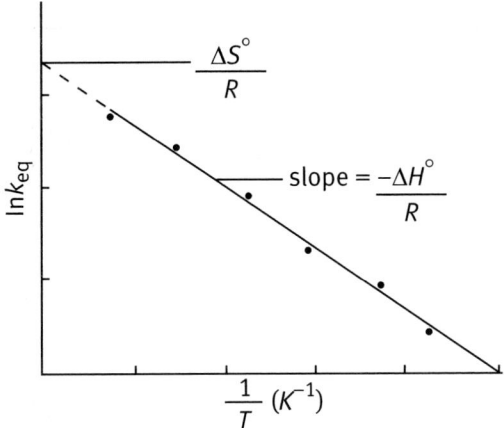

Figure 6.6. A typical van't Hoff plot of ln K_{eq} vs $1/T$.

Equation (6.11) implies that lnK_{eq} is inversely proportional to the absolute temperature. If H and S are not greatly influenced by a small change in temperature, often the case over the relatively small biological temperature range, a plot of lnK_{eq} vs $1/T$ should be linear, with a slope equal to $-\Delta H^0/R$ and an intercept equal to $\Delta S^0/R$.

This is called a van' Hoff plot, after J.H. van't Hoff' (1852–1911) a Dutch physical chemist, the first Nobel laureate in chemistry, awarded in 1901 (Figure 6.6).

The value of such a plot is that it can yield values for ΔH^0 and ΔS^0, so long as K_{eq} can be measured experimentally over a suitable temperature range (Haynie 2001).

Equation (6.11) also shows that the equilibrium constant depends on $\Delta H^0_{reaction}$. If the reaction is endothermic ($\Delta H^0_{reaction}$ is positive) then $(-\Delta H^0/RT)$ will be negative, so as T is increased, K will increase.

We can therefore conclude that:

a) for an endothermic reaction, increasing the temperature will increase K, that is will shift the equilibrium in favour of the products
b) if the reaction is exothermic, increasing the temperature will, for similar reasons, cause a decrease in K and shift the equilibrium to favour the reactants.

Both conclusions are probably familiar from elementary chemistry as examples of Le Chatelier's principle.

Secondly, let us consider the influence of concentration on the position of equilibrium. For the reaction:

$$A + B \rightleftharpoons C + D$$

$$K_{eq} = \frac{[C][D]}{[A][B]}$$

Suppose fresh A and B atoms were to be added to those already present at equilibrium (without changing the volume). What will occur? This amounts to an increase in concentration of A and B, which will increase the forward reaction rate to form new C plus D, until the new C plus new D molecules reach a concentration such that the reverse reaction rate once again is equal to the forward rate, and equilibrium will be re-established—another example of Le Chatelier's principle at work. During the many processes involved in metabolism in a biological system, we seldom have the simple system described above. Usually there are other reactions feeding in to form A and B, while C and D are usually removed by further reactions in the metabolic pathway. In living organisms, there obviously needs to be some control over the production of A and B, and over the removal of C and D.

The *how far* of the chapter title also concerns how far towards or how far from equilibrium living organisms exist, and how they achieve and maintain these positions. Supposing we have a situation inside a cell in which the equilibrium constant of a reaction is small. Its ΔG is small and positive, meaning that there is very little tendency for the reaction to proceed spontaneously in the direction written.

Now suppose the cell is in need of the products C and D, or perhaps needs to get rid of A and B. How can this be achieved? This is a real situation for cells, an example being the oxidation of lactate to pyruvate at pH 7. Lactate (lactic acid) builds up in muscle cells when they work vigorously, say during a sprint race, as aerobic respiration can't keep up with energy demand. This can only be tolerated for a short time. We experience pain as lactate builds up, and eventually are forced to stop, pant and use the gulped air to remove the lactate from our aching muscle cells. The cells use the gulped air to oxidize the lactate to pyruvate, which then enters aerobic metabolic pathways and is oxidized further. The oxidation reaction requires the involvement of an enzyme, lactate dehydrogenase (LDH), and a coenzyme, NAD. An enzyme doesn't alter the equilibrium constant of the reaction it catalyses; it increases the rate at which the reaction occurs. A coenzyme is usually a small molecule which, as its name suggests, helps some enzymes to achieve their purpose. Not all enzymes require a coenzyme.

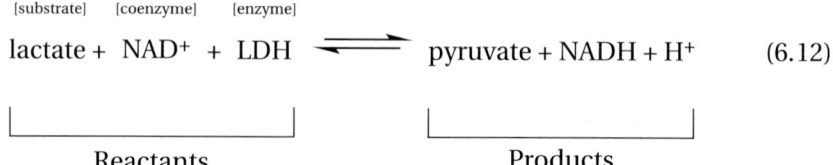

$$\text{lactate} + \text{NAD}^+ + \text{LDH} \rightleftharpoons \text{pyruvate} + \text{NADH} + \text{H}^+ \qquad (6.12)$$

Reactants — Products

At pH 7 and 37°C, ΔG^0 is positive, and the oxidation of lactate is not possible under physiological conditions. However, if the concentrations of lactate and NAD^+ are in great excess over pyruvate and NADH (which is the case after enough vigorous exercise) the overall ΔG actually becomes negative and the reaction will proceed spontaneously in the direction written. Actually it is the concentration ratio

$$\frac{[\text{pyruvate}][\text{NADH}][\text{H}^+]}{[\text{lactate}][\text{NAD}^+]}$$

rather than the absolute concentrations, that are the key factors determining the direction of the reaction. To keep the reaction running, it is necessary to keep the above concentration ratio small, that is much less than 1.

This can be achieved by removing the products, pyruvate and NADH, as rapidly as they are made. This has important consequences for an energetically unfavourable reaction embedded in a metabolic pathway. The statement can be made:

It is possible to keep energetically unfavourable reactions running by continuously removing the products.

In the above example, pyruvate is removed by immediate conversion to another compound, acetyl coenzyme A, in a manner that will be described later (Chapter 11). Everyone at some time has surely felt the discomfort of lactate build up in the muscles after strenuous exercise and knows that it takes a while for this pain to ease. Once again a knowledge of thermodynamics, this time as applied to chemical equilibria, has helped us to explain the workings of part of a biological system with which we are all familiar.

Let us consider at this point some more of the requirements for effective metabolism in a living organism.

All metabolic reactions are governed by the same laws of chemistry described above. There is nothing special or mystical about the basic chemistry of life, but there are many things very special about the way this chemistry is arranged, regulated, and coordinated. If we take humans as the example organism, all the enormous variety of reactions needs to occur at about body temperature, 37°C. This is a considerable constraint, or at least would be for traditional synthetic chemists. In addition, the reactions occur mainly in an aqueous environment, they must usually avoid strongly acidic or strongly alkaline conditions (which are damaging to our tissues), and be protected against too much oxygen. Such constraints would send most synthetic chemists crazy, for they usually need to spend a great deal of time and effort to prevent getting water, and often oxygen, into their reactions. The fact that so many biological reactions are able to take place under mild aqueous conditions is a tribute to the power of biological evolution, which has, over time, been able to hone the workings of life with such exquisite subtlety that even we, arguably its cleverest product, can't come close to matching them.

Why should there be so many reactions needed by a living organism, even a simple one? One reason is that to obey the laws of thermodynamics, under mild conditions and in water surroundings, some rather round-about pathways need to be followed. Suppose an organism needs endothermic reactions to take place, for example a plant needs to produce cellulose to strengthen it or an animal needs to produce collagen, a protein in tendons and connective tissues. Putting glucose molecules together to form cellulose requires an input of energy; so does putting amino acids together to form proteins and two-carbon units to synthesize fatty acids.

What are the sources of energy available to the plant or animal? The reactants can't be heated to boiling, as a chemist might be able to do, so where can the energy be accessed? The answer is deceptively simple.

The energy released by a spontaneous, exothermic chemical reaction ($\Delta G < 0$) may be used by the plant or animal or any living system to 'drive' an endothermic reaction ($\Delta G > 0$) of whatever type may be needed. To achieve this in a controlled manner requires the two reactions to be 'coupled' in a specific way. This coupling of reactions is used by living organisms for a number of purposes:

1) to allow the synthesis of molecules needed for various structures such as cellulose or proteins
2) to maintain the temperature of warm-blooded creatures
3) to synthesize a range of essential biological molecules which cannot be provided directly by the diet
4) for movement
5) for transport of molecules around the body and through biological membranes.

All of these processes are endothermic: they require an input of energy to proceed. The use of coupled chemical reactions is one of the ingenious means by which biology provides such specific energy requirements.

As an example of energy coupling, firstly of a non-chemical nature, let us look at a mechanical analogy. Suppose a builder wants to raise a block of stone, mass 50 kg, from the ground to a ledge 5 m above. This is an energy-requiring, endothermic process, and will certainly never proceed spontaneously. Suppose the builder arranges to have a pulley system attached to a beam about 7 m above ground, with one end of a rope tied around the block of stone, running over the pulley to the other end, which is held by a 80-kg man standing on the 5 m ledge. If the 80-kg man were to jump from the ledge, his gravitational potential energy would become available, through the rope, to lift the block. The tendency of the man to fall to ground would be his weight, 80-kg, and the tendency to stop him falling would be the weight of the block, 50-kg weight. The net tendency of the block to move upwards would be about 30 kg, that is $\Delta G = -30$ units for the upwards direction. The 30 kg should be plenty to overcome the small amount of friction in the pulley. The block would move up and the man would fall to the ground, but more slowly (and safely) than if he had jumped freely. Thus an exothermic process, an 80-kg man falling under gravity, will have been coupled to an endothermic process, the block being raised against gravity, in such a way that the block is raised successfully. The whole process will occur spontaneously once the man jumps free of the ledge. In our example of mechanical work, the intensity factor, gravitational potential, is the height the weight was lifted, and the capacity factor is the mass of the block.

So much for the mechanical analogy. Now let's consider an actual example of energy coupling for a biochemical reaction. The amide group of the common amino acid glutamine is a major donor of nitrogen in the biosynthesis of compounds such as purines, pyrimidines, and some other amino acids. Glutamine is formed from glutamate and ammonium ion, catalysed by the enzyme glutamine synthetase. The reaction is ATP-dependent, and requires Mg^{2+} ion, as many ATP reactions do. It is a central control point in nitrogen metabolism. Glutamine is the amino group donor in many biosyntheses and is a storage form of ammonia.

The two reactions involved are:

$$\text{glutamate} + \text{NH}_3 \longrightarrow \text{glutamine} + \text{H}_2\text{O} \quad \Delta G^{0\prime} = +14.2 \text{ kJ mol}^{-1} \quad (6.13)$$

$$\text{ATP} + \text{H}_2\text{O} \longrightarrow \text{ADP} + \text{phosphate} \quad \Delta G^{0\prime} = -30.5 \text{ kJ mol}^{-1} \quad (6.14)$$

Reaction (6.13) is endothermic, and won't proceed spontaneously. Reaction (6.14) is exothermic and if coupled to (6.13) the overall reaction should be exothermic by $(14.2 - 30.5) = -16.3$ kJ mol^{-1} and spontaneous.

Adding eqns (6.13) and (6.14) leads to:

$$\text{glutamate} + \text{NH}_3 + \text{ATP} \longrightarrow \text{glutamine} + \text{ADP} + \text{phosphate}$$
$$\Delta G^{0\prime} = -16.3 \text{ kJ mol}^{-1} \quad (6.15)$$

The reaction mediated by glutamine synthetase, and coupled via ATP, becomes:

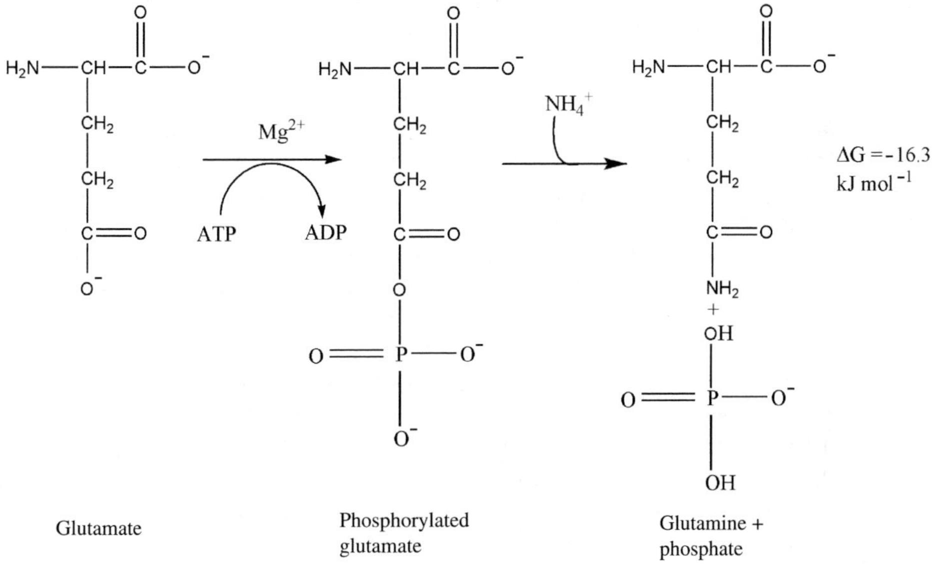

[Reaction scheme: Glutamate → (Mg²⁺, ATP → ADP) → Phosphorylated glutamate → (NH₄⁺) → Glutamine + phosphate; $\Delta G = -16.3$ kJ mol^{-1}]

This example reinforces what we have shown above—it is possible to predict whether or not pairs of coupled reactions will occur spontaneously simply by summing the Gibbs energy changes for each reaction and looking at the sign of the resulting ΔG. Let us take as a second example adenosine triphosphate, ATP itself. ATP is used to drive many metabolic reactions, but then it must be regenerated from the 'lower-energy' compound such as adenosine diphosphate (ADP) formed as a result of the reaction.

The energy-producing reaction is:

$$\text{ATP} + \text{H}_2\text{O} \longrightarrow \text{ADP} + \text{phosphate} \quad \Delta G^{0\prime} = -30.5 \text{ kJ mol}^{-1} \quad (6.16)$$

The reverse, or regeneration, step is:

$$\text{ADP} + \text{phosphate} \longrightarrow \text{ATP} + \text{H}_2\text{O} \quad \Delta G^{0\prime} = +30.5 \text{ kJ mol}^{-1} \quad (6.17)$$

Reaction (6.17) is obviously an endothermic reaction that will not proceed without some energy input. One way in which the required energy is obtained is by coupling to an exothermic reaction, such as that involving the conversion of another 'high-energy' compound, phospho[enol]pyruvate (PEP) to pyruvate:

$$\text{PEP} + \text{H}_2\text{O} \longrightarrow \text{pyruvate} + \text{phosphate} \quad \Delta G^{0\prime} = -62.2 \text{ kJ mol}^{-1} \quad (6.18)$$

Adding eqns (6.17) and (6.18) we obtain:

$$\text{PEP} + \text{ADP} \longrightarrow \text{ATP} + \text{pyruvate} \quad \Delta G^{0\prime} = 30.5 - 62.2 = -31.7 \text{ kJ mol}^{-1} \quad (6.19)$$

As a shorthand way of expressing what is taking place, it is legitimate to eliminate phosphate and water from eqn (6.19), as each occurs on both sides when we add eqns (6.17) and (6.18), thus cancelling one another out. In this sense, we are treating eqn (6.19) as the algebraic sum of eqns (6.17) and (6.18). The resulting coupled reaction (6.19) is exothermic and will occur spontaneously from left to right as written. The reaction is catalysed by pyruvate kinase, an important reaction in glycolysis, that we will return to in Chapter 11.

This involvement of an enzyme is a vital aspect of coupled biochemical reactions. Reactions (6.17) and (6.18) would not be coupled if the reagents were merely mixed together in a test-tube, just as the block in the example above would not move upwards unless it was connected via the rope to the man when he jumped. Many ATP-coupled reactions involve the phosphorylation of an intermediate such as the phosphorylated glutamate above. The presence of the enzyme doesn't alter the overall thermodynamics of eqns (6.15) and (6.19), which on the basis of their negative ΔG values we would predict to occur spontaneously.

Just how great an effect can ATP coupling have on the equilibrium position of a reaction? Suppose we wish to know quantitatively by how much the equilibrium and the product/reactant ratio change for an ATP-coupled *in vivo* reaction. We saw in the LDH example above the importance of concentration ratios in keeping an energetically unfavourable reaction running. Here, we must include biologically realistic concentrations of ATP and other reactants in the calculation. In a cell, the concentrations of metabolites, ions, coenzymes, etc., vary with time and location. As pointed out above, the standard ΔG^0 or $\Delta G^{0\prime}$ values we have been using never occur in real cells. These values are used in many textbook examples because the real cellular concentrations of many species are difficult to determine experimentally. The standard values give a useful picture, but sometimes more realistic ones must be used.

Before answering the question at the beginning of the last paragraph, let's look at an example involving 'real' values of ΔG for physiological conditions.

For the hydrolysis of ATP:

$$\text{ATP} + \text{H}_2\text{O} \rightleftharpoons \text{ADP} + \text{P}_i \quad \Delta G^{0\prime} = -30.5 \text{ kJ mol}^{-1} \quad (6.20)$$

$$K_{eq} = \frac{[ADP][P_i]}{[ATP]}$$

$$\Delta G = \Delta G^{0\prime} + RT \ln \left[\frac{[ADP][P_i]}{[ATP]}\right]$$

Taking some 'typical' cell values:

$$[ATP] = 3.0 \text{mM}, [ADP] = 0.8 \text{mM}, \text{ and}$$

$$P_i = 4.0 \text{ mM at } 37°C \text{ (Voet } et\ al.\ 2002)$$

This reduces to $\Delta G = -30.5 - 17.6 = -48.1$ kJmol^{-1}

This value of -48.1 is much more exothermic than the standard state one of -30.5 kJ mol^{-1}.

Other values quoted are (i) -47.6 kJ mol^{-1} for a bacterial cell, using [ATP], [ADP], and [P_i] values of 8 mM, 8 mM, and 1 mM, respectively, at 25°C (Garrett and Grisham 2010), and (ii) -55 kJ mol^{-1} in human erythrocytes (Garrett and Grisham 1999). The results reinforce the effect of conditions on real ΔG values.

To our question. Using the values provided by Garrett and Grisham (2010) for a bacterial cell; ATP is typically present in 1000-fold excess over its hydrolysis products:

$$\frac{[ADP][P_i]}{[ATP]} = 10^{-3} \text{ approx.}$$

We can take any hypothetical equation having a $\Delta G^{0\prime}$ of our own choosing:

$$X \rightleftharpoons Y \qquad \Delta G^{0\prime} = +20.9 \text{ kJ mol}^{-1}$$

Then from the equation

$$\Delta G^{0\prime} = -RT \ln K_{eq}$$
$$20\,900 = -(8.31)(298) \ln K_{eq}$$
$$\ln K_{eq} = -3.6642$$
$$K_{eq} = 2.26 \times 10^{-4} = [Y_{eq}]/[X_{eq}]$$

This is an energetically unfavourable reaction.

Coupling the same reaction to the hydrolysis of ATP, we obtain the overall equation:

$$X + ATP \rightleftharpoons Y + ADP + P_i$$

This is the same as the sum of the two uncoupled reactions:

$$X \rightleftharpoons Y \qquad \underline{\Delta G}^{0\prime} = +20.9 \text{ kJ mol}^{-1}$$
$$ATP + H_2O \rightleftharpoons ADP + P_i \qquad \Delta G^{0\prime} = -30.5 \text{ kJ mol}^{-1}$$

Overall $X + ATP + H_2O \rightleftharpoons Y + ADP + P_i \quad \Delta G^{0\prime} = -9.6 \text{ kJ mol}^{-1}$

So
$$-9600 = -RT \ln K_{eq}$$

$$\ln K_{eq} = \frac{9600}{2476} = 3.87$$

$$K_{eq} = 48$$

By definition:

$$K_{eq} = \frac{[Y_{eq}][ADP][P_i]}{[X_{eq}][ATP]}$$

$$48 = \frac{[Y_{eq}][8 \times 10^{-3}][10^{-3}]}{[X_{eq}][8 \times 10^{-3}]}$$

$$\frac{[Y_{eq}]}{[X_{eq}]} = 48000$$

For the uncoupled reaction $\frac{[Y_{eq}]}{[X_{eq}]} = 2.26 \times 10^{-4}$

So the ratio $\frac{K_{coupled}}{K_{uncoupled}} = \frac{4.8 \times 10^4}{2.26 \times 10^{-4}} = 2.12 \times 10^8$

The equilibrium ratio of products:reactants in the coupled reaction is about 2×10^8 times that of the uncoupled one, turning an energetically unfavourable reaction into one that will proceed spontaneously to completion.

The above example illustrates a general principle. It reveals why coupling of reactions to ATP hydrolysis is utilized in so many different ways in living cells. The reaction coupled can be of any type, such as formation of a covalent bond in biosynthesis, ion transport across a membrane against a concentration gradient, or a change in conformation of a protein, as in muscle contraction or assisted protein folding.

A few more words on 'high-energy' biomolecules are appropriate here. Although most of the energy for life comes from the Sun, not all living organisms can process sunlight directly. Those which can are able to capture and store the light energy in various chemical compounds, such as carbohydrates. Examples are algae and green plants, which store mainly the carbohydrates starch or sucrose. Other organisms, such as mammals, including humans, are able to feed on these stored compounds, releasing the energy in reactions usually involving oxygen (recall the 'controlled burning' of sugars mentioned earlier) during the process of respiration. Despite some differences, both types of organism share mechanisms for generating useful forms of chemical energy. One of these is the use of high-energy phosphate compounds, such as ATP, GTP, ADP, PEP, acetyl phosphate, glucose-1-phosphate, and creatine phosphate. There are 'high-energy' compounds other than phosphates that are also used (see Box 11.1). Compounds are classified as being high energy if they have a ΔG of hydrolysis more negative than about -25 kJ mol^{-1}. Equation (6.19) illustrates this: $\Delta G^{0\prime} = -31.7$ kJ mol^{-1}.

Hydrolysis is the name given to reactions involving compounds that are 'split' (lysed) with the involvement of water. In the case of phosphorylated compounds, during hydrolysis the phosphate group is transferred to water. It is important to distinguish between these high-energy phosphate compounds and the long-term energy storage compounds such as starch in plants and glycogen and fat in animals. The ATP type of high-energy compounds are transient forms of energy storage, molecules which carry energy from point to point for the moment-to-moment running of the cell. The high-energy compounds themselves must be regenerated for further use when the demand arises later. ATP in eukaryotes, for example, is largely synthesized from ADP during the process of respiration, in a key series of reactions called oxidative phosphorylation, details of which are discussed in Chapter 10.

This completes the discussion on the way in which coupled reactions may be used to utilize special high-energy compounds to form essential biomolecules. The reasons why some compounds have a large negative ΔG of hydrolysis are discussed in Box 6.1.

So far in this chapter we have seen two ways in which energetically unfavourable reactions can be made to proceed:

1) by removing the products of the reaction as fast as they are produced – see eqn (6.12)
2) by coupling of reactions involving, for example, ATP (eqns (6.15) and (6.19)). As explained in Box 6.1, this latter process is classified as a group transfer reaction. In these cases the essential group was phosphate.

Another mechanism of energy transfer, termed electron transfer, is important in biological energy processes and is discussed in Chapter 10. We have discussed endothermic and exothermic reactions and processes, and the fact that exothermic reactions and processes proceed spontaneously if there is a suitable pathway available. We have rationalized this by saying that spontaneous chemical reactions occur because the products are at a lower Gibbs energy than the initial reactants (ΔG is negative).

Reversible reactions are very useful in metabolism as they can be controlled to go one way under one set of conditions and in the opposite direction under a different set of conditions. This property is important in regulation of metabolic pathways (Chapter 11). We have stated that isolated systems, which don't exchange matter and energy with their surroundings, if left to their own devices will eventually reach a state of thermodynamic equilibrium.

However, in living systems, reversible reactions do not necessarily reach thermodynamic equilibrium as described above; in fact they mainly do not. A living organism that fits our definition of a system is an example of an open system, which may exchange matter and energy with its surroundings and is not at thermodynamic equilibrium. Such a system normally exists in a state of dynamic energy balance, sometimes called a steady state, but not in a state of thermodynamic equilibrium.

128 Introducing Biological Energetics

Box 6.1 ATP: the energy currency of cells

ATP is the canonical 'high-energy' compound of the biological world. Almost all cellular chemical energy use can be traced back to it. We have defined high-energy compounds as those having a standard Gibbs energy of hydrolysis, $\Delta G^{0\prime}$, more negative than about -25 kJ mol^{-1}. Thus:

$$\text{ATP} + \text{H}_2\text{O} \longrightarrow \text{ADP} + \text{phosphate} \quad \Delta G^{0\prime} = -30.5 \text{ kJ mol}^{-1} \quad (6.20)$$

The structure and roles of ATP are worthy of further attention.

Coupled reactions using ATP and other high-energy compounds take advantage of their phosphoryl group transfer potential. Phosphoryl group transfer potential is a measure of the tendency of such compounds to transfer their phosphoryl groups to water, expressed in order of their $\Delta G^{0\prime}$ values (Table 6B1).

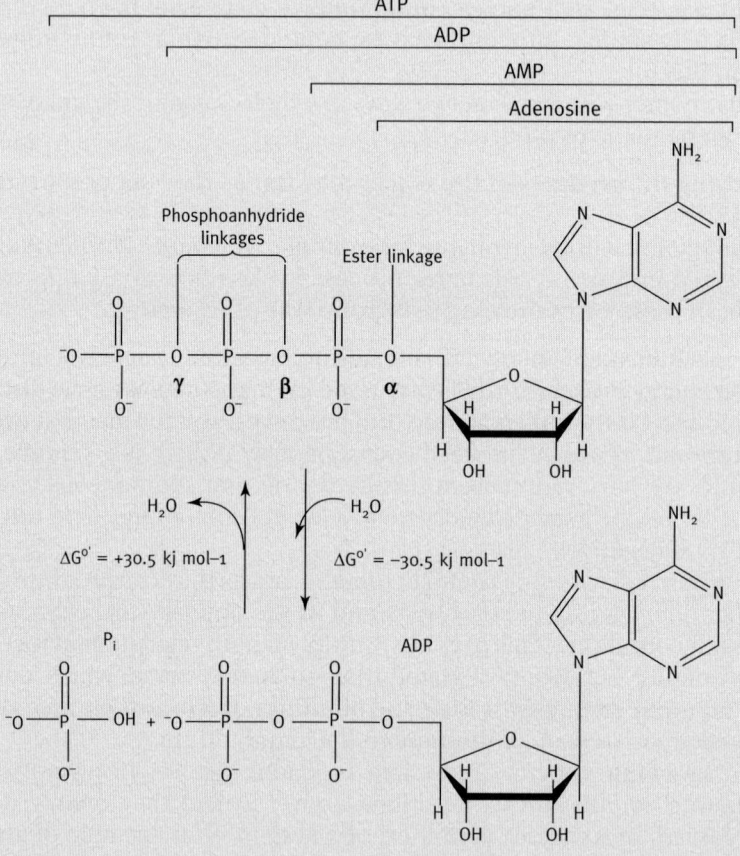

Figure 6.B1. ATP and related structures, showing the energetics of hydrolysis to ADP and inorganic phosphate (P$_i$) and the reverse reaction. The free energy of hydrolysis of ATP to AMP and inorganic pyrophosphate (PP$_i$) is also widely utilized in biological systems.

Box 6.1 (continued)

Table 6.B1. Some standard free energies of hydrolysis.

Compound	$\Delta G^{0\prime}$ (kJ mol^{-1})
Phosphoenolpyruvate	−61.9
1, 3-Bisphosphoglycerate	−49.4
Acetyl phosphate	−43.1
Phosphocreatine	−43.1
PP$_i$	−33.5
ATP ($\rightarrow$ AMP + PP$_i$)	−32.2
ATP ($\rightarrow$ ADP + P$_i$)	−30.5
Glucose-1-phosphate	−20.9
Fructose-6-phosphate	−13.8
Glucose-6-phosphate	−13.8
Glycerol-3-phosphate	−9.2

Source: Jencks, W.P., in Fasman, G.D. (ed.) (1976) *Handbook of Biochemistry and Molecular Biology*, (3rd edn.), Physical and Chemical Data, Vol. I, pp. 296–304, CRC. Press.

Table 6.B1 shows that ATP is in about the middle range. The compounds above ATP in the table can, with suitable enzyme coupling, form ATP from ADP and P$_i$ by substrate level phosphorylation. This can be used as an alternative to oxidative phosphorylation that takes place in mitochondria.

Why should ATP and other compounds have such exothermic $\Delta G^{0\prime}$ of hydrolysis values?

Bear in mind that, in general, $\Delta G^{0\prime}$ values depend on concentrations of reactants and products. In addition, in the case of ATP and its hydrolysis products, there is a dependence on pH and ionic strength, as ions are involved. Let us look specifically at ATP and other phosphorylated compounds.

The bonds/linkages involved are labelled β- and γ- in the ATP structure in Figure 6.B1. They are phosphoanhydride bonds, in contrast to the phosphate ester bond labelled $\tilde{\alpha}$.

The 'high energy' involved here has nothing to do with bond energy, which is a measure of the energy needed to break a covalent bond. ATP-coupled reactions involve the energy liberated during the process of hydrolysis. The explanation is in what happens *after* hydrolysis, relative to what happens before. We are looking for reasons why, energywise, the components of ATP would 'prefer' the energy state after hydrolysis to that before hydrolysis. There are several proposed contributions:

1. ATP is destabilized by electrostatic repulsion between relatively close positive charges. The electronegative O atoms polarize the P–O bonds as shown below, increasing the electrostatic repulsion between them.

Box 6.1 (continued)

Hydrolysis relieves this source of instability by separation of the charges.

2. Entropic factors: There is an increase in the number of molecules after hydrolysis, allowing for an increase in entropy.
3. Smaller solvation energy of ATP relative to ADP + P_i. In the latter there are two smaller molecules with five negative charges to be solvated by water. ATP has four negative charges spread over a larger molecule. The magnitude of this effect is difficult to assess, but is probably quite significant.
4. To help with this explanation, we need to deal with an aspect of the stabilization of molecular species that was not mentioned in Chapter 5. This is the concept of resonance.

Some molecules can exist in two or more forms by rearrangement of their electrons. This can only happen to any significant extent if the electron orbitals involved are in 'allowed' orientations, with little or no energy barrier between them.

Thus, acetate ions can be written in two equi-energy forms:

As these forms are equal in energy, then entropy will ensure that both are equally likely to occur. Entropy favours the distribution of energy over as many states as possible.

In the case of the inorganic phosphate ion from the hydrolysis of ATP, there are at least three resonance forms capable of existing simultaneously:

inorganic phosphate (P_i) resonance forms

Box 6.1 (continued)

We could imagine more, formed by locating the proton on other oxygen atoms. This can be illustrated with a diagram of what is known as a resonance hybrid:

$$\left[O^{\delta-} \cdots P \cdots O^{\delta-} \atop {O^{\delta-} \atop O^{\delta-}} \right]^{3-} H^+$$

This shows that the H^+ and double bond are equally likely to involve any one of the four oxygen atoms.

After hydrolysis, there are at least three viable resonance forms for the remaining ADP:

ADP resonance forms

By contrast, in the case of ATP, there will be three forms for the end (γ) phosphate group, corresponding to the three for the β-phosphate in ADP above. There will be two competing forms, (a) and (b) below, that can exist alternately, but not simultaneously. This limits ATP to about four viable resonance-stabilized forms.

ATP resonance

Overall, the resonance stabilization for ATP is much less than for its hydrolysis products ADP and P_i.

All these influences add up to produce a more favourable, that is lower overall, energy distribution for the hydrolysis products over the unhydrolyzed ATP. Similar arguments can be made of other 'high-energy' compounds.

132 Introducing Biological Energetics

> **Box 6.1** (continued)
>
> The above discussion is for the thermodynamics of group-transfer reactions, which are vitally important in biology. Group transfer potential is discussed in terms of free energy of hydrolysis, but in most biological examples water doesn't even enter the equation. Consider the general case of the biosynthesis of a molecule that involves the condensation of two compounds (de Duve 2005):
>
> $$\text{XOH} + \text{HY} \longrightarrow \text{X–Y} + \text{H}_2\text{O} \qquad (6.\text{B}1)$$
>
> If this is carried out by a coupling reaction with ATP as the energy source, the steps involved are:
>
> $$\text{ATP} + \text{XOH} \longrightarrow \text{ADP} + \text{X–O–P} \qquad (6.\text{B}2)$$
>
> In the second step, X is transferred to Y–H:
>
> $$\text{X–O–P} + \text{HY} \longrightarrow \text{X–Y} + \text{P}_i \qquad (6.\text{B}3)$$
>
> Adding (6.B2) and (6.B3):
>
> $$\text{ATP} + \text{XOH} + \text{HY} \longrightarrow \text{X–Y} + \text{ADP} + \text{P}_i \qquad (6.\text{B}4)$$
>
> No water in its free form is involved. It is transferred via the XOP intermediate from XOH to P_i. Instead of the energy of ATP hydrolysis being dissipated as heat, this vitally important mechanism results in the synthesis of a covalent bond. The step in (6.B2) to form X–O–P is called the activation step, as XOH is activated chemically by the phosphoryl transfer to form XOP, which is capable of reacting with Y–H.
>
> As pointed out by de Duve, many important biochemical processes involve the activation of X–O–H by transfer of inorganic pyrophosphate, PP_i, which involves splitting of the inner (β-) phosphoanhydride linkage rather than the outer (γ-) linkage (Figure 6.B1).
>
> This summarizes to:
>
> $$\text{ATP} + \text{X–O–H} \longrightarrow \text{AMP} + \text{X–O–PP}$$
>
> $$\text{X–O–PP} + \text{HY} \longrightarrow \text{X–Y} + \text{PP}_i$$
>
> Overall $\quad \text{ATP} + \text{X–O–H} + \text{H–Y} \longrightarrow \text{X–Y} + \text{AMP} + \text{PP}_i$
>
> Aerobic cells maintain a high level of ATP via photosynthesis or respiration to ensure that a ready supply of energy is available.

For living systems an overall approach too close to equilibrium is an approach to death. Why do I make such a dramatic statement?

All living systems must be able to carry out work of various types to stay alive. Systems at equilibrium cannot carry out work, which is reflected in the fact that $\Delta G = 0$ in such systems. In one sense, life can be considered as a struggle against reaching equilibrium. When death occurs, the organism ceases its normal energy-driven processes and the inevitable 'downward' path of chemical reactions in its body towards equilibrium commences. Constantly during its lifetime, each beetle, each eagle, each whale, and each of the 6.8 billion, and rising, humans on Earth, every last bacterium of the billions seething about in your gut is 'battling' against thermodynamics to survive, each an improbable island of order in a sea of chaos. Each can only succeed in this by processing the energy in its food to maintain an ordered state. All are non-equilibrium systems, engaged in dissipating some of the concentrated energy of the universe. They take in 'high-energy' chemical compounds and degrade them to 'low-energy' compounds, making use of the released potential energy for their own survival.

Suppose we look at a living organism in terms of the basic processes involved in its existence. In the broadest sense, a living organism is a transducer of energy. By transducer I mean that an organism has the ability to take energy in one form and to transform at least some of that energy to other forms, with the 'objective' of its own survival. The energy that is transformed goes, in part, towards maintaining the high degree of organization necessary to sustain all the life functions of the organism. We have discussed many of those functions, and in most cases it is easy to recognize why a supply of energy is necessary to carry them out. Maintenance of body temperature, growth, and muscular activity all need a source of energy. In addition, there is the less obvious yet crucial energy function of maintaining the degree of order we associate with a living organism. All its organs and systems are specifically constructed and integrated in such a way as to function as a whole. Failure of one or more vital organs, such as a kidney, the liver, the brain, or the heart, can lead to the death of a mammal. After death, the ability of an animal or plant to maintain its ordered structure is lost, as it can no longer act as a successful transducer of energy. Processes of decay commence as bacteria and other biological consumers take control and convert whatever energy is stored in the body of the deceased animal to their own advantage. Most of this residual energy is recycled in such a way because the living world is adept at survival and the death of one creature usually means life for others. The ultimate fate of the energy of life is low-intensity heat.

In order to maintain the dynamic energy balance mentioned in Chapter 1, we humans must achieve an overall daily balance of energy intake, anabolism, catabolism, and energy output. The maintenance of this balance is the role of metabolism, assisted by our own conscious contribution in the form of a suitable diet and lifestyle.

As I have stressed previously, the simple maintenance of our energy balance is not the whole story.

Most biological molecules are 'turned over'; they are replaced periodically. Most of us are familiar with the fact that most of the cells in our bodies are replaced a number of times during our lifetime; nerve cells seem to be a notable exception, although even the consensus on this is changing. The rates of molecular turnover may vary greatly. Enzymes and some other proteins have a half-life of 2–8 days. Structural proteins such as collagen in connective tissue and cartilage last much longer, often having a half-life of greater than 200 days. We shed hundreds of thousands of dead skin cells every day. Much of the 'dust' that gathers so annoyingly in our bedrooms and bathrooms consists of skin cells. Red blood cells work hard, being squeezed through narrow capillaries as well as being pushed through the major blood vessels at a punishing rate, and as a result they suffer much wear and tear. They are replaced after about 120 days.

Similarly, most of our carbohydrates, fats, and nucleic acids are being continuously replaced, although we don't notice this process at all.

As a result of all this replacement activity, each of us is literally not the person he or she used to be. Each of us is a temporary, reasonably organized arrangement of chemical compounds, but over our lifetime we do not even consist of the same molecules. Suppose it were to become possible to label each individual molecule in a newborn female baby. After 15 years, how many labels would we be likely to find in her body (assuming the ethics of the time allowed us to look)?

Apart from entirely new sets of molecules such as those arising at puberty, the same types of molecules—same proteins, same enzymes, same fats, same sugars—would be found. However, most—I can't hazard a guess at the percentage, but it would be low—would be unlabelled because of turnover. Turnover occurs from the molecular to the cellular level. Over time, most of the original labels will disappear, but our girl as an individual organism will continue to exist. The more widely she travelled, the broader the spread of her molecular labels would be. She would probably pick up a few of her own labels as they were recycled. Is this the origin of the saying 'getting your own back'?

Many of the individual atoms that once made up your body may well be spread widely through the neighbouring biosphere. Some might well be inspiring the leaps of your favourite sporting hero or fuelling the gyrations of a pop star. This turnover of molecules is one of the unique characteristics of living organisms. In contrast, inanimate objects such as motorcar engines simply wear out and need to be replaced. Your shoes will stretch and wear out, but the cells of your feet will replace themselves.

What are the origins of the order of life? Order and organization are consumers of energy, but how is it that out of the chaos and disorder of non-living matter, life has managed to arise? From a purely thermodynamic point of view, a living creature is a most improbable occurrence. So long as it remains alive, a plant or animal is able to convert some of the energy it consumes to resist the tendency to become disordered and fall apart. As soon as its normal energy flow ceases, the resistance disappears, the plant wilts, the animal collapses, and the inexorable degradations characteristic of death are set in motion. The highly structured leaves and flowers, the stems and roots, the legs, the paws, the heart and the brain all too soon lose their familiar form. Soon only the harder parts, the teeth and the skull, the skin, and the trunk of a tree, will be left to view. Of these, a tiny fraction may later be preserved as fossils, of which an even smaller fraction may become exposed by scouring wind or water, or by the probing tools of humans. Energy is 'running down' for these examples of ours and in the universe as a whole, as it expands, cools, and approaches maximum entropy. The stars are constantly generating and radiating vast amounts of energy, but will eventually 'burn out'. They pass through a series of evolutionary stages that are quite well understood. Our own Sun is burning away its mass at about 4,674,000 tonnes per second as it converts hydrogen to helium by nuclear fusion, with the generation of a vast amount of energy that radiates away into space in all directions. We alive today need not be concerned with this seemingly wasteful dissipation of our major source of energy. The Sun contains over 99.8% of all the mass in the solar system and has been estimated to have more than half of its predicted lifetime of 10 billion years remaining. As far as we know, Earth is the only place in the solar system where life has evolved. Life as we are privileged to observe it may well be unique to our galaxy, if not to the universe itself.

Apart from a few glitches, such as the major extinctions, life on Earth has not only survived for 3.5 billion years, overall it has positively thrived. Its complexity, diversity, and sheer extravagance fascinate all who take the time to wonder at its workings. Life's underlying order is another of its intriguing assets. The maintenance of order brings us back to the inescapable laws of thermodynamics. Any real-world process involves energy changes. There are macroscopic forms of energy, involving the properties of the system as a whole, such as the kinetic and potential energies of rocks, cars, bullets, or people. Microscopic forms of energy are related to molecular structure and molecular activity, and are essentially independent of what is happening outside. This microscopic energy includes the motion of molecules as they vibrate, rotate, stretch, break, and reform their bonds. The sum of all these microscopic forms of energy is called the internal energy. The internal energy associated with chemical bonds is called chemical or bond energy. During a typical chemical reaction some bonds are destroyed and new ones are formed. The internal energy changes and some chemical energy may be stored in a 'concentrated' form for later access. In order to have access to reactions with a large negative ΔG, living organisms need to have access to such compounds with concentrated, high-intensity energy. We can argue that the quality/intensity of this chemical potential energy, largely stored in food, is the ultimate result of the 'intensification' of solar energy by the process of photosynthesis.

The Second Law says that in any spontaneous process the overall entropy of the universe will increase. At the same time it does not exclude the possibility that locally the entropy may actually decrease during a process. This aspect is one that has confused some who have tried to justify the existence of God on thermodynamic grounds. Such people assert that as living organisms exhibit order, that is the origins of their highly organized bodily systems are characterized by a decrease in entropy (as when a large protein is synthesized from a number of small amino acids), they defy the Second Law and only a God can do that. The misconception arises because the whole picture is not being considered. Only a local part (you) of the whole system (the universe) is being considered in the argument. Consider what has really happened to yourself during your lifetime so far. You as an organized system have probably existed essentially unchanged since reaching maturity. That is only part of the story. As we have seen, most of our body cells will be replaced a number of times during our lifetime. What has happened to all the food you have consumed and all the water you have drunk? The individual atoms in all the chemicals in the mountain of faeces and the small lake of urine you have generated could be scattered almost anywhere on Earth, especially if you are well travelled. The same applies to the water vapour you have breathed out or perspired into the air. The carbon dioxide you have exhaled, millions of litres of it probably, could have been through plants and recycled to animals which yourself or almost anyone else could have eaten. Some of your carbon dioxide molecules may have reached the tables of people as far away as

America or Africa. If that isn't an example of an increase in overall disorder and spreading out of the energy involved, I don't know what is.

Other statements of the Second Law include:

'It is impossible to turn all heat into work.' (Sadi Carnot)

'Heat can never pass spontaneously from a body at lower temperature to one at a higher temperature.' (Clausius)

'No process is possible whose only result is the abstraction of heat from a single heat reservoir and the performance of an exactly equivalent amount of work.' (Kelvin-Planck)

Systems tend to proceed from ordered (low entropy or low probability) states to disordered (high entropy or high probability) states.

All naturally occurring processes proceed towards equilibrium, that is towards a state of minimum potential energy.

The total amount of entropy in the Universe is increasing.

There are others. The simplest I know is 'things wear out'. The story of Humpty Dumpty is another. Machines begin to wear out spontaneously immediately they start to be used. They never spontaneously repair themselves.

Why are there so many statements of the Second Law? One reason is that the Second Law has wide applicability, so the wording can be expressed in many ways. Another is that the pioneers of thermodynamics were not always clear in their own minds about entropy. Let's look at a few more everyday examples.

We all have experience of the tendency of things in our daily lives to proceed, annoyingly in most cases, spontaneously towards disorder. My teenage son's room is an example. Any handyman is soon amazed at the amount of disorder he creates in a short time. Tools become scattered, bits of timber lie everywhere, wood shavings, nails, and sandpaper clutter the workplace. All this is a result of a redistribution of some of the handyman's concentrated, high-intensity energy into a more disordered form. In the process, the mess was made and the handyman's muscles produced a lot of heat. Another familiar example is the incredible mess created during the cooking of a three-course meal. All these need an input of our energy to reorganize. Unfortunately, the utensils will never spontaneously reorganize themselves into storage. We don't really expect this, for it would be against all experience. Nevertheless, it is the result of the Second Law. Similarly, improbable arrangements don't occur spontaneously. Tossing a handful of childrens' blocks into the air will not result in a model medieval castle.

Irreversible processes are sometimes described as examples of 'time's arrow' and are characterized by an increase in entropy.

Many of the things worth achieving in life seem to involve creating order—building a complex structure or piece of machinery, creating the ordered beauty of a work of art, writing a novel by ordering your thoughts into a sequence then typing them out, composing a symphony. I can't recall an example of a noteworthy human achievement that has occurred without considerable organizational effort occurring at some level.

There are examples of good fortune, such as a hole in one at golf or a spectacularly successful gambling win. Statistically, these little-effort events are possible, but rare, and don't really count as achievements in the way I mean.

They are highly improbable, which is reflected in the very low frequency with which they occur.

Although perhaps not everyone would express it this way, real achievements are admired because most people appreciate the difficulty and effort involved in overcoming the natural tendency towards disorder and in bringing about a decrease in entropy, even at a local level.

The fundamental role of thermodynamics in living organisms was lucidly described by the physicist Erwin Schrodinger in his masterful little book *What is life?* First published in 1944, 9 years before the structure of DNA was determined, it throws a little more light on the nature of entropy (Schrodinger 1967). From Chapter 6:

'It is by avoiding the rapid decay into the inert state of equilibrium that an organism appears so enigmatic; so much so, that from the earliest of times of human thought some special non-physical or supernatural force was claimed to be operative in the organism... How does a living organism avoid decay? The obvious answer is: by eating, drinking, breathing, and (in the case of plants) assimilating. The technical term is metabolism... Every process, event, happening—call it what you will; in a word, everything that is going on in Nature means an increase of the entropy of the part of the world where it is going on. Thus a living organism continually increases its entropy... and thus tends to approach the dangerous state of maximum entropy, which is death. It can only keep aloof from it, i.e. alive, by continually drawing from its environment negative entropy—which is something very positive as we shall immediately see. Or to put it less paradoxically, the essential thing in metabolism is that the organism succeeds in freeing itself from all the entropy it cannot help producing while alive.... The device by which an organism maintains itself stationary at a fairly high level of orderliness (fairly low level of entropy) really consists in continually sucking orderliness from its environment. This conclusion is less paradoxical than it appears at first sight... in the case of higher animals we know the kind of orderliness they feed upon well enough, *viz*, the extremely well-ordered state of matter in... complicated organic compounds, which serve them as foodstuffs. After utilizing it they return it in a very much degraded form—not entirely degraded however, for plants can still make use of it.'

In the last few sentences you will recognize echoes of what I mentioned in Chapter 1, where I discussed the utilization of available energy in the food we consume. We have now been able to deal not only with the relatively simple concept of energy release from our food, but also with the less obvious and more fundamental considerations of entropy compensation and its role in the maintenance of life.

Finally, it is interesting to contemplate two great evolutionary concepts of the nineteenth century: entropy and evolution by natural selection. Both are evolutionary in the sense that their results and predictions unfold with the passage of time. Darwin claimed that the living world was not created as it is now, but rather that living organisms are and always have been undergoing constant change and adaptation, evolving over time into many different forms. Life is characterized by complexity (high degrees of order and a multiplicity of chemical reactions), diversity (of species and of form), and randomness (of mutations). On the other hand, entropy predicts simplicity, maximum disorder, lack of motion, and the same ultimate thermodynamic fate for everything in the universe.

A burning question in the nineteenth century was, could these two concepts be reconciled? It seemed an impossible task in those times, and as a result the biological and physical sciences continued to follow their own 'laws' and to develop different ways of thinking about the world. Some scientists believed for a time that biology was based on a number of yet-undiscovered laws of nature. This was understandable but incorrect. As we have just seen, thermodynamics and biology are interrelated, but this was not widely recognized until about the 1960s.

We are getting closer to an explanation of our dynamic energy balance in terms of fundamental chemical and physical laws. Here I stress again that I am not implying that living organisms, either their workings or their behaviour, are fully explicable by the simple application of purely physical and chemical laws. There is more to life than this, but whatever processes are involved in the workings of biology, they do not and cannot violate those laws.

When developments in physics and chemistry led to the elucidation by Watson and Crick of the structure of DNA in 1953, the science of molecular biology became possible. The determination of the chemical structure of the genetic material allowed application of the laws and techniques of the physical scientist to the most fundamental biological questions and processes. When it became possible to determine the sequence of the components of DNA, genes were 'mapped' at the molecular level. Definite sequences of DNA eventually became identified with specific genes on specific chromosomes. Further developments in X-ray crystallography and nuclear magnetic resonance spectroscopy led to the determination of high-resolution, three-dimensional structures of proteins and many other biomolecules, and the era of genetic engineering was just around the corner.

GENERAL REFERENCES

Keeler, J. and Wothers, P. (2003) *Why Chemical Reactions Happen*. Oxford University Press, Oxford.

REFERENCES

De Duve, C. (2005) *Singularities*. Cambridge University Press, New York, pp. 29–35.
Garrett, R.H. and Grisham, C.M. (2010) *Biochemistry*, 4th edn. Brooks/Cole, Cengage Learning, Boston, p. 67.
Garrett, R.H. and Grisham, C.M. (1999) *Biochemistry*, 2nd edn. Saunders College Publishing, Fort Worth, pp. 66, 860–1.
Haynie, D.T. (2001) *Biological Thermodynamics*. Cambridge University Press, Cambridge, pp. 98–9.
Schrodinger, E. (1967) *What is Life*. Cambridge University Press, Cambridge, pp. 70–1.
Streitwieser, A. and Heathcock, C.H. (1981) *Introduction to Organic Chemistry*, 2nd edn. Macmillan Publishing Co. Inc., New York, p. 52.
Voet, D., Voet, J.G., and Pratt, C.W. (2002) *Biochemistry*, upgrade edn. John Wiley & Sons Inc., New York, pp. 395, 404.

7
The Strange Story of Water and Oil

Please consider the message of this chapter thoroughly. The differences in properties between water and water-loving molecules, and oily/greasy substances have allowed living cells to evolve, and thus to provide life with a safe and selective environment. Energy and information are transferred into and out of cells, allowing interaction and communication between individuals, be they unicellular bacteria or human beings. Essential chemical gradients are set up and maintained by the expenditure of energy.

All the biological reactions we have discussed must take place at some specific location in some specific medium. We haven't discussed this explicitly, but now is the time to do so. It is easy to forget, when looking at the equations representing such biochemical reactions, the conditions under which they occur. Although thermodynamics may favour a reaction proceeding from reagents to products, as written on a page, the conditions required for it to take place may not be obvious. Biochemical reactions in general occur under aqueous conditions, although the enzymes involved could be in free solution, bound to membranes or to other cell structures such as ribosomes and DNA. The actual reactive sites where reactions take place may be located in the so-called 'hydrophobic pockets' in the enzyme surface, where water is excluded. Enzymic reactions occur overwhelmingly under quite mild conditions of temperature, pressure, pH, and salt concentration.

In this, biochemical reactions differ markedly from many reactions carried out by synthetic chemists. Chemists often need to resort to high temperatures, long reaction times, and catalysts to achieve reasonable yields of product. In a large number of instances, they also have to avoid the presence of water. The 'drying' of organic solvents to remove traces of moisture, and the need to protect reactions from even the relatively small levels of water in the air, requires significant amounts of time and effort. Failures to reach the expected yield of product often have their roots in careless or insufficient drying of solvents.

On the other hand, biological systems are capable of highly specific, high-yielding, rapid reactions in the presence of water. Chemists lag far behind in these capabilities, but biotechnologists have begun to use the lessons of biology to improve the situation. One important result of the use of biotechnology could be a reduction in the levels of chemical pollutants that have given much of the traditional chemical industry such a poor reputation. Such a result is not guaranteed, as there will always be some by-products to dispose of. With imaginative science, backed by sensible legislation, it should be possible to design biochemical processes with little negative environmental impact. Although biotechnology on a wide scale is a relatively recent industry, we at least have the chance and the knowledge to ensure that the errors of the past are not repeated. Even so, it will not be easy in these days when economic rationalism and globalization are powerful influences. Although science and technology can solve almost any problem, economics and politics often enforce a compromise.

Water is my favourite substance and, as we'll see, it has some remarkable and unique properties. Even in this age of the hyperbole, it is no exaggeration to state that without the availability of the water molecule, life on Earth would never have developed and survived. Water in its many aspects has always been a source of fascination to humans. From the purely aesthetic through to the practical and the essential, it is part of the very fabric of our existence. Besides covering some 70% of the surface of the planet to an average depth of about 3.8 km, water pervades the senses, permeates our bodies, inspires artists and poets, provides transport, and is at the heart of many natural disasters. It is the basis of some of our most enjoyable recreations, from swimming to sailing, from snow-skiing to shooting rapids. Water provides all that and more. As the main component of coffee, tea, and alcoholic beverages it helps stimulate a jaded humanity. It may occupy all three common states of matter—solid, liquid, and gas—depending on the amount of energy it possesses. Energy-dependent transformations between these states determine our climate and drive the weather patterns of the world.

Most water on Earth is found in the oceans and is salty, or saline. Fresh water is one of the factors limiting the development of civilizations, and is one of humanity's most precious resources. An enormous proportion of the fresh water on Earth is bound in polar, mainly the Antarctic, ice-caps. On a human timescale these remain almost constant in size, but none the less continual change is occurring on the geological time-scale. Fresh snowfalls carry pollution that reflects the state of the atmosphere, and have always done so. Cores drilled from the ice-caps provide a history of the atmosphere going back thousands of years. Polar ice-caps spread outwards under the pressure of layer upon layer of new ice and snow. Pieces, some many kilometres square, break off the edges and may drift away as icebergs, eventually to melt. Glaciers continue to wear the rocks below and around them. Ice ages have come and gone as changes in atmospheric energy balances have moved Earth's climate one way then another. As areas of snow and ice reflect solar radiation rather than absorb it, the lower amount of retained energy probably increased the length of the ice ages.

Over the oceans and lakes of the Earth, water is evaporated by the energy of the Sun and is transported by the wind. As it rises and cools, water vapour condenses to form clouds, and the resulting rainfall incessantly scours the landscape, while streams pour silt and soluble minerals into the oceans. This hydrological cycle has continued since the earliest cooling of the Earth's crust allowed the formation of liquid water on a vast scale. Dissolved oxygen plus minerals help to sustain much of aquatic life today. Countless watery niches have been and continue to be created, allowing the evolution of unique species, the development of populations, and the emergence of whole ecosystems. Even a small rock pool teems with life, much of it visible to our curious eyes. Whether or not life began in such a pool is not known. Perhaps it started in the depths of the ocean, in darkness near some smoking, sulphurous hydrothermal vent. Wherever and whenever the spark was ignited, we can be certain that water was intimately involved.

Biologically, water is much more than a simple solvent in which the molecules of life react. It is an enormous contributor itself to the very success of many processes. One way water does this is by providing microenvironments of differing pH and salt strength. The pH of an aqueous solution is a measure of its concentration of protons, H^+, or, more strictly, its concentration of H_3O^+, because $H^+ + H_2O \rightarrow H_3O^+$ almost exclusively in the presence of water. Once again this is dictated by energy considerations. A naked H^+ floating about in a sea of polar water molecules just doesn't happen, and just like any other ionic species will have a 'shell' of water molecules surrounding it. The H^+ is deficient in electrons, and the lone pair of electrons in the water molecule is literally an attractive solution to the problem.

The pH value is a more familiar measure of the acidity of an aqueous solution. Formally, $pH = -\log[H^+]$, where $[H^+]$ is the concentration of hydrogen ions in moles per litre. The pH scale ranges from 0 to 14, with $pH = 7$ being called neutral. This arises from the fact that pure water dissociates thus:

$$H_2O \rightleftharpoons H^+ + OH^- \tag{7.1}$$

By definition

$$K_a = \frac{[H^+][OH^-]}{[H_2O]} \tag{7.2}$$

As the molar concentration of pure water is usually very large compared with that of a solute, it can be regarded as constant. The molarity of pure water is 1 L/mol. wt water $= 1000/18 = 55.5$ molar.

Thus

$$K_a = \frac{[H^+][OH^-]}{[55.5]}$$

So

$$K_a \times 55.5 = [H^+][OH^-] = K_w$$

where K_w is defined as the ion product constant of water.

At 25°C in pure water it is found that

$$[H^+] = 10^{-7} = [OH^-]$$

So

$$K_w = [H^+][OH] = (10^{-7})(10^{-7}) = 10^{-14}$$

Taking the negative logarithm:

$$pK_w = pH + pOH = 14.0 \tag{7.3}$$

This relationship is valid for any aqueous solution, whether acidic, basic, or neutral.

Any pH below 7 is said to be acidic, while a pH above 7 is said to be alkaline or basic. Strong acids such as sulphuric, hydrochloric, or phosphoric provide protons readily and produce low pH values. Strong alkalis (bases) such as sodium hydroxide (caustic soda) produce high pH values in solution. There are so-called 'weaker' acids, for example acetic acid found in vinegar, and weaker bases, for example ammonia, sodium bicarbonate (baking soda), and sodium carbonate (washing soda).

A strong acid such as hydrochloric will dissociate completely into ions in water, generating a large number of H^+ ions:

$$HCl(aq) + H_2O(aq) \longrightarrow H_3O^+(aq) + Cl^-(aq) \tag{7.4}$$

(From now I'll omit the 'aq' subscripts for clarity, and use H^+ instead of H_3O^+.)

The single arrow indicates complete dissociation into ions.

Thus 0.1 M HCl has a $[H^+]$ of 0.1 or 10^{-1} molar.

So

$$pH = -\log[H^+] = -\log 10^{-1} = 1$$

A weak acid such as acetic acid dissolved in water occurs only partially as ions, that is its dissociation into ions is a reversible reaction and an equilibrium is set up. This is indicated by the familiar double arrow:

$$CH_3COOH \rightleftharpoons H^+ + CH_3COO^- \tag{7.5}$$

(acetic acid)   (acetate ion)

Weak acids don't give such a low pH as strong ones of the same concentration. Thus 0.1 M acetic acid in water gives a pH of 2.88.

To simplify the argument, let HA be any weak acid. Then in aqueous solution

$$HA \rightleftharpoons H^+ + A^-$$

A^- is called the conjugate base of the acid HA.

For this ionization, the equilibrium constant is called the acidity constant or acid ionization constant, K_a. We define this in the usual way (see Chapter 6)

$$K_a = \frac{[H^+][A^-]}{[HA]} \tag{7.6}$$

Acidity constant values are often tabulated as their pK_a, where $pK_a = -\log K_a$.

The strength of the acid is related to the magnitude of K_a. The stronger the acid, the larger the value of K_a and the lower the value of pK_a. As a weak acid is never fully ionized, there will always be some H^+, A^-, and HA present in the solution.

An important application of chemical equilibrium in biological systems is the common ion effect, a special case of Le Chatelier's principle (see Chapter 6).

Suppose we have a solution of a weak acid, acetic acid, to which we add sodium acetate.

Both will dissociate and form acetate ions:

$$CH_3COOH \rightleftarrows H^+ + CH_3COO^- \text{ (acetate ion)} \tag{7.7}$$

$$CH_3COONa \longrightarrow Na^+ + CH_3COO^- \text{ (acetate ion)} \tag{7.8}$$

The acetate ion is correctly called the conjugate base of acetic acid. It is often also called the salt, as it is derived from the salt, sodium acetate. In the general expression for an acid, HA, the A^- ion may be called either the conjugate base *or* the salt. Both are acceptable and are used interchangeably. The sodium acetate is a strong electrolyte and dissociates fully, hence the single arrow.

According to Le Chatelier's principle, addition of acetate ions from CH_3COONa will disturb the equilibrium of the CH_3COOH, as the extra CH_3COO^- will 'force' reaction (7.7) from right to left. This will decrease the hydrogen ion concentration $[H^+]$ and the pH will increase. The common ion, CH_3COO^-, caused the effect. Stating this in general terms:

The shift in the position of equilibrium caused by the addition of a substance having an ion in common with the dissolved substance is called the common ion effect. (Chang 1994)

The common ion effect plays an important role in determining the pH of a solution. If we take eqn (7.6)

$$K_a = \frac{[H^+][A^-]}{[HA]}$$

Rearranging gives

$$[H^+] = \frac{K_a[HA]}{[A^-]}$$

Taking the negative log of both sides gives

$$-\log[H^+] = -\log K_a - \log\frac{[HA]}{[A^-]}$$

or

$$-\log[H^+] = -\log K_a + \log\frac{[A^-]}{[HA]}$$

Thus

$$pH = pK_a + \log\frac{[A^-]}{[HA]} \tag{7.9}$$

Where

$$pK_a = -\log K_a$$

Equation (7.9) is called the Henderson–Hasselbalch equation.

If we know pK_a and the concentrations of the acid [HA] and the conjugate base $[A^-]$ we can calculate the pH of the solution. The equation is valid regardless of the source of A^-, that is it can come from both the acid HA and the salt A^- (e.g. from CH_3COOH(HA) and CH_3COO^-(A^-) in eqns (7.7) and (7.8)).

An important application of the above theory is that of buffers. A buffer is a solution that has the ability to resist changes of pH if small amounts of acid or base are added. Physiological or biological buffers are used by organisms to maintain the pH of various regions of cell or tissues within quite narrow limits to maintain the optimum performance of enzymes, and the optimum electrical charge on other molecules. For example, blood has a pH range of 7.35–7.45, stomach is approximately pH 2, and the small intestine is above pH 8.

How do buffers work? A buffer consists of a solution of (a) a weak acid and (b) the salt of the acid. Again using the example of acetic acid and sodium acetate, suppose we have a solution 0.2 M in acetic acid and 0.3 M in sodium acetate:

$$CH_3COOH \rightleftarrows H^+ + CH_3COO^- \text{ (acetate ion)} \quad (7.7)$$

$$CH_3COONa \longrightarrow Na^+ + CH_3COO^- \text{ (acetate ion)} \quad (7.8)$$

What will happen if we add a small amount of some acid (H^+) to the solution? The equilibrium of eqn (7.7) will change (Le Chatelier's principle) to minimize the effect of added acid. The added H^+ will react with some CH_3COO^- to form CH_3COOH.

Effectively, this means that the added H^+ will be removed and the pH will remain relatively constant. We have a buffer, as long as we don't add too much acid and overwhelm the buffering capacity of the system.

Similarly, what will happen if we add some base, as OH^- for example? The OH^- will react with some H^+ to form water. This will reduce the $[H^+]$ in eqn (7.7), causing the system to compensate by making more H^+ available via the dissociation of some CH_3COOH (which we know will be present). Again, the pH will remain relatively unchanged, for small amounts of added OH^-.

Let's do some quantitative calculations to determine the pH of our buffer solution. Assume that $[CH_3COOH] = 0.2$ M and $[CH_3COO^-] = 0.3$ M. We can justify these assumptions because CH_3COOH is a weak acid. The presence of the common ion CH_3COO^- further suppresses the ionization of CH_3COOH, so its concentration is essentially 0.2 M. Also, CH_3COONa is a strong electrolyte and is fully dissociated, so $[CH_3COO^-] = 0.3$ M.

The K_a for acetic acid $= 1.8 \times 10^{-5}$.

Rearranging eqn (7.6) we obtain:

$$[H^+] = \frac{K_a[HA]}{[A^-]}$$

$$= \frac{(1.8 \times 10^{-5})(0.2)}{0.3}$$

$$= 1.2 \times 10^{-5} \text{ M}$$

So

$$pH = -\log[H^+] = -\log(1.2 \times 10^{-5})$$

$$= 4.92$$

We can also find the pH using the Henderson–Hasselbalch equation, first converting the K_a for acetic acid of 1.8×10^{-5} to pK_a:

$$pK_a = -\log K_a = -\log(1.8 \times 10^{-5}) = 4.74$$

Then

$$pH = pK_a + \log \frac{[CH_3COO^-]}{[CH_3COOH]}$$
$$= 4.74 + \log \frac{0.3}{0.2}$$
$$= 4.92$$

What would the pH of 0.2 M acetic acid have been if we had not added the 0.3 M sodium acetate? I won't calculate it here, but this turns out to be a pH of 2.72.

This is considerably less than 4.92 and shows that, as expected, the presence of the common ion CH_3COO^- considerably suppresses the ionization of CH_3COOH.

It can be shown by similar calculations to those above that a solution 1 M in acetic acid and 1 M in sodium acetate has a pH of 4.74. To test its effectiveness as a buffer, suppose we were to add 0.1 mole equivalents of HCl (say as a gas, so as not to change the volume of our solution and so make the calculation simpler). This addition of HCl will force reaction (7.7) to the left, reducing $[CH_3COO^-]$ and increasing $[CH_3COOH]$ to make the final concentrations:

$$[CH_3COO^-] = 0.9 \text{ M and } [CH_3COOH] = 1.1 \text{ M}$$

Using

$$[H^+] = \frac{K_a[HA]}{[A^-]}$$

We have

$$[H^+] = \frac{((1.8 \times 10^{-5})1.1)}{0.9}$$
$$[H^+] = 2.2 \times 10^{-5} \text{ M}$$
$$pH = 4.66$$

The addition of 0.1 moles of HCl has caused the pH to fall by only (4.74–4.66) or 0.08 pH units. This corresponds to a change in $[H^+]$ by a factor of only 1.2. If the 0.1 moles of HCl had been added to pure water, we would have $[H^+] = 0.1$ M.

For pure water:

$$H_2O \rightleftharpoons H^+ + OH^-$$
$$K_w = [H^+][OH^-] = 1.0 \times 10^{-14}$$

So

$$[H^+] = [OH^-] = 1.0 \times 10^{-7} \text{ M}$$

This is a change in [H⁺] by a factor equal to

$$= \frac{[H^+]_{\text{after HCl addition}}}{[H^+]_{\text{before HCl addition}}}$$

$$= \frac{0.1}{1.0 \times 10^{-7}}$$

$$= 1.0 \times 10^6 \text{ (This number is a dimensionless ratio.)}$$

This is a millionfold increase compared with a 1.2-fold increase in the presence of acetate buffer, graphically illustrating (i) the buffering effect, (ii) Le Chatelier's principle, and (iii) the common ion effect on chemical equilibria.

Here is another point about the effectiveness of buffers. When the ratio

$$\frac{[A^-]}{[HA]} = 1$$

Then

$$\log \frac{[A^-]}{[HA]} = 0$$

So the Henderson–Hasselbalch equation leads to

$$\text{pH} = pK_a$$

This means that when the salt/acid ratio is

$$\frac{[A^-]}{[HA]} = \frac{1}{1}$$

the pH of the solution is equal to the pK_a of the acid involved. This pH is also where the buffer exhibits its maximum buffering capacity, as $[A^-]$ and $[HA]$ are both high, and can resist added base or acid to the same (maximum) extent. Thus, a plot of the pH of a buffer system titrated with base shows that over the range $pK_a \pm 1.0$ unit, the change is quite small (Figure 7.1).

A useful rule of thumb for the buffering range, or the pH range over which a buffer is useful, is

$$\text{pH range} = pK_a \pm 1.0 \text{ pH unit}$$

This allows us to select a suitable buffer for a particular pH range. Choose an acid with a pK_a as close as possible to the pH required, and adjust the pH with acid or base to the required value.

As the ionization of acids is a process involving changes in energy, thermodynamics is involved. The pK_a is related to the magnitude of the standard state free energy of ionization. Using eqn (6.9) we can say in general:

$$\Delta G^0 = -RT \ln K_{eq} = -2.303 \, RT \log K_{eq}$$

In this case the equilibrium constant is K_a, the equilibrium constant for the ionization of HA, and ΔG^0 is the standard free energy of ionization.

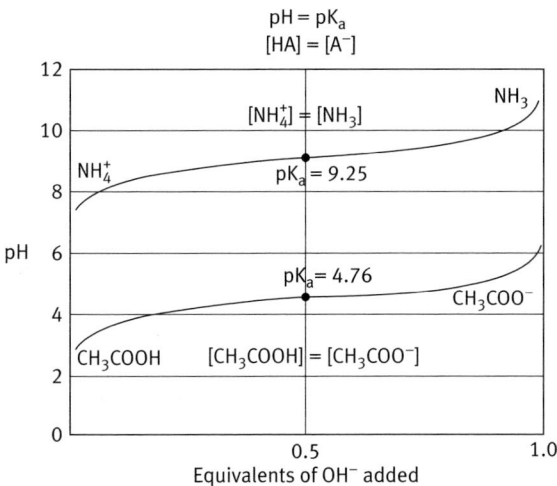

Figure 7.1. Titration curves of acetate buffer and ammonia by the addition of sodium hydroxide solution. Buffers in general are effective over the pH range $pK_a \pm 1$ unit.

Also, as $pK_a = -\log K_a$ it follows that

$$\Delta G^0 = 2.303\, RT pK_a \qquad (7.10)$$

So

$$pK_a = \Delta G^0/(2.303 RT)$$

Equation (7.10) shows that the standard free energy of ionization is proportional to the pK_a. Is this reasonable? For acid dissociation in general we can write

$$HA \rightleftharpoons H^+ + A^-$$

The equilibrium constant for this is given by eqn (7.6)

$$K_a = \frac{[H^+][A^-]}{[HA]}$$

For a very strong acid, this equilibrium will be far to the right, and K_a will be large, for example suppose $K_a = 10^4$ for a strong acid at 25°C. Then $\log K_a = 4$, and

$$-\log K_a = -4 = pK_a$$

Substituting in eqn (7.10) gives

$$\begin{aligned}\Delta G^0 &= 2.303\, RT(-4) \text{ J mol}^{-1}\\ &= -(2.303 \times 8.31 \times 298 \times 4)\\ &= -22{,}812 \text{ J mol}^{-1}\\ &= -22.8 \text{ kJ mol}^{-1}\end{aligned}$$

Table 7.1. Dissociation constants and pK_a values for some biological electrolytes.

	K_a	pK_a
Acetic acid	1.74×10^{-5}	4.76
Lactic acid	1.38×10^{-4}	3.86
Glutamic acid (COOH in proteins)	(~4.5)	
Phosphoric acid pK_{a1}	7.08×10^{-3}	2.15
Phosphoric acid pK_{a2}	6.31×10^{-8}	7.2
Phosphoric acid pK_{a3}	3.98×10^{-13}	12.4
Imidazole (histidine in proteins)	(5.6–7.0)	
Cysteine (SH in proteins)	(9.1–10.8)	
Tyrosine (phenolic OH in proteins)	(9.8×10^{-4})	
Carbonic acid (H_2CO_3) pK_{a1}	4.3×10^{-7}	6.35
Bicarbonate (HCO_3^-) pK_{a2}	5.7×10^{-11}	10.33
Ammonium ion (NH_4^+)	5.62×10^{-10}	9.24

Data in brackets are from Haynie (2001), p. 101. Other data are from Aylward & Findlay (1994).

Thus, the ΔG^0 for ionization is large and negative, consistent with our proposal for a strong acid, essentially fully ionized, with the equilibrium greatly in favour of $H^+ + A^-$.

Similarly, a weak acid having $K_a = 10^{-3}$ has p$K_a = 3$.

Substituting p$K_a = 3$ in eqn (7.10) yields $\Delta G^0 = 17.1$ kJ mol^{-1}, a large positive value indicating that HA predominates, and we have a very weakly dissociated acid.

There's thermodynamics in everything, even buffers!

Let's look at some biological examples involving buffers. Buffering in biological systems can be achieved by several means. About 80% of the buffering capacity of blood is due to proteins, mainly haemoglobin and serum albumin.

Side-chains of the amino acids in proteins have a variety of acid and basic groups capable of forming a buffer system (Table 7.1).

Physiological pH in most organisms is close to 7, and though the phosphate system seems suitable (p$K_{a2} = 7.2$) it is not used in the case of human blood (pH ~7.4) because the phosphate ion concentration in blood is too low. Bicarbonate is used in blood, while phosphate is the buffer system in intracellular fluids.

Another of the reasons why pH is so important in biology is that the reactivity of important molecules such as enzymes is highly pH dependent. Digestive enzymes such as trypsin and chymotrypsin occur in the stomach, which maintains a strongly acidic pH of about 1.2–2.5. These enzymes are quite active and reasonably stable in the stomach, but many other enzymes would be inactivated by such conditions. Proteins, including enzymes, are in general active and their 'native' three-dimensional shape (native conformation) is stable within certain pH limits only, and the appropriate pH is usually maintained in their region of the cell. Even if an enzyme is not irreversibly inactivated by a pH change, it will almost certainly require a certain pH for optimum activity. Sudden changes in pH are generally speaking to be avoided, and cells take advantage of the properties of weak acids and/or weak bases to achieve pH stability through buffering.

Let's now look at some other properties of water molecules, concentrating on 'bulk water' properties instead of individual water molecules. We'll also look at some other molecular types which behave quite differently from water. We have seen that water molecules are highly polar, and that they form a three-dimensional network of hydrogen bonds. The partially-positive ($\delta+$) hydrogen atoms arrange themselves to be close to the partially-negative ($\delta-$) oxygen atoms of neighbouring water molecules. This is a relatively low-energy arrangement, where the two opposite charges tend to cancel one another, but remember that the hydrogen bonds are constantly forming and breaking and reforming with other neighbours. In liquid water at room temperature about 85% of the possible hydrogen bonds are in place at any moment. Liquid water is a dynamic system. Hydrogen bonds have an average lifetime of about 10 picoseconds (1 picosecond = 10^{-12} s) and water has a kind of temporary, fluttering 'structure'. Locally, forms called clusters containing about 100 water molecules may occur.

These clusters, as well as the bulk hydrogen bonds in water, take energy to disrupt. One way to disrupt them is to raise the temperature, and the resulting increase in heat energy causes the water molecules to move faster and their increased kinetic energy rapidly disrupts the weak hydrogen bonds. The 'structure' of water decreases as the temperature rises. Eventually, at the boiling point, the water molecules possess so much energy that the hydrogen-bonding attractive forces are no longer sufficient to hold them together. In the body of the liquid, bubbles of water vapour may form, and molecules at the surface fly out of the liquid state and eventually become a gas, steam. Steam itself is invisible, and the clouds we see near the nozzle of a boiling kettle are droplets of water vapour, from steam that has condensed as it strikes the cooler air.

On the other hand, as the temperature of water is lowered, hydrogen bonding is enhanced, more clusters of water molecules form, and near 0°C the proportion of these clusters increases until the water begins to crystallize and the water freezes to form ice. Quite a large amount of energy must be taken from liquid water at 0°C before it will form ice at 0°C. This is called the latent heat of fusion, or enthalpy of fusion, ΔH_{fusion}, which for water is 6.0 kJ mol^{-1}. In ice, most of the water molecules are fixed into a hexagonal lattice, an arrangement different from the temporary, tetrahedral structure of liquid water (Figure 7.2). Depending on the temperature and pressure, there are six crystalline forms of ice. The form we normally observe is called ice 1. Figure 7.2 includes the two orbitals of water that contain lone pairs of electrons (Chapter 5). As they contain two electrons and occupy space, lone pair orbitals contribute to the shape and properties of a molecule. In a water molecule, if one includes the two lone-pair orbitals, the shape is close to tetrahedral. This is reflected in the arrangement of hydrogen bonds in liquid water (Figure 7.2(b)).

In ice 1, which is fully hydrogen bonded, the average distance between the individual molecules is actually greater than in the liquid state, giving the ice a more open structure than water. As a result, ice has a lower density than liquid water (about 0.92 instead of 1). This is one of the fascinating properties of water and is the reason that ice floats. The peculiarity has important consequences for the survival of life. Ice forms on the surface of water first and is a good insulator, so lakes and streams seldom freeze to the bottom in winter. Fish, otters, seals, and other animals survive under the protecting layer of ice. The individual water molecules in ice still vibrate (as they would do right down to absolute zero, 0 K (-273.15°C)) but their average position in the solid ice remains the same, locked into the constraining lattice.

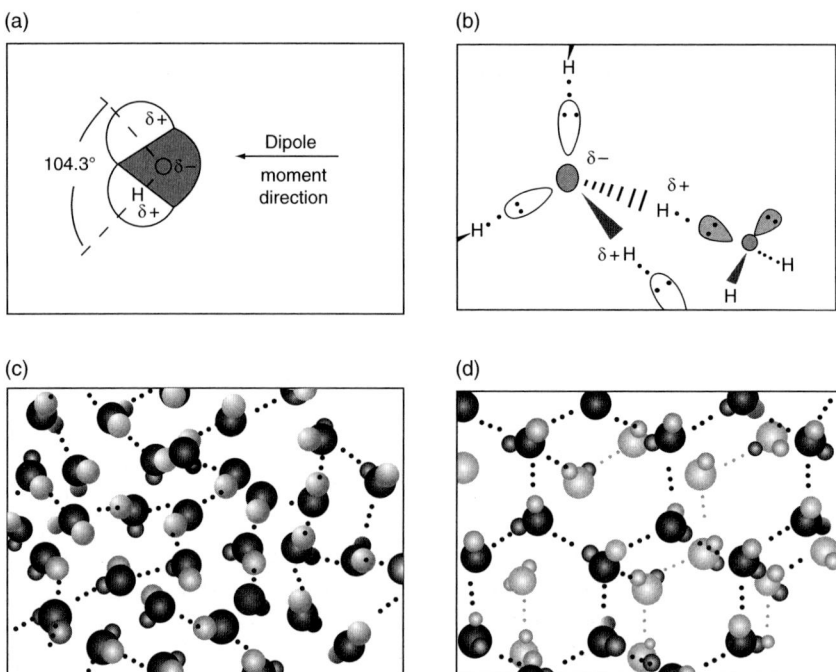

Figure 7.2. The structures of water and ice. (a) Space-filling structure of water. The direction of the dipole moment is from negative to positive. (b) The structure of a water molecule showing the two non-bonding oxygen orbitals. Each orbital contains a lone pair of electrons that can be involved (as acceptors) in hydrogen bonding to the hydrogen atoms of neighbouring water molecules. (c) The instantaneous structure of water. Average hydrogen bond lifetime is 10 picoseconds, and water molecules are in constant motion. (d) The structure of ice. Hydrogen bonding forms a hexagonal-based lattice, with the water molecules fixed in a more open, less dense (0.92) arrangement than in liquid water (1.0).

On the other hand, the non-polar, non-hydrogen-bonding methane CH_4, with a molecular weight of 16, is close in size and shape to water (if we consider the lone pairs in the latter) but is a gas, with a boiling point of $-162°C$ and zero polarity ($p = 0$). These two molecules differ in boiling point by a staggering 262°C. Chloroform, $CHCl_3$, has a similar shape, a molecular weight of 119.5, a medium polarity ($p = 3.4$), boils at just 61.7°C, and does not mix with water (Figure 7.3).

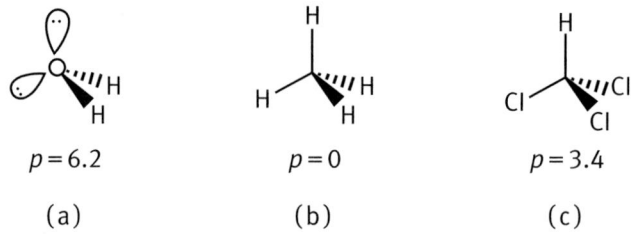

Figure 7.3. Polarities (p) of (a) water, (b) methane, and (c) chloroform. The three molecules are similar in size and shape, but their physical properties differ greatly.

Structure and bonding determine properties in the molecular world. Differences in electronegativity between the atoms can have an enormous effect on the physical properties of the molecules.

Water is an excellent solvent. It is the most effective liquid we know for its ability to dissolve solids and to mix with other liquids such as alcohols and organic acids. Water does not, however, mix well with substances such as fats and oils, and this fact allows the formation of some molecular structures that are essential for life.

What characterizes substances that dissolve in water?

Polar compounds often dissolve in water and are said to be hydrophilic (water loving).

Ethyl alcohol (ethanol) is a polar liquid.

Compare ethanol structure with that of water:

$$\overset{\delta-\ \delta+}{CH_3-CH_2-O--H} \quad \overset{\delta+\ \delta-\ \delta+}{H--O--H},$$
$$p = 5.6 \quad p = 6.2$$

Note that the two structures have an $-OH$ group in common. The $-OH$, or hydroxyl group, is characteristic of alcohols, is polar as shown, and forms hydrogen bonds (see Chapter 5). Thus, ethyl alcohol ($p = 5.6$) forms hydrogen bonds to water and will mix with it in all proportions. Ethanol boils at 78.3°C, well below the boiling point of water, even though its molecular weight of 46 is about 2.5 times that of water. Alcohol molecules hydrogen bond to one another, but this is not as effective as in water. Having a non-polar CH_3-CH_2- group present is part of the reason for this. Only the OH hydrogen can act as a hydrogen-bond donor, whereas water has two such donors. As in water, the oxygen atom can act as a hydrogen bond acceptor.

Similarly, glycerol (glycerine), with the structure

$$\begin{array}{l} CH_2OH \\ | \\ CHOH \\ | \\ CH_2OH \end{array}$$

has three polar hydroxyl groups and is also fully miscible with water. Glycerol forms multiple hydrogen bonds very effectively to itself. As you probably know, glycerol feels quite thick and viscous, and although its molecular weight is only 92, it boils at an astonishing 290°C, demonstrating again the profound effects of hydrogen bonding on physical properties.

The water solubility of non-ionic polar compounds such as sugars is related to the ability of their OH groups to form strong hydrogen bonds to water. For a sugar to dissolve spontaneously, the ΔG of sugar–water hydrogen bond formation, plus the increase in entropy brought about by the change from organized solid lattice to widely dispersed molecules in solution, must overcome the lattice enthalpy of the solid sugar.

Hydrogen bond formation is not restricted to hydroxyl groups. Other polar bonds to hydrogen may provide hydrogen bond donors, and polar groups other than OH may act as acceptors and donors. For proteins in water, hydrogen bond formation may occur as follows:

Hydroxyl to hydroxyl in water:

$$\overset{H}{\underset{|}{}} \quad \overset{\delta+\delta-}{}$$
$$O-H\text{---}O-H$$
$$|$$
$$H$$

Carbonyl to hydroxyl:

$$\overset{\delta-\delta+}{R-C=O\text{---}H-O-R_2}$$
$$|$$
$$R_1$$

Amide to hydroxyl:

$$\overset{H}{\underset{|}{}} \quad \overset{\delta-\delta+}{}$$
$$R-N-C=O\text{---}H-O-R_2$$
$$|$$
$$R_1$$

Amide to carbonyl:

$$\overset{R}{\underset{|}{}} \quad \overset{\delta-\delta+}{}$$
$$HN-C=O\text{---}C=O$$
$$||$$
$$R_1R_2$$

R and R_1 represent rest of the protein joined by the amide linkage $-CONH-$ (see Chapter 8).

Hydrogen bonding is also vital in the specific interactions in the DNA double helix and also to the protein α-helix, where nitrogen as well as oxygen atoms are involved as donors and acceptors.

In Chapter 5 we saw that common salt, sodium chloride, was an ionic compound, best shown by the formula Na^+Cl^- rather than by a covalent formula Na–Cl. Similarly Na^+OH^- better represents sodium hydroxide. In the solid salt, sodium ions and chloride ions are packed into a regular lattice, which takes a significant amount of energy to disrupt.

What happens when the ionic substance salt is put in contact with water? Both water and salt are very polar substances, so the respective positive and negative regions will tend to attract one another and thus minimize the overall energy. What happens to the Na^+ and Cl^- ions? At the surface of the solid, both ions rapidly become surrounded by water molecules, which form hydration shells. All ions in aqueous solution are hydrated, that is they are surrounded by oriented shells of water. The number of water molecules and the strength of their interaction with the ions are dependent on the size and the charge on the ion involved. Small, highly charged ions have tightly bound hydration shells, sometimes with multiple layers of water molecules. The hydration shells are dynamic entities, with individual molecules changing place continuously. In the electrical field surrounding an ion, the water molecules orient themselves according to the charge on the ion (Figure 7.4).

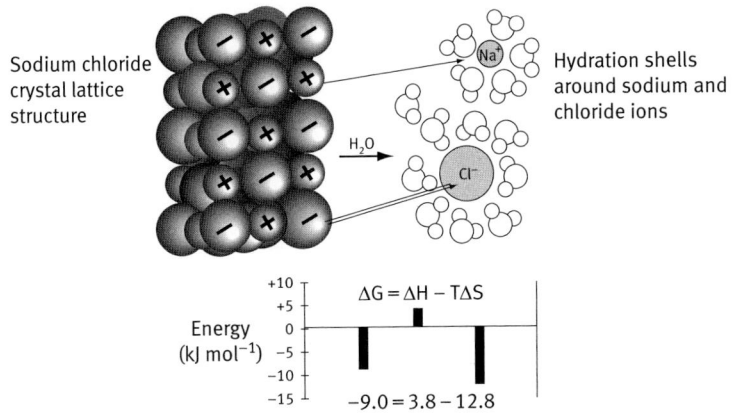

Figure 7.4. Energetics of the dissolution of sodium chloride in water. Energy is absorbed from the surrounding water to disrupt the ionic lattice of salt crystals (ΔH is positive) so the temperature of the solution falls. The process is spontaneous because the large increase in entropy makes the term $-T\Delta S$ large and negative, such that ΔG becomes negative. Water molecules in the solvation shells orient themselves according to the ionic charges: $-\delta H^+$ close to Cl^- and δO^- close to Na^+.

As water has a high dielectric constant of 78, the hydration shells will shield or insulate the Na^+ and Cl^- ions from one another, and greatly reduce the forces of attraction between them (to about 1/78 compared with that in air). The hydrated ions are thus able to move from their fixed lattice positions, and the resulting increase in mobility and hence entropy favours the spontaneous formation of a solution, despite the fact that energy is required to disrupt the ionic lattice in solid salt. The overall process is exothermic and will proceed spontaneously.

Differences in the size and hydration of ions are vitally important in biological systems, as we'll see later. Such differences allow, for example, the plasma membranes in animal cells to maintain a concentration difference between sodium (Na^+) and potassium (K^+) ions; a concentration gradient is set up. A typical mammalian cell has an extracellular sodium ion concentration of about 145 millimolar (mM), while its intracellular concentration is only 5–15 mM. Potassium ion is the opposite at about 140 mM (intracellular) and about 5 mM (extracellular). This imbalance is maintained by the operation of the transport protein Na^+/K^+ exchanging ATPase. It requires energy, supplied by ATP, to maintain the ion imbalance because the natural (thermodynamic) tendency is for an equilibrium to be set up with roughly equal concentrations on both sides of the membrane. The ATPase acts as a pump, selective for Na^+ and K^+.

This gradient involving the two ions is essential in nerve transmission. The pump can be selective for Na^+ and K^+ as the two ions differ quite significantly in size and charge distribution, especially considering the respective volumes of their hydration shells. The first layer of water in the hydration shell is the most tightly 'bound' to the central ion, although its individual water molecules are rapidly exchanging places with surrounding water molecules.

In general, water is a good solvent for ions, but other liquids, such as oils, kerosene, and alcohols, are not. Non-polar substances such as oils don't mix

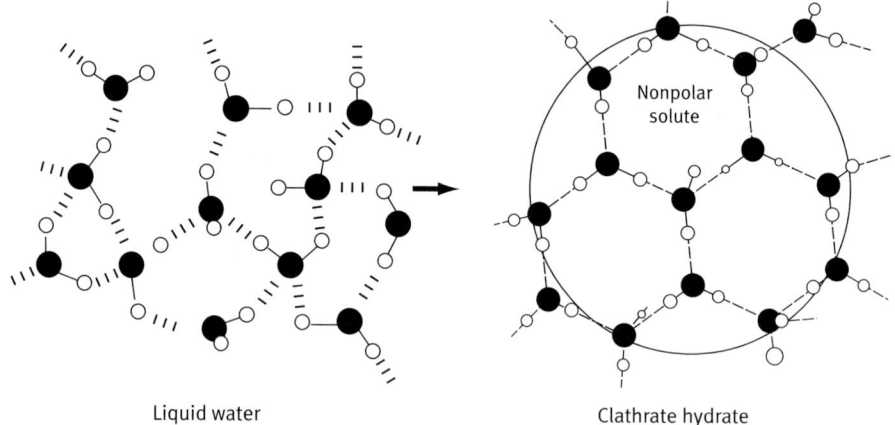

Figure 7.5. Formation of a clathrate hydrate between water and a non-polar solute.

with water and are said to be hydrophobic (water hating). Everyone is familiar with the saying 'Oil and water don't mix'. It is basically true, but why should this be so? The simple answer is that the process is energetically unfavourable. This raises a further question—why?

The answer lies largely in the differences in polarity between water molecules and oil molecules. Firstly, water and oil don't react chemically under normal conditions, so no new compounds are formed when we shake an oil/water mixture.

Initially, the energy delivered by our vigorous mixing will break up the two phases (oil and water) into small oil droplets dispersed in the water. Maintenance of the mixed state is energetically unfavourable. Consider the insertion of a single oil molecule into a large volume of water. The oil will disrupt the hydrogen-bonding network of water molecules in its immediate surroundings, without being able to hydrogen bond to them. The affected water molecules will rearrange themselves to re-establish and maximize their hydrogen bond network. This results in the formation of a cagelike structure of water molecules around the oil molecule, called a clathrate (Figure 7.5).

The hydrogen-bond-stabilized clathrate restricts some water molecules to a 'fixed' orientation; they become more ordered compared to 'free' water molecules further away from the oil. The water molecules in the clathrate cage have fewer orientational options because of the presence of the oil molecule. Clathrate formation therefore involves a reduction in entropy. Some clathrates under favourable conditions are reasonably stable, for example methane (see below), but in general clathrate formation is limited. In a situation where many oil molecules are dispersed in water, what can the system 'do' to minimize the effect of the decrease in entropy? We are familiar with the answer. Many small droplets spontaneously coalesce to form larger droplets. This is referred to as the 'oil-drop effect' (Figure 7.6).

In some cases the two phases might separate completely, with the oil floating in a layer on top of the water. When several small droplets come together, the

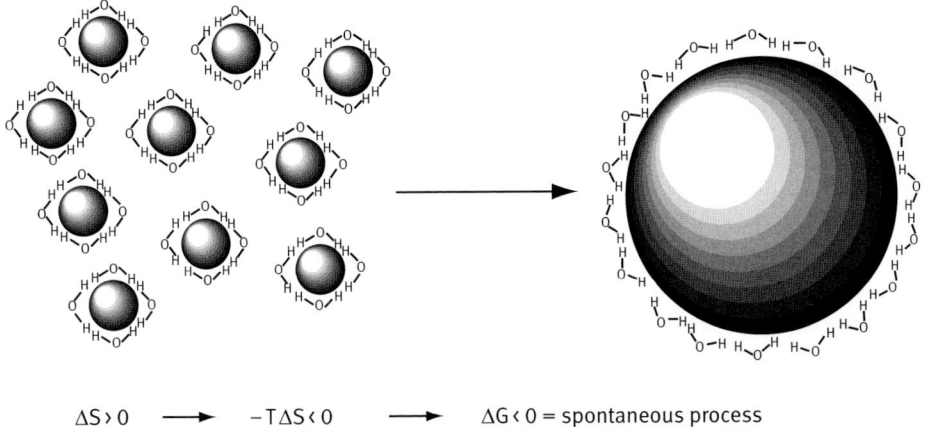

$\Delta S > 0 \longrightarrow -T\Delta S < 0 \longrightarrow \Delta G < 0$ = spontaneous process

Figure 7.6. The oil drop effect. The volume and surface area of a single large drop are less than those of the sum of the individual drops. This imposes less overall order on the surrounding water molecules, causing an increase in entropy that favours the formation of the large drop.

solvation cage must rearrange. The larger drop has a lower surface area than the sum of the individual drops, and imposes less overall order on the surrounding water than the separated droplets in their individual cages. There are no actual attractive forces between the droplets, rather the process is said to be 'entropy driven'. This may seem strange, but recall the large entropy effect in the dissolution of salt (Figure 7.4), although the circumstances were somewhat different. Entropy effects are real and are often quantitatively important.

The discussion above is the basis for understanding the hydrophobic effect and what are called hydrophobic interactions between non-polar molecules or non-polar parts of molecules such as between aromatic compounds or some non-polar side-chains in proteins. The term 'hydrophobic bond' should be avoided because of the variable nature and stoichiometry of such interactions, and the fact that they don't fit the usual concept of bonds between specific atoms.

The boiling point of any liquid is largely determined by the size and shape of its molecules, and by the polarity of those molecules. Polar molecules will bond to one another and a polar liquid will have a higher boiling point than a non-polar liquid of a similar molecular mass, as discussed above. An example of a typical oil is a hydrocarbon produced from petroleum or coal. The simplest hydrocarbon is methane, CH_4 (b.p. $-161.7°C$). Methane is soluble in water to some extent (3.7 mL/100 mL water at $17°C$) even though it is non-polar. Methane forms a number of clathrates or methane clathrate hydrates. These are not particularly stable solids, which at atmospheric pressure begin to decompose into water and methane gas at about $0°C$. Large deposits (corresponding to 500–2500 gigatonnes carbon) of methane clathrate have been found under sediments on some ocean floors. Suitable conditions for their formation are found in polar continental sedimentary rocks where surface temperatures are less than $0°C$, in oceanic sediment at water depths greater than 300 m, where the water temperature is around $2°C$, or in deep lakes. Methane clathrate hydrates are potential sources of fuel, and the possibility of their sudden release of methane is of concern to some climate change scientists.

Methane is followed by ethane (C_2H_6, $-88.6°C$), propane (C_3H_8, $-42.1°C$), n-butane (C_4H_{10}, $-0.5°C$), n-pentane (C_5H_{12} $36.1°C$), and so on. Propane is compressed in cylinders and sold in this form as bottled gas for barbecues, etc. Butane, often the principal hydrocarbon in liquified petroleum gas (LPG), can be liquified readily by compression. When compressed, the molecules are forced closer together and the weak, short-range attractive forces between them are then sufficient to allow the butane to condense to form a liquid. When the pressure is released, however, the butane (b.p. $-0.5°C$) will form a gas suitable to run a combustion engine, as in a motor car. The first hydrocarbon of the series that is a liquid at room temperature and atmospheric pressure is n-pentane (C_5H_{12}), boiling at $36.1°C$ which, being below human body temperature, indicates that it is very volatile, even though its molecular mass is 72 units or four times that of water. The total attractive forces between pentane molecules are obviously quite low, in keeping with their non-polar structure. As we go to higher numbers of carbon atoms in the hydrocarbon series, the boiling points increase and the properties progress from 'light' liquids such as hexane to 'heavier' ones, which are oily, through margarine-like and butter-like greases, to soft then hard waxes, which appear quite solid at room temperature. All these hydrocarbons will not dissolve in or mix with water to any significant extent. The solid-looking ones such as paraffin wax or candle wax will melt readily when heated and 'solidify' on cooling. Hydrocarbons don't dissolve in water, but they do dissolve in non-polar solvents such as ether and chloroform, and of course other hydrocarbon liquids such as petrol. You have probably had enough of hydrocarbon properties by now, so where is all this leading?

We have established so far that oil and water don't mix for energetic reasons. I have shown by use of specific examples that the polar properties of water on one hand and non-polar hydrocarbons on the other hand are very different. Polar solvents tend to dissolve other polar substances, while non-polar solvents tend to dissolve non-polar substances. This is the basis of the very useful chemistry rule of thumb, 'like dissolves like'. Having considered the behaviour in water of substances at the two extremes of polarity, ionic and polar materials on the one hand and hydrophobic oils on the other, what should we expect to happen when both polar and non-polar groups occur in the same molecule? Would we, for instance, expect a mixture of properties? This is in fact what is found. Molecules containing both polar and non-polar groups are called amphiphilic or amphiphatic (from the Greek *amphi* meaning 'both', *philos* meaning 'loving', and *pathos* meaning 'passion'). A prime example of amphiphatic molecules are the fatty acids, so-called because they are major constituents of animal fats. A typical fatty acid molecule has a polar carboxylic acid group, $-COOH$, at one end, connected to a non-polar hydrocarbon chain, represented as $CH_3(CH_2)_n-$, where n is a number from 4 to about 20 in biological systems.

Saturated fatty acids have the general formula $CH_3(CH_2)_nCOOH$ or, more simply, RCOOH. A common way to represent linear hydrocarbon chains is the zig-zag conformation. Thus, ⋀⋀⋀⋀⋀⋀COOH is a short way of representing

$CH_3-CH_2-CH_2-CH_2-CH_2-CH_2-CH_2-CH_2-CH_2-CH_2-CH_2-CH_2-CH_2-CH_2-CH_2-CH_2-CH_2-COOH$. Each bend represents a $-CH_2-$ unit. Another way is $CH_3(CH_2)_{16}-COOH$. All these represent stearic acid, which is a C_{18} acid, so-called because of the number of carbon atoms in its chain.

The carboxyl group is often referred to as the head, and the hydrocarbon or R– end as the tail of the fatty acid. As with the common hydrochloric and sulphuric acids, fatty acids can exist in the form of salts, for example the sodium salt $CH_3(CH_2)_n COO^- Na^+$ (or $RCOO^- Na^+$). As the salts are ionic, they are more soluble in water than the respective free acids, RCOOH. The solubility of a fatty acid depends on the length of the hydrocarbon chain. Longer tails make the molecules as a whole larger, so they disrupt the water structure more, and their solubility is reduced due to the increased hydrophobic effect described above.

What happens when an amphiphatic compound such as a fatty acid is added to water and shaken? The polar heads become readily hydrated, but the hydrocarbon tails, for the energetic reasons explained above, 'avoid' the water and reduce their area of contact with it by forming clusters with themselves, as described in the oil drop effect. Above a certain concentration, fatty acids tend to form structures called micelles. The concentration at which a particular fatty acid forms micelles, called the critical micellar concentration (CMC), depends on the hydrocarbon chain length—the longer the tail, the lower the CMC.

The hydrophobic tails cluster together and form a spherical 'oily' phase, surrounded by the polar head groups, which are happy to be in contact with water. This is a lower-energy arrangement than having the individual fatty acid molecules floating around in solution disrupting the hydrogen-bonded pattern of many more water molecules. The clustering together of the tails is another example of hydrophobic interactions between non-polar groups in the presence of water. Hydrophobic interactions may occur between molecules such as hydrocarbons, fatty acids, long-chain alcohols, and cholesterol, and are critical in the formation of biological membranes which surround cells and form the structures and compartments within them, such as mitochondria, lysosomes, endoplasmic reticulum, and the nuclear membrane.

Fatty acid molecules will also arrange themselves at the surface (interface) between water and air. The polar heads remain embedded in the water, while the tails poke out into the air. This strange arrangement is reminiscent of ticks embedded in a dog's ear (Figure 7.7).

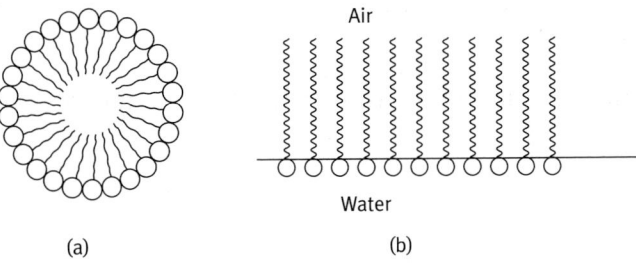

Figure 7.7. (a) A micelle formed from fatty acids in water. (b) Fatty acids form a monolayer at the water/air interface. Circles represent polar COOH head groups.

In nature, fatty acids are usually combined with glycerol to form glycerides, commonly known as fats. The fats constitute part of a larger group of non-polar compounds known as lipids. The lipids are a quite diverse group chemically, the common features being their solubility in organic solvents such as chloroform, ether, or acetone, and their insolubility in water. In mammals, fatty acids are typically stored as triglycerides such as glyceryl tristearate (below) in cells called adipocytes.

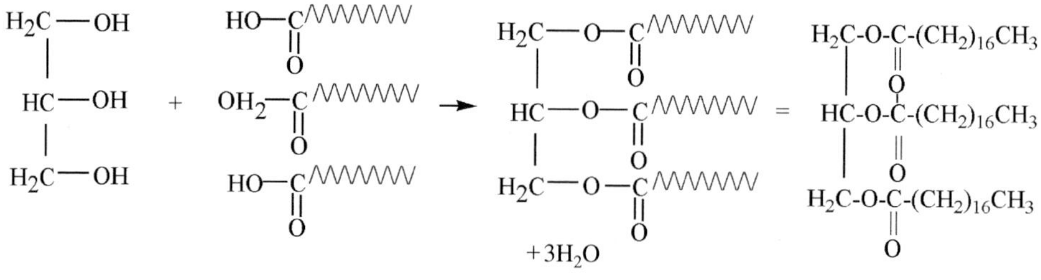

$$H_2C-OH \quad HO-\underset{O}{\overset{O}{C}}-\!\!\sim\!\!\sim \quad H_2C-O-\underset{O}{\overset{O}{C}}-\!\!\sim\!\!\sim \quad H_2C-O-C-(CH_2)_{16}CH_3$$
$$HC-OH \;+\; OH_2-\underset{O}{\overset{O}{C}}-\!\!\sim\!\!\sim \;\rightarrow\; HC-O-\underset{O}{\overset{O}{C}}-\!\!\sim\!\!\sim \;=\; HC-O-C-(CH_2)_{16}CH_3$$
$$H_2C-OH \quad HO-\underset{O}{\overset{O}{C}}-\!\!\sim\!\!\sim \quad H_2C-O-\underset{O}{\overset{O}{C}}-\!\!\sim\!\!\sim \quad H_2C-O-C-(CH_2)_{16}CH_3$$

$$+\,3H_2O$$

glycerol   fatty acid   triglyceride

An important group biologically are the phospholipids, major components of cell membranes. These amphiphatic molecules readily form lipid bilayers or double membranes. The basic structure is shown in Figure 7.8(b), with the hydrophobic tails intercalating slightly, and the hydrophilic heads facing the water phase.

If the ends of such a membrane fold around such that they come in contact and fuse together, a vesicle is formed, whose contents are isolated from the environment. At some stage in the evolution of life from prebiotic chemicals, the encapsulation of replicating molecules must have occurred, probably in a vesicle of the type described above. Once inside a protective vesicle, the advantages for survival of essential and delicate molecules are obvious, and it is not surprising that such an arrangement became the universal model for living cells.

The point to remember from Figure 7.8 is that these structures form because of the bifunctional nature of the molecules—hydrophilic and hydrophobic properties in the same structure. Their behaviour in water reflects the universal

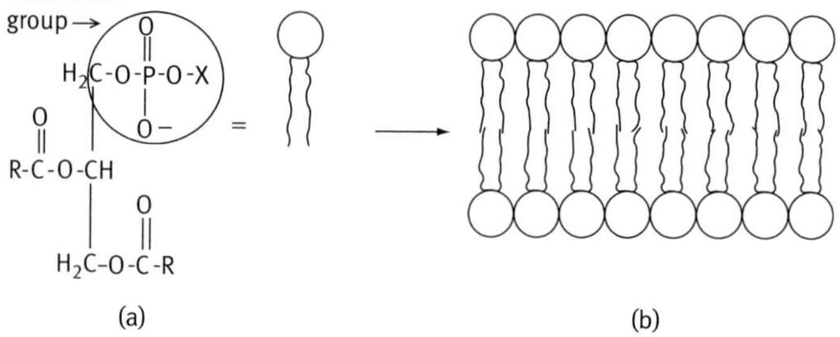

Figure 7.8. (a) Chemical structure of a glycerophospholipid. R = a fatty acid hydrocarbon chain. X is typically choline, serine, or ethanolamine. (b) A phospholipid bilayer.

tendency of systems to achieve the lowest energy possible under the prevailing conditions.

Existing cells have much more complex membranes than the simple model described above. Membrane lipids come in a variety of structures, each type having some useful function. Proteins are nearly always present in membranes, especially those of multicellular organisms, and carbohydrates also commonly occur. Depending on the organism or tissue, the cell membrane may be embedded with special proteins for assisting the passage of specific molecules (pumps and channels) with receptors for chemical messengers such as hormones and neurotransmitters, with recognition molecules, which determine cell–cell interactions and blood-group specificity, and with cholesterol molecules. Carbohydrate chains are present as membrane components in many cells, usually attached to proteins or lipids on the outside of the cell.

Despite the complexity of many eukaryotic cell membranes, the basic reason for their stability remains as simple in principle as the vesicle model above. The membrane of a cell is not rigid, like actual cell walls possessed by some organisms. The components of cell membranes are to an extent mobile. Individual lipid components, proteins, and receptors are able to move laterally in the membrane, and some such movements are believed to be important in their function. A widely accepted model for biological membranes is the fluid mosaic model (Figure 7.9).

This embodies the above ideas that the components exist side by side, and that lateral movement of membrane molecules is allowed. Solidification of the cell membrane could in fact be disastrous, as its normal functions would be impaired, for example transport of some essential components through the membrane would not be possible. Inserted cholesterol molecules, and phospholipid tails with kinks due to the presence of double bonds, disrupt the molecular ordering in the bilayer and maintain its fluidity. In addition to the simple separation of the cell contents from the outside environment, there are four major functions of membranes: their catalytic properties, transport of substances into and out of a cell or organelle, their receptor properties, and the formation of boundaries that allow the setting up of essential gradients—pH, electrical potential, ionic, osmotic. The membranes of both mitochondria and chloroplasts are the sites of a number of enzyme-catalysed reactions. Some enzymes are bound tightly in the membrane and the reactions take place there. Closely coupled enzyme reactions are often found to have the enzymes physically bound close by one another for ease of substrate or electron transfer between them. Not all enzymes are membrane bound. Some exist 'free' in the cytoplasm or in vesicles, such as lysosomes, from which they are released when required.

Transport of substances across membranes is a study in itself, but I'll only mention a few points here. In the process of passive transport a substance moves across the membrane from higher concentration to lower concentration, that is down a concentration gradient. No energy needs to be expended by the cell. On the other hand, in active transport the substance is moved against a concentration gradient, a process requiring the expenditure of cellular energy. This is analogous to pumping water uphill; important examples are the sodium–potassium ion pump mentioned above, and the Ca^{2+}-ATPase involved in maintaining low levels of calcium ion in the cell cytosol relative to

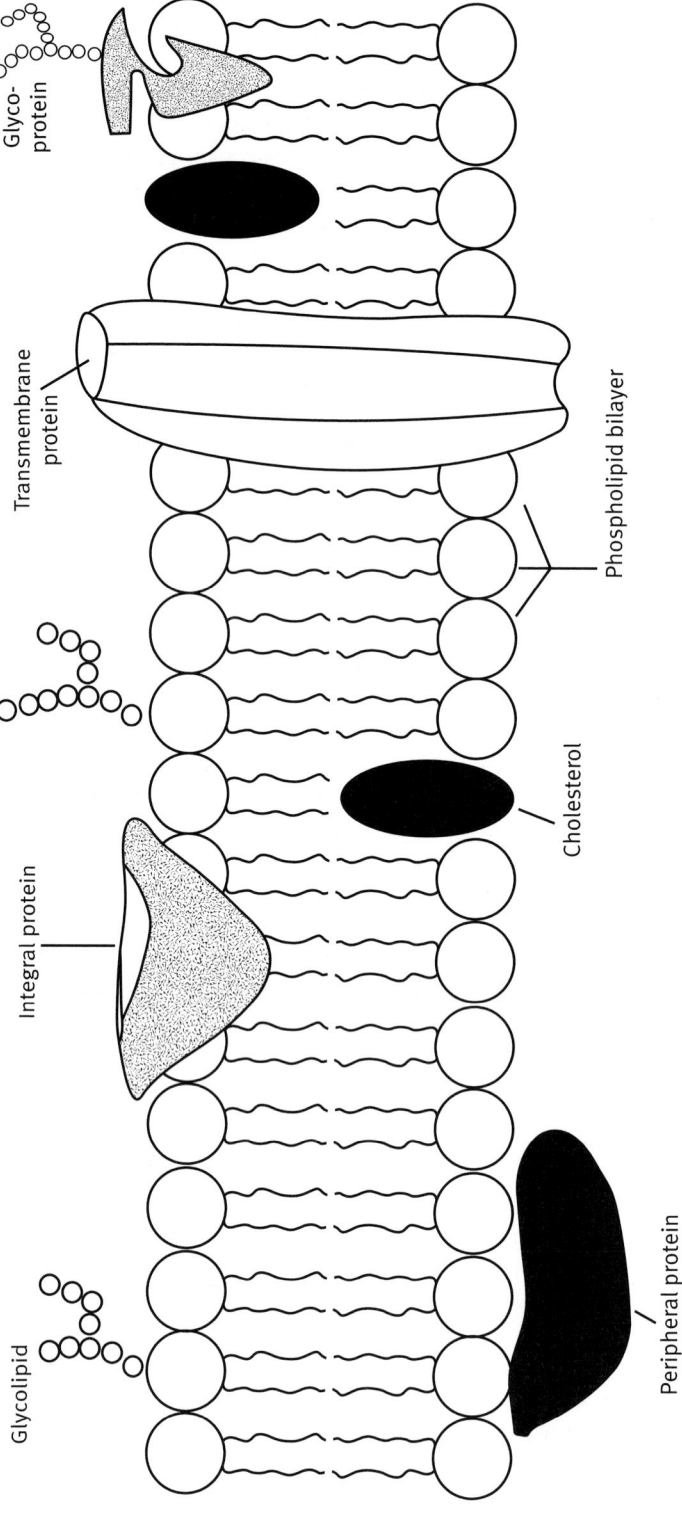

Figure 7.9. The fluid mosaic model of a biological membrane. Recent work has shown that there are many more membrane-associated proteins than originally thought.

that in the extracellular space. The Ca^{2+}-ATPase pumps calcium ion out of the cell, at the cost of ATP hydrolysis. There are many other examples of transmembrane transport, some of which we'll encounter later.

Hydrophobic interactions are also involved in the maintenance of the all-important three-dimensional structure of many proteins. Globular proteins, which include many enzymes, tend to have most of their hydrophobic amino acid side-chains clustered inside the globular structure, with many of the hydrophilic side-chains located on the outer surface, in contact with the polar surrounding water. Disruption of the native three-dimensional structure of such proteins, for example by heat, mechanical shear, or extremes of pH, will often eliminate their natural function, for example as enzymes, and render them useless. This process is called denaturation. The covalent bonds linking the individual amino acids need not be broken for proteins to become denatured. As mentioned in Chapter 5, the non-covalent bonds that maintain the three-dimensional shapes of proteins, the secondary bonds, individually require relatively small amounts of energy to disrupt. They include hydrogen bonds, ionic attractions between charged groups such as $-COO^-$ and $-NH_4^+$, hydrophobic interactions, and very weak, close-range van der Waals forces. The cooking of proteins such as those in meat, fish, or eggs is sufficient to disrupt secondary bonds and denature proteins, but, happily for consumers, improving their texture and taste in the process. The properties of secondary bonds, individually involving only small amounts of energy, have been recruited over evolutionary time to provide subtle but extremely important fine tuning of the properties of biological macromolecules. Small alterations in the conformation of the subunits of haemoglobin account for its greatly enhanced affinity for oxygen in the lungs, and for its subsequent release in oxygen-depleted cells. The haemoglobin can also reversibly bind CO_2 and protons, processes that affect its oxygen-binding properties and buffering capacity. The flexibility and generally dynamic nature of haemoglobin or enzyme molecules allow the binding to occur and, just as importantly, to be reversed. The entire complex process is a result of ubiquitous, although individually weak, secondary bonds. Might is not always right in the natural world, as we'll see in the next chapter.

GENERAL REFERENCES

Campbell, M.K. and Farrell, S.O. (2006) *Biochemistry*, 5th edn. Thomson Brooks/Cole, Belmont, CA.
Garrett, R.H. and Grisham, C.M. (2010) *Biochemistry*, 4th edn. Brooks/Cole, Cengage Learning, Boston.
Haynie, D.T. (2001) *Biological Thermodynamics*. Cambridge University Press, Cambridge.

REFERENCES

Chang, R. (1994) *Chemistry*, 5th edn. McGraw-Hill Inc., New York, pp. 653, 913–4.
Voet, D., Voet, J.G., and Pratt, C.W. (2002) *Fundamentals of Biochemistry*, upgrade edn. Wiley, New York, p. 248.

8
Size Matters: Proteins and Enzymes

8.1 Principles of protein structure

Big molecules are special. Modern life has been transformed by large, man-made molecules, the synthetic polymers, such as plastics, fibres, adhesives, paints, and elastomers (rubbers). Their unique properties are related to the fact that the molecules are large. Small molecules simply cannot generate the same range of properties. Polymers are made up of smaller subunits, or monomers, which are joined by covalent bonds into long chains. The properties of the monomers will naturally influence the properties of the polymer, as will the number of each monomer in the polymer chain, because this will determine the final size of the polymer.

The ingenuity displayed by modern chemists to produce polymers with useful properties is impressive, and since the 1930s an enormous array of these useful materials has been produced. The end is not yet in sight. New polymers are being introduced on a regular basis because the possible combination of available monomers, plus the ways in which they may be arranged, is almost unlimited. Clever as it may be, all this endeavour is relatively insignificant compared with the products of biological evolution.

Big biological molecules have unique properties that allowed life on Earth to begin and to flourish. The ones that concern us most are the nucleic acids, the proteins, and the polysaccharides. These large biological molecules are built up in a living organism from smaller ones, and are best described as modular. All are made by the bonding of monomers-simple sugars to form polysaccharides, amino acids to form peptides and proteins, and nucleotides to form DNA and RNA. Although the monomers differ in chemical structure between the different types, there are features common to biopolymers worthy of mention:

- They are modular.
- They are formed biochemically by removal of water between the individual monomers, a process called dehydration or condensation. The condensation process is endothermic, using ATP as a source of energy.
- ATP requirements per monomer added: DNA/RNA 2, proteins 4, polysaccharides 2.
- They can be hydrolysed by enzymes back to the monomers.
- Each one has a characteristic three-dimensional architecture.
- Weak forces (secondary bonds) maintain their characteristic structure and help determine biomolecular interactions.

- Macromolecules and their monomers have a 'sense' or directionality, that is they do not 'read' the same in both directions.
- Macromolecules are informational, for example DNA and RNA.

Some of these properties are illustrated below.

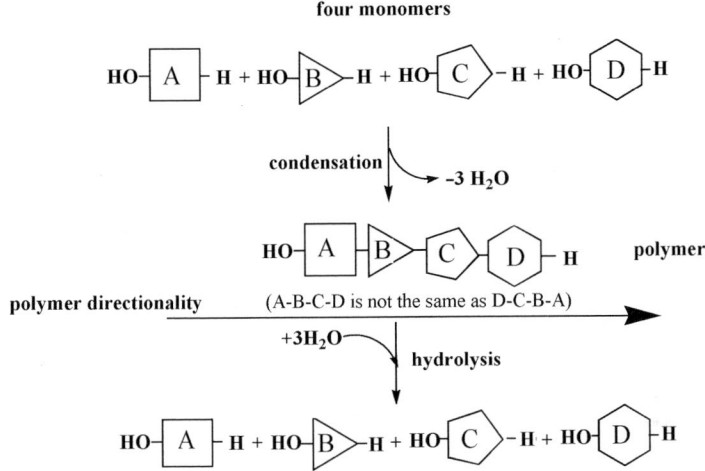

Many monomers have molecular masses up to about that of glucose (180 Da), which is still quite small compared with the molecular mass of a protein or a DNA molecule, which may range from several thousand to several hundred thousand daltons. Haemoglobin, the oxygen-carrying protein in red blood cells has a relative molecular mass of about 258,000 Da. One of the components of starch, the multiple-branched amylopectin, is believed to be the largest naturally occurring molecule and, depending on the plant source, may have a relative molecular mass up to 10 million Da.

All biopolymers have a property called directionality, meaning that they are not the same when 'read' from different ends. This property will become obvious in the following discussion. The main concern of this chapter is the proteins, and especially the enzymes.

The word protein is derived from the Greek *proteios*, meaning 'primary', as it has long been considered that these large molecules were essential components of all life. They were almost certainly not, however, the first potentially biological macromolecules to be formed on Earth. This role probably fell to the nucleic acids, most likely RNA, but this will be dealt with later. I use the word 'potentially' as the formation of many kinds of molecules preceded the emergence of life. Some of these precursor molecules were the relatively simple, stable monomers that eventually became joined to form the first polymers.

Proteins have a number of roles in living organisms, including structural (collagen in connective tissue and tendons, keratin in hair and feathers), transport (haemoglobin carries oxygen, some lipoproteins carry fats and cholesterol), messenger (certain peptide hormones act as molecular signals), catalytic (enzymes), regulatory (insulin controls entry of glucose into cells), protective (immunoglobulins of the immune system), contractile (actin and myosin of muscle), and storage (albumin in egg, casein in milk).

Our major focus in this chapter will be the enzymes, which illustrate many of the amazing variety of properties which proteins may exhibit, and which are intimately involved in metabolism and its control.

The monomers that are strung together to form proteins are the amino acids. There are about 20 amino acids that appear commonly in proteins, plus a few others that are not so widely used in nature. It is possible to draw a common structural formula for the type of amino acids found in proteins:

$$H_2N-\underset{\underset{H}{|}}{\overset{\overset{R}{|}}{C}}-\underset{O}{\overset{\|}{C}}-O-H$$

The H_2N- is an α-amino group (it is on the carbon α to the COOH) and the COOH is an acid (carboxyl) group, hence the name α-amino acid. The R in this case stands for any one of the 20 side-chains that make each amino acid different from the rest. I won't give the structures for all of these here, but refer you to Appendix E.

Box 8.1 Some properties of α-amino acids

At physiological pH values, the amino acids occur not in the form illustrated above, but as dipolar ions, called zwitterions:

$$\overset{+}{H_3N}-\underset{\underset{H}{|}}{\overset{\overset{R}{|}}{C}}-\underset{O}{\overset{\|}{C}}-O^-$$

The zwitterion arrangement occurs because the pK values of carboxyl groups occur in a small range around 2.2, and the pK values of the amino group occur around 9.4. This means that at pH ~7, both groups are more than two pH units away from their pKs, and will be effectively fully ionized. The pK values given in Appendix E refer to the pK_1 (for COOH groups) and pK_2 (for NH_2 groups). Other ionizable side-chains also have their pK values listed. Amino acids can act as both acids and bases. In free form, or in peptides and proteins, they are important in maintaining the pH in their particular cellular environment. The actual pK of an ionizable group will depend on its surroundings. In a protein, for example, amino acid side-chains that lie far apart in the primary sequence may be close neighbours in space due to folding of the protein backbone. Thus $-COO^-$ and $-H_3N^+$ in close proximity can have their pK values moved by one pH unit or more. Other close neighbour groups can also exert an influence. The ability of the side-chains of amino acids to gain or lose charge, for example by proton exchange, explains their presence in the active sites of enzymes, where they often act as acid/base catalysts (see Figure 8.15).

I'll mention certain ones where necessary. The side-chains vary in their chemical and physical properties, from acidic through neutral to basic, and from hydrophilic to hydrophobic. This range of side-chain properties plus the fact that they can be arranged in any order, provides for a vast array of protein properties.

The simplest side-chain is a hydrogen atom, so the simplest amino acid, glycine (abbreviation: Gly) has the structure

$$H_2N-\underset{\underset{H}{|}}{\overset{\overset{H}{|}}{C}}-COOH$$

Two amino acids may be covalently linked by joining the amino end of one to the carboxyl end of the other. For two glycines, it is possible to 'eliminate' a water molecule, in a typical condensation reaction, to yield the two glycines linked as follows:

$$H_2N-CH_2-COOH + H_2N-CH_2-COOH \longrightarrow$$
$$H_2N-CH_2-\mathbf{CONH}-CH_2-COOH + \mathbf{H_2O}$$

The $-\mathbf{CONH}-$ part of the new molecule links the two glycines together by a peptide linkage, one of the most important bonding types in biochemistry. The new molecule is called a dipeptide as it consists of two amino acids. Its specific name is glycylglycine, reflecting its amino acid composition. In similar fashion we may have tripeptides, tetrapeptides, and so on, up to hundreds of amino acids. Above 10 or so amino acids, the term polypeptide is used. In nature, peptide bonds are formed by enzymes, on the cellular structures called ribosomes, as outlined in Chapter 9. Peptide bond formation is endothermic, the energy being provided by ATP via various coupling reactions. The chemical and physical nature of the side-chains represented by R varies widely, leading to a variety of properties when the amino acids are joined to make a protein chain. This ensures that a plethora of useful properties is available in proteins. The amino acids may be joined in any gene-coded sequence, so with 20 possibilities to choose from at any position in a protein chain, there is an enormous number of unique sequences available for the formation of proteins. As an illustration let's consider joining any two of the 20 amino acids to form a dipeptide.

The first position of the dipeptide can be taken by any of the 20 amino acids, and so can the second position. The number of possible dipeptides is thus $20^2 = 400$. There are 20^3 or $20 \times 20 \times 20 = 8000$ possible tripeptides, and so on. The number of possible peptides soon becomes mind boggling. There is almost an unlimited number of protein chains possible, each with a unique amino acid sequence. For example, there are in theory, 20^{50} different peptide chains 50 amino acids long, but nowhere near this many types actually exist.

It is the sequence of amino acids in the chain, plus the number of amino acids, which *ultimately* determine shape and the properties of an individual protein. There is more to this statement than meets the eye because functional proteins

166 Introducing Biological Energetics

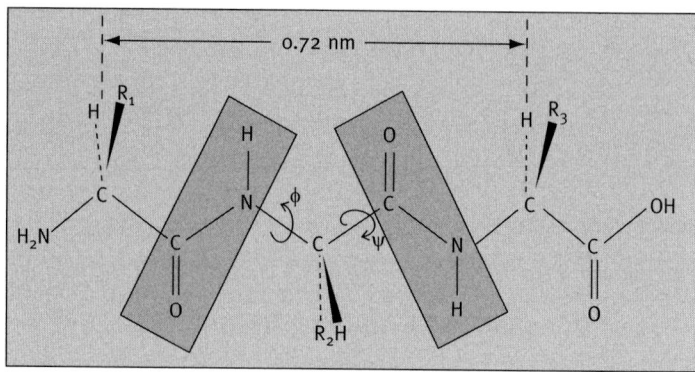

Figure 8.1. A tripeptide formed from amino acid residues having side-chains R_1, R_2, and R_3. Note the rigid, planar peptide linkages (shading) and the bonds ϕ and ψ that allow relatively free rotation. The fully extended β-conformation is shown.

are not merely long chains of amino acids strung together like beads on a string—far from it. They can adopt a large variety of three-dimensional shapes, or conformations. Figure 8.1 shows the bonds that confer flexibility on peptide and protein chains. Depending on the side-chains involved, there is relatively free rotation allowed around the bonds marked ϕ and ψ. There is a ϕ and ψ angle for each amino acid in a protein chain.

There are four well-recognized levels of complexity in proteins: the primary, secondary, tertiary, and quaternary structures (Figure 8.2).

The primary structure of proteins is the amino acid sequence, for example Gly–Val–Asp–Glu–Tyr–Cys–Ser.

Secondary structure refers to the folding of the primary chain into a number of shapes. There are two major regular shapes, or ordered conformations. The best-known ordered conformations in protein secondary structure are the alpha helix (α-helix) and the beta-strand (β-strand) (Figure 8.2(b)). The latter is also termed the β-sheet or β-pleated sheet.

Tertiary structure involves further folding of the protein chain, which when completed may have some parts of its length in the α-helix and/or some in the β-strand form, all folded into a three-dimensional shape (Figure 8.2(c)). The muscle protein myoglobin in its native form is folded into a compact globular shape, and this fact is reflected in the name (Figure 8.3). Myoglobin is very similar in overall shape to the individual chains of haemoglobin.

Myoglobin contains eight α-helical regions (A to H in Figure 8.3) separated by bends or kinks in the polypeptide chain (AB, BC, etc.) but no β-pleated sheets. About 75% of the chain is α-helical. Myoglobin is extremely compact, there being very little 'free space' inside, much of which is occupied by non-polar side-chains, in accordance with the hydrophobic effect.

Proteins may also adopt a quite linear (elongated) overall ordered conformation, such as that found in the fibrous silk protein, fibroin, which consists largely of β-sheets (Figure 8.2(b)). The arrangement of protein chains in fibroin is stacks of antiparallel β-sheets. The amino acid side-chains (–R) protrude alternately above and below the plane of the sheet. In fibroin, there are many

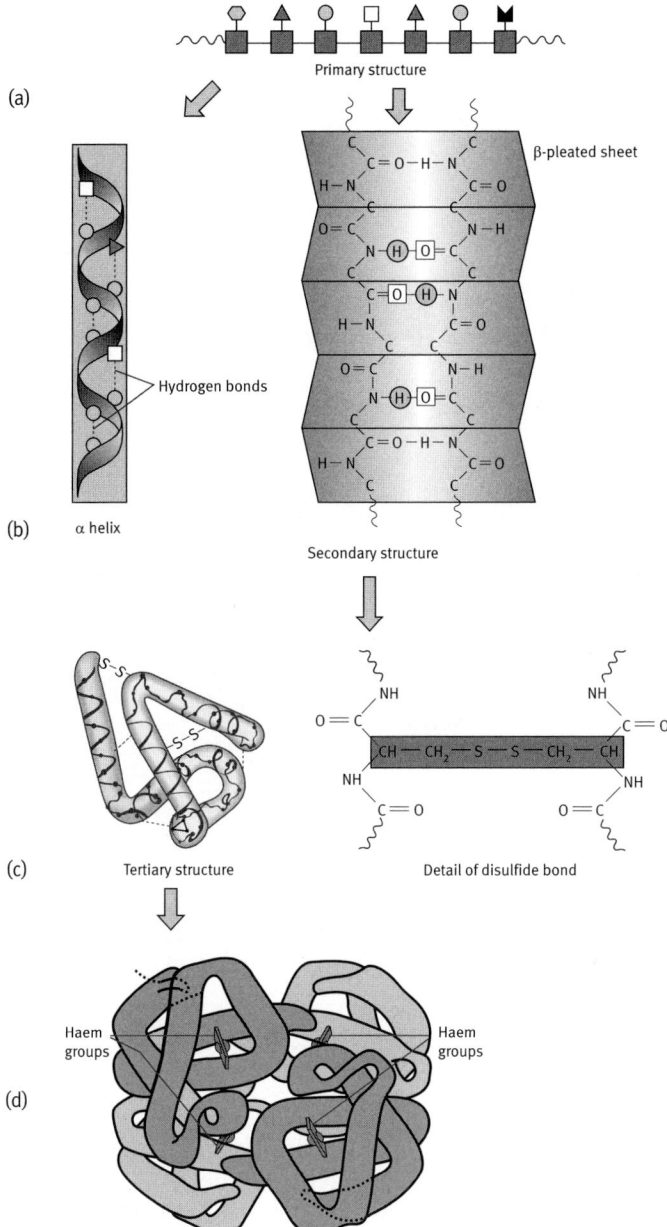

Figure 8.2. The four levels of protein structure, illustrated by stages in the formation of haemoglobin. (a) Primary structure is the amino acid sequence. (b) Secondary structure involves the ordered conformations of the α-helix and the β-pleated sheet. Secondary structure is stabilized by hydrogen bonds and other secondary bond types. (c) Further folding of secondary structures leads to the tertiary structure, which is often stabilized by disulphide bonds. One of the four subunits of haemoglobin is shown. (d) Haemoglobin formation is completed by the association of four subunits to form the active quaternary structure. (Illustrations (a) to (c) from Talaro, K. and Talaro, A. *Foundations in Microbiology*, 2nd edn, p. 47. © 1996 Times Mirror Higher Education group, Inc. Illustration (d) from Irving Geis. Image from Irving Geis Collection, Howard Hughes Medical Institute. Rights owned by HHMI.)

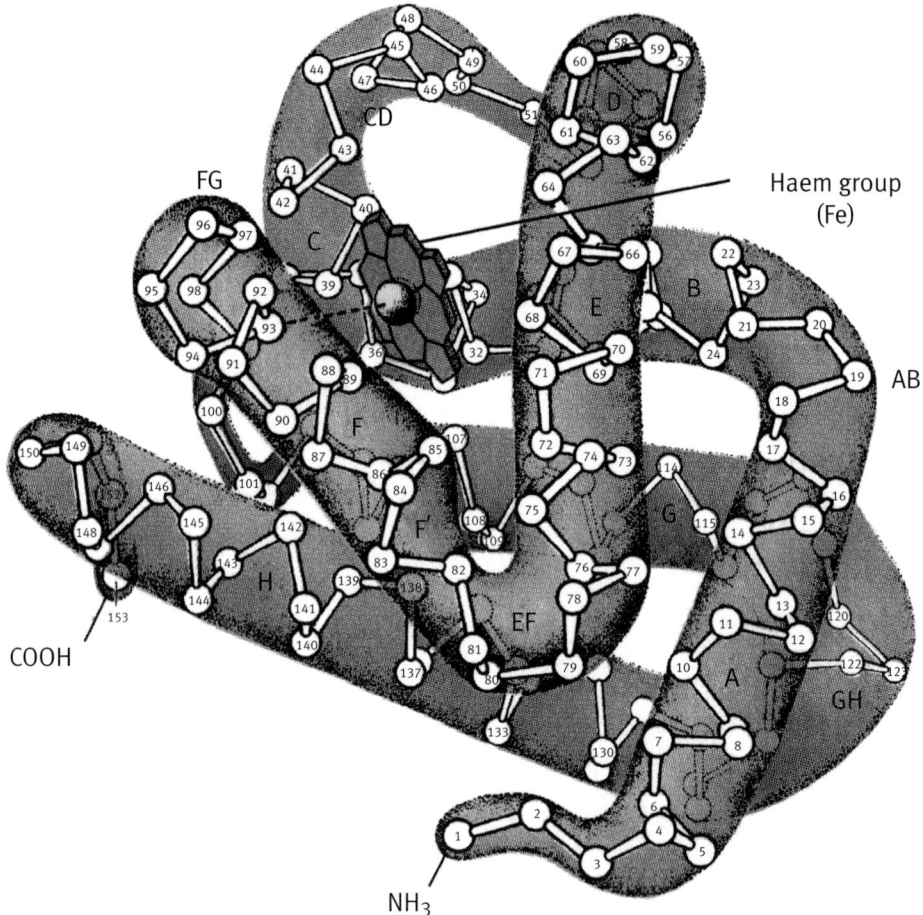

Figure 8.3. Structure of myoglobin, showing the secondary and tertiary structure (Illustration from Irving Geis. Image from Irving Geis Collection, Howard Hughes Medical Institute. Rights owned by HHMI.)

regions where glycine (R = H) alternates with either alanine (R = CH$_3$) or serine (R = CH$_2$OH). The glycines are arranged on one side of the sheet and the alanines/serines on the other, allowing two sheets to pack closely together— glycine to glycine, or alanine/serine to alanine/serine. About 80% of fibroin consists of glycine, alanine, and serine (Figure 8.4).

The keratins of bird feathers are also made up of stacked β-sheets. Such a regular, tightly packed arrangement allows stabilizing secondary bond forces to operate maximally. The result in silk is a fibre with high tensile strength and resilience.

Another fibrous protein, collagen, found in cartilage and connective tissue, consists of three α-helical chains, each formed into a superhelix and intertwined (Figure 8.5). Collagen has an unusual amino acid composition, made up of multiple repetitions of the sequence glycine, proline, and hydroxyproline.

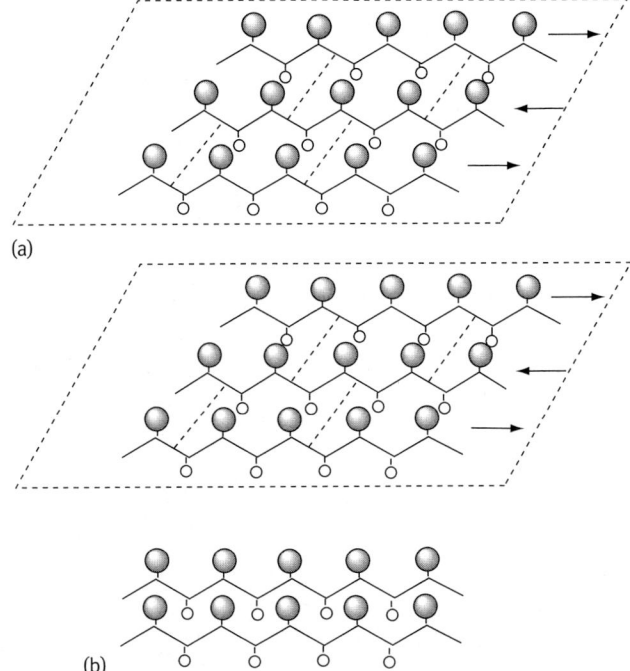

Figure 8.4. The structure of silk fibroin. (a) Two layers of fibroin showing antiparallel β-sheet arrangement of the chains. The vertical distance between sheets 1 and 2 is greatly exaggerated for clarity. Interchain hydrogen bonds between −C=O and −HN− groups in the protein backbone are also shown. (b) Front-on view of layers 1 and 2 showing the intercalation of large sidechains ◯ (alanine and serine.) with the H sidechains of glycine (◌).

Hydrogen bonds occur between −OH groups on side-chains of amino acids in the helices. Low hydroxylation can lead to scurvy. Adequate hydroxylation levels in mammals require an adequate supply of vitamin C in the diet. Covalent bonds (not shown) help to hold the three superhelices together.

Other proteins consist of several individual chains, or subunits, held together by secondary forces to form quaternary structures. The haemoglobin (Hb) molecule is an example. Each protein has an overall structure suited to its biological function. As proteins come off the ribosome in a linear fashion, they must be somehow folded into this active shape or conformation. Figure 8.2 illustrates this process schematically for haemoglobin. Protein folding and its energetics are discussed later in the chapter.

At each of the structural levels described above, maintenance of that structure is important for the function carried out by the protein.

Some obvious questions arise:

1) What determines the three-dimensional shapes a given protein chain may adopt?
2) What causes it to adopt such a shape?
3) What maintains the shape once it has been achieved?

170 Introducing Biological Energetics

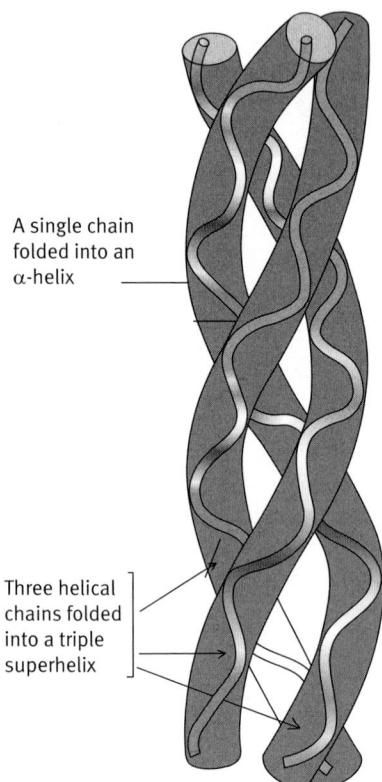

Figure 8.5. Collagen: a triple superhelical protein found in skin, bone, tendon, cartilage, and eye lens. (Illustration from Irving Geis. Image from Irving Geis Collection, Howard Hughes Medical Institute. Rights owned by HHMI.)

The answer to each of these questions involves energy and includes a number of principles already discussed. The discussion below is restricted mainly to the proteins that form enzymes or enzyme systems.

1) The amino acid sequence (primary structure) of a protein is the major determinant of the shapes it may adopt by folding of the protein chain. Also involved will be the precise environment within the cell, for example whether the protein will be free in solution, inserted into a lipid membrane, or be part of a structure such as a ribosome or a gene.

2) What causes the protein chain to adopt a particular shape? How is energy involved to bring the folding into being, and to maintain it?

Consider the synthesis of the protein chain that will become a globular myoglobin molecule. It comes off the ribosome in a linear fashion, like a string of beads, and (eventually) adopts its secondary structure with a high proportion of α-helical regions. The α-helical regions are joined by folds and bends, rather like a long sausage folding into a heap, and finally form into the compact globular tertiary structure. The millions of myoglobin molecules in mammalian muscle all fold unerringly into the same shape. The process must therefore be thermodynamically favourable. A net driving energy (negative Gibbs energy, ΔG) is involved overall, even though there is a decrease in entropy as the folding of the chain makes it more ordered. What are the origins of this negative ΔG?

For the overall process of folding

$$\Delta G = \Delta H - T\Delta S$$

A contribution to ΔG that favours folding comes from the formation of hydrogen bonds between adjacent parts of the polypeptide chain during the formation of α-helices, which are apparently formed early in the folding process (see the α-helix in Figure 8.2(b)). Some favourable energy derives from the association between oppositely charged $-R$ groups, such as $-COO^-$ and $-NH_3^+$, or coordination of metal ions such as Mg^{2+} or Ca^{2+} to negatively charged side-chains.

Calculations have shown that for the polar side-chains, the entropy and enthalpy terms approximately cancel out, so that $\Delta G_{folding} = \sim 0$. For the non-polar side-chain contribution, the ΔH and $T\Delta S$ terms are positive, and are thus unfavourable for folding. The solvent water now plays an important part in folding energetics. Large numbers of water molecules around non-polar groups in the unfolded state are restricted in motion (see Chapter 7). When these are 'liberated' as the hydrophobic groups fold inside, there is a large increase in entropy, which compensates for the repulsive energy and makes the overall folding process favourable. The largest contribution to the stability of a folded protein derives from this entropy change. Interestingly, the ΔG for folding is typically not large, in the -20 to -40 kJ mol^{-1} range for the macromolecule, leading to the conclusion that a folded protein is barely stable, that is its native conformation is somewhat flexible and is often easily denatured. The flexibility has important implications for protein dynamics, especially in the case of enzymes, as discussed below. The propensity for denaturation needs to be tolerated, and could be considered as the price paid for the subtleties of protein regulation. Even natural selection can't win all the time. (In hyperthermophiles, natural selection has, once more, done very well. The amino acid sequences 'chosen' for their enzymes are such that high temperatures (100°C plus) can be tolerated.)

A fully folded, water-soluble globular protein usually has the majority of its polar side-chains facing out towards the aqueous environment, with the hydrophobic side-chains buried deep inside, well shielded from being able to disrupt the structure of the surrounding water. Despite this general principle, there are many examples of water molecules also being found inside globular proteins. They may have some specific role in forming bridging hydrogen bonds between polar groups or be involved in proton-catalyzed reactions. The globular structure minimizes the volume in contact with the water environment, thus maximizing the number of energetically favourable hydrogen bonds in the surrounding water. Disulphide bonds between two cysteine residues in different parts of the protein chain may also be involved. These are covalent bonds that, when in place, restrict the folding capability of the protein chain, so are therefore important in determining, stabilizing, and maintaining its overall three-dimensional shape. The $-R$ group in the amino acid cysteine is −CH$_2$SH. Schematically, the formation of a disulphide bond between two cysteine groups in a protein chain can be represented as:

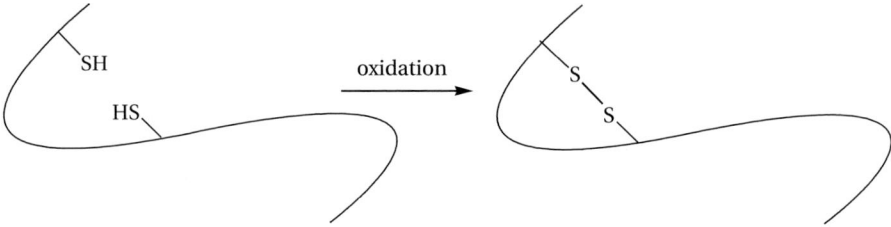

In vivo, the oxidation of the two −SH side-chains is usually carried out by a specific enzyme. A recent review points out the importance of disulphide bond formation in mitochondria and the endoplasmic reticulum system (Riemer *et al.* 2009). The above description is a simplified account of protein folding. In principle, there are many tertiary conformations that a linear molecule can adopt. The only requirement is that the ΔG for the entire folding process be negative. There can be many conformations of roughly comparable energy level that satisfy this criterion. This leaves plenty of scope for some proteins to adopt a non-biological conformation. Inside cells, protein folding processes can be complex. As the correct folding of a protein chain is so important, considerable research effort has been put into studying this process, and three protein-folding mechanisms have been identified (Figure 8.6).

Some proteins fold without assistance—they spontaneously adopt the correct conformation on thermodynamic grounds alone (Figure 8.6(a)). Others may fold

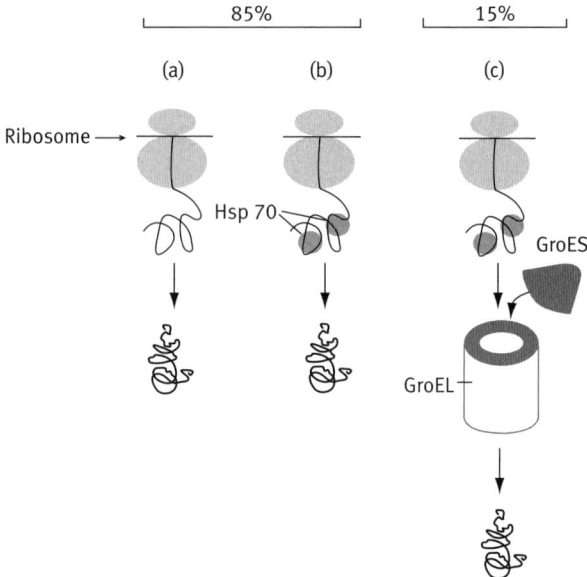

Figure 8.6. Mechanisms of protein folding. (Adapted from Netzer, W.J. and Hartl, F.U. (1998) Protein folding in the cytosol: Chaperonin-dependent and -independent mechanisms. *Trends in Biochemical Sciences* **23**, 68–73.)

incorrectly unless they interact with a chaperone, such as Hsp70 (heat-shock protein 70) (Figure 8.6(b)). The process is driven by ATP hydrolysis. About 85% of proteins fold as shown in Figure 8.6(a) and (b). The rest are folded with the aid of other proteins called chaperonins (Figure 8.6(c)). Chaperonins, large structures consisting of protein subunits, form a cage around each newly formed protein chain. Protein folding, which takes place inside the cage, is ATP dependent. Up to 100 ATP molecules must be hydrolyzed to complete protein folding. This is a large expenditure of energy. For the cell to invest such an amount of energy in this process (plus the associated genes, their control, and synthesis of the subunit proteins) it must be very important. As we have seen, natural selection is parsimonious and tends to waste nothing. In fact, there is good evidence that if a correctly functioning, properly folded protein somehow loses its native conformation, the result for the organism can be disastrous. Prions (proteinaceous infectious particles) are proposed to be particles consisting of proteins that have adopted an abnormal conformation. They appear to be a prime cause of some diseases of the central nervous system. In humans, Creutzfeldt–Jacob disease and kuru are fatal degenerative diseases. In other mammals the diseases scrapie in sheep and BSE/mad cow disease, have similar origins. Although apparently inherited, according to some evidence, no nucleic acid involvement has been shown. Prions may be acquired by infection, perhaps by ingesting food from an infected animal. The prions are believed to differ from the natural proteins only in their secondary and tertiary structures. One proposal is that the natural form (PrP^c) has a large proportion of α-helices, whereas the the scrapie form (PrP^{sc}) consists of both α-helices and β-strands. Apparently, the presence of PrP^{sc} can induce the normal PrP^c to change to the pathogenic conformation, forming plaques of ever-increasing size that eventually develop into the disease. The plaques ultimately destroy tissue in the central nervous system.

In the normal course of protein maturation, the folded, active protein structure is maintained by the bonds and forces that stabilize the tertiary and quaternary structure (Figure 8.7).

Hydrogen bonding was described in some detail in Chapters 5 and 7 as an example of secondary bonds, the weak interactions that are important in biomolecules. Hydrophobic interactions were also covered in Chapter 7. Secondary bond strengths relative to some covalent bonds are quite low, but their influence can be large, as many secondary bonds can act cooperatively:

Type	Strength (kJ mol^{-1})	Comments on strength
Covalent bonds	Range 200–900	Strong primary bonds
Hydrogen bonds	0–30	Varies with bond polarity; directional
Van der Waals	0.4–4.0	Close contact/large area enhances strength
Ionic bonds	20	Depends on polarity of the + and − species
Hydrophobic interactions	4–12	Highly dependent on environment, especially the effect on water structure

174 Introducing Biological Energetics

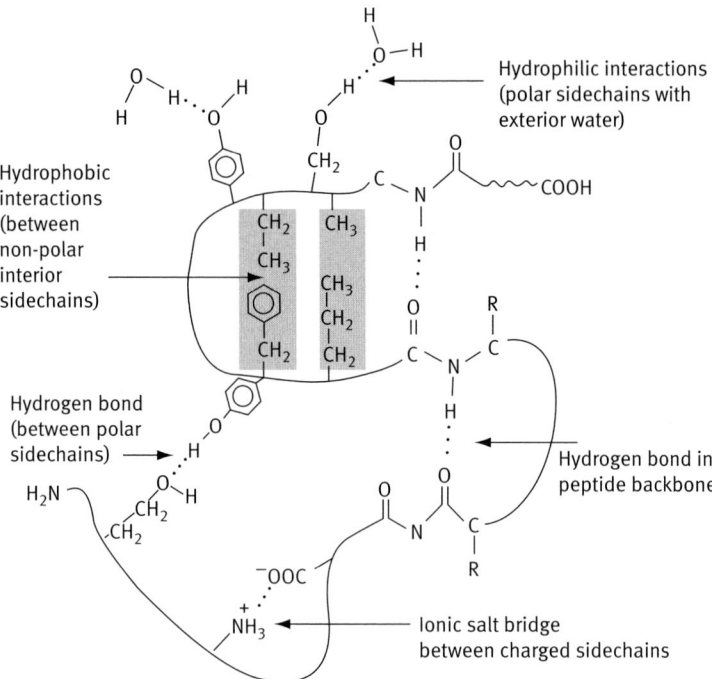

Figure 8.7. Secondary forces that stabilize protein structure. (Illustration adapted from Crowe, J., Bradshaw, T., and Monk, P. *Chemistry for the Biosciences*, p. 106. © 2006 Oxford University Press.)

Van der Waals interactions occur between all molecules, and are very short range. They are the result of induced electrical dipoles involving electrons of molecules in close contact. Their strength is enhanced by complementarity of shape and area of contact between the molecules involved. These conditions can occur when substrates bind to enzymes. An example is the binding of the enzyme lysozyme to a carbohydrate substrate, where the attractive van der Waals energy has been calculated to be about 60 kJ mol^{-1}, a considerable value (Garrett and Grisham 1999).

Many proteins, especially enzymes, undergo slight but functionally important conformational changes when carrying out their biological roles. They can change shape. Haemoglobin is one of the most fascinating examples of this. Each of the four subunits of haemoglobin has a globular shape similar to that of a single myoglobin molecule. Both molecules possess a haem prosthetic group, a structure containing a ferrous iron (Fe^{2+}) ion that binds molecular oxygen, O$_2$. The four haem groups in haemoglobin allow it to act as an oxygen carrier in red blood cells. As venous blood, low in oxygen and high in carbon dioxide collected from working cells, circulates in the alveoli of the lungs, carbon dioxide is lost from the red cells and is exhaled. As each new breath of air is taken, the partial pressure of oxygen in the lungs becomes quite high and oxygen is taken up by the haemoglobin of red cells passing through capillaries very close by. The blood continues to the heart, to be transported around the body. As the oxygenated blood passes 'hard-working' cells where metabolic activity has depleted the oxygen during the processes of

Proteins and Enzymes 175

Haem groups

Haem groups

Figure 8.8. Quaternary structure of haemoglobin. (Illustration from Irving Geis. Image from Irving Geis Collection, Howard Hughes Medical Institute. Rights owned by HHMI.)

respiration, the haemoglobin releases most of its oxygen to the cells and picks up carbon dioxide in the form of bicarbonate (HCO_3^-) (Figure 8.8).

The relative positions of the four chains alter when oxygen binds, indeed crystals of haemoglobin shatter when exposed to oxygen, a dramatic demonstration of the effect. Studies by X-ray crystallography have revealed that oxyhaemoglobin and deoxyhaemoglobin differ significantly in their quaternary structures. The precise movements and their distances are known in detail. Why should haemoglobin behave in this way? The biological significance is great. The role of haemoglobin in mammals is to efficiently bind oxygen from the lungs, transport it to the tissues, then efficiently release it. I emphasize the binding and release processes, as these appear to be contradictory properties in the same molecule. If oxygen is efficiently bound by haemoglobin, how can it also be released efficiently later? The answer is by making use of the allosteric effect. Put concisely, the movement of Hb subunits when the first oxygen molecule binds, assists the binding of the second and subsequent oxygen molecules (haemoglobin can bind four oxygen molecules, one per subunit). This process is a form of cooperative binding, and is typical of allosteric effects. Allosteric means 'in another place' ('other space' in Greek). In haemoglobin and in the case of allosteric enzymes, the term allosteric indicates that the effects are caused by molecules or ions binding away from the active site.

When the fully oxygenated haemoglobin (HbO$_2$) reaches working tissues, the environment is quite different from that in the lungs. The pH, particularly in active muscle cells, is lower as lactic acid is produced. Protons have an affinity for, and bind to, the HbO$_2$, altering the spatial relationship between the α- and β-chains. This movement causes a shape change that lowers the affinity of HbO$_2$ for oxygen and assists in its release, leaving protons bound to Hb.

Also, as an end-product of aerobic metabolism, carbon dioxide will be present in high concentration in the active tissue. Because the pH of blood is about 7.4, most CO$_2$ will be in the form of bicarbonate, HCO$_3^-$, because at pH 7.4 the reaction below is driven towards the right:

$$CO_2 + H_2O \rightleftharpoons H_2CO_3 \rightleftharpoons HCO_3^- + H^+$$

Protons formed in this way further assist the release of oxygen from HbO$_2$. The HCO$_3^-$ is transported in the blood back to the lungs. When our oxygen-depleted Hb arrives back in the oxygen-filled lungs, its bound protons are released and Hb combines with fresh oxygen. In the reverse of the reaction shown in the equation above, the released H$^+$ combines with HCO$_3^-$ to form H$_2$CO$_3$, which dissociates into water plus CO$_2$, both of which are exhaled in the breath. Removal of the CO$_2$ in this manner helps to drive the reversible reaction above to the left. A small amount of CO$_2$ binds to Hb directly and is transported thereby to the lungs and released.

To complete the picture of this remarkable molecule, another metabolite bisphosphoglycerate (BPG, see Chapter 11) also has a marked effect on the oxygen-carrying capabilities of Hb. When oxygen-laden Hb reaches the cells, it needs to release the four bound O$_2$ molecules efficiently. The highly charged BPG is present in about a 1:1 ratio with Hb, and binds strongly to its allosteric site, in the interior between the two β-chains, stabilizing the deoxy form and promoting the release of O$_2$. BPG plays an important role in supplying oxygen to a growing mammalian foetus. Foetal Hb (HbF) binds BPG less efficiently than adult Hb, with the result that HbF has a greater affinity for oxygen than the mother's Hb. This ensures that oxygen transfer from mother to foetus is favoured. Why should foetal Hb bind BPG less strongly? In maternal Hb, the negatively charged BPG binds to the positively charged histidine side-chains at position 143 in the β-chains. Foetal Hb possesses a different equivalent to the adult β-chains, called γ-chains. These γ-chains have serine at position 143 instead of the basic histidine. The serine side-chain is $-CH_2OH$, is neutral, and does not become positively charged, hence removing two positive charges for binding to BPG. The result is that foetal Hb still binds oxygen at low partial pressures of O$_2$, whereas adult Hb will have lost it.

The behaviour of Hb is known in great detail, much of which has not been dealt with above. All I wish to add is that the precise mechanism of oxygen uptake and binding, plus the movements of the four subunits, are known down to the exact amino acid side-chains involved, including the location of those involved in important hydrogen bonds, the location of salt bridges (some of which involve chloride ions Cl$^-$), and the location of others which form between $-H_3N^+$ and $-COO^-$ groups of amino acids in the chains. The disruption of some of these secondary bonds during oxygen uptake is energetically favourable overall. Although individually small in magnitude, they are collectively and essentially involved in the binding and subsequent release of oxygen. The energy contributions of all these secondary bonds are vital to the function of haemoglobin, and are thus vital to our own survival. The quaternary structure of Hb, with its four subunits, has evolved to allow the emergence of a property, cooperativity, of which a single polypeptide chain is incapable—another example of the art of the possible. The allosteric effect is a major mechanism by which enzyme activity is regulated (see below).

I have treated the principles of protein three-dimensional structure in some detail because of its close relationship to function. In general, if a protein is 'denatured', that is if the native conformation is somehow destroyed, its natural function is also likely to be destroyed. This is particularly the case with enzymes. Most of the forces holding proteins in their native conformations are relatively weak secondary forces, and as such are individually easy to disrupt. If enough of its hundreds of hydrogen bonds, electrostatic forces, and van der Waals forces are disrupted, a typical protein will be forced to change its tertiary structure, often drastically. Familiar examples include the cooking of an egg or a piece of steak: enzymes become useless, the white and yolk solidify, collagen in the steak shrinks and loses water. The changes in properties of these high-protein foods brought about by cooking is obvious to the naked eye. It is possible to demonstrate that under carefully controlled conditions denaturation is not necessarily irreversible. During experiments to demonstrate that the native shape of a protein is determined by its primary structure, Christian Anfinsen was the first to demonstrate that the refolding of a denatured protein (the enzyme ribonuclease) in vitro could be a spontaneous process. Put briefly, after denaturation and allowing some time for renaturation under appropriate conditions, the ribonuclease was found to have its enzymic activity restored. When in the denatured form, it was completely inactive.

How do proteins become denatured? Heat is the most familiar cause. Heat energy, for example that involved in cooking, is sufficient to disrupt the secondary bonds in most proteins. Changes in pH and ionic strength, plus treatment with detergents, may also cause denaturation by changing the patterns of hydrogen bonds and ionic bonds. Some proteins in solution are denatured by vigorous mixing. Denatured proteins may collect at the air/water interface as froth or foam. Egg white when beaten forms a protein foam that is stabilized by denatured ovalbumin. Cooking causes further denaturation, which hardens the surface of the foam and the result (for a skilful cook) is a light shell used for the delicious dessert known in Australia as Pavlova. Further examples of deliberate denaturation of proteins for purposes of customer appeal include beer pouring and cappuccino coffee making. Blasted into existence by a jet of steam, cappuccino froth consists largely of heat-denatured milk proteins. Some addicts almost swoon with delight at the structures and odours so generated. Beer drinkers can become passionate about the 'head' of their favourite brew, although it is merely a foam stabilized by a little denatured protein (and perhaps an additive) produced when pressurized carbon dioxide is suddenly released. Try telling that to a crowded bar full of Guinness drinkers on a Friday night!

8.2 Some very special molecules: enzymes

As we have seen in Chapter 5, a catalyst is a substance that increases the rate of a chemical reaction by altering the reaction pathway, thus lowering the overall activation energy. In biological systems, some reactions would be impossibly slow, and cause a log-jam in metabolic pathways, without the presence of enzymes to speed them up. In some inherited metabolic diseases, a crucial enzyme is missing (or present in low amounts), leading to just such log-jams. An example is the genetic disease phenylketonuria, which I will discuss later in the chapter. Like all catalysts, enzymes change the rate of a reaction without being changed permanently themselves. This latter property is important, as a single enzyme molecule may be used many times over. Enzymes are usually highly specific for the reaction they catalyze, and their activity can be regulated. Essentially, all enzymes are proteins. Some enzymes are able to carry out their catalytic functions using their protein structure alone. Others need the help of cofactors. Cofactors can be metal ions or organic molecules called coenzymes, such as the nicotinamide adenine dinucleotide (NAD) or nicotinamide adenine dinucleotide phosphate (NADP) required in electron transfer reactions. We know many of the precursors of coenzymes as vitamins, about which more will be said later. Other enzymes may need the presence of a specific metal ion, such as Ca^{2+}, Mg^{2+}, or Zn^{2+}, to be active. A cofactor that is covalently linked to the protein is called a prosthetic group. Although not enzymes in the accepted sense of the word, some catalytically active RNA molecules have been identified. These might, and I stress *might*, have had a role in the early development of life (Luisi 2006).

Enzymes certainly increase reaction rates, but they do much more as well. A particular enzyme is in general highly specific for its particular substrate and usually will not catalyze reactions with other substrates. For biochemical reactions, embedded in biochemical pathways which need to be strictly controlled, this is a matter of necessity.

A suite of enzymes exerts control over a coordinated series of reactions making up a particular pathway, allowing the pathway to adapt to changing conditions. If there is a demand for more ATP, as during heavy exercise, the pathways involved in respiration, glycolysis, the TCA cycle, and the electron transport chain (ETC) of oxidative phosphorylation are brought into high activity. When demand falls, feedback mechanisms, mediated by key enzymes in the pathways, reduce the overall activity back to normal levels.

There are some important exceptions to the specificity rule, such as a number of enzymes found in mammalian liver. The liver is the site of detoxification of foreign substances potentially harmful to the body. Enzymes that catalyze oxidations, reductions, and desulphurations are examples of the first phase of detoxification reactions in the liver. Many of the oxidations are performed by cytochrome P450 systems, which have a broad substrate specificity. Obviously the liver doesn't 'know' what unwanted compounds are likely to be ingested by the mammal, so the ability to cope with a range of possibilities is a distinct advantage.

How do enzymes perform all the functions we have mentioned? Let's take them one at a time, starting with the catalytic properties. How is it that a protein, present in low concentrations, can increase the rate of a reaction by a factor of a staggering 10^{12} or more? Such increases in reaction rate are far more than man-made, synthetic catalysts can achieve. Let's take the familiar biological oxidation of glucose as an example:

$$C_6H_{12}O_6 + 6CO_2 \longrightarrow 6CO_2 + 6H_2O \quad \Delta G^{0\prime} = -2871 \text{ kJ mol}^{-1}$$

This is a large amount of energy, but we are considering thermodynamics alone. The thermodynamics tells us nothing about the rate at which the reaction will occur. One could leave glucose (or just about any carbohydrate) lying around at room temperature, in contact with the oxygen in the air, for months and it would remain essentially unchanged. That situation is obviously unacceptable for living cells, which may need the energy immediately. The series of enzymes involved in glucose metabolism allows the energy to be released more rapidly in a controlled way that allows the regeneration of ATP, the major immediate source of cellular energy. Enzymes give the cell kinetic control over the thermodynamic potential of glucose.

What actually happens when an enzyme becomes involved in a reaction such as (8.1)?

$$W + X \longrightarrow Y + Z \quad (8.1)$$

In an uncatalysed biochemical reaction the reactants, molecules W and X, are surrounded by a hydration shell of water molecules, and they move about in solution in a random fashion. Only rarely will W and X collide with sufficient energy, and in the correct orientation, for a collision complex (X–W) to occur (Figure 8.9).

Before the products, molecules Y and Z, can form, the collision complex must pass through a transition state, which can be thought of as orienting W and X in such a way and with sufficient energy to facilitate any bond making/bond making necessary to form Y and Z. It has been suggested that one important aspect of enzyme catalysis is the ability of the reactive site to organize the substrate into a near-attack conformation.

The near-attack conformation puts the atoms that will react in close contact, in the correct orientation for bond breaking and making. It has been estimated that near-attack conformations leading to the transition state occur about 0.0001% of the time in uncatalysed reactions (Bruice 2002). Formation of the transition state requires a relatively large activation energy (ΔG^*). In the uncatalysed reaction, very

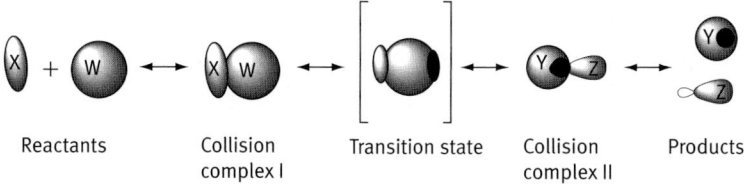

Figure 8.9. Mechanism of an uncatalysed reaction. Collisions leading to a viable transition state are rare. Near-attack conformations leading to the transition state occur about 0.0001% of the time (Bruice 2002).

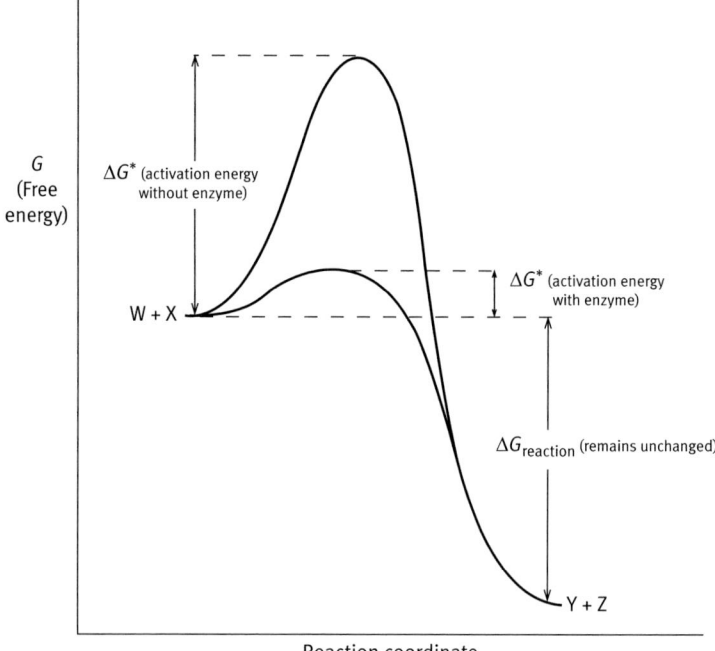

Figure 8.10. The Gibbs energies of activation, ΔG^*, of an uncatalysed and an idealized enzyme-catalysed reaction. The overall free energy change ΔG of the reaction is not altered by the presence of the enzyme. The reaction is spontaneous and will proceed at a faster rate at the same temperature in the presence of the enzyme.

few of the W−X complexes that happen to form will possess this amount of energy, so the reaction rate to form Y + Z will be slow (even though $\Delta G_{\text{reaction}}$ may be negative). In aqueous solution, a significant amount of the activation energy can be required to remove the hydration shells of W and X. Also there will be energy involved in rearrangement(s) of electronic charge within W and/or X during formation of the transition state. The reaction profiles of an uncatalysed and a simple idealized enzyme-catalyzed reaction (Figure 8.10) show that the enzyme has the effect of lowering the activation energy ΔG^* so that more W and X molecules achieve the transition state and the reaction will proceed at a faster rate.

To be useful, the study of enzyme activities must be quantitative. Any proposed mechanisms must be consistent with the experimental results. This is the province of the enzyme kinetics, a topic that concerns the rate at which enzyme-catalyzed reactions occur, and which provides information about enzyme mechanisms, inhibition, and regulation of activity. The Michaelis–Menten model of enzyme kinetics was introduced in 1913 and, despite many modifications, it is still the basic model for many enzymes.

$$E + S \underset{k_{-1}}{\overset{k_1}{\rightleftharpoons}} ES \overset{k_2}{\longrightarrow} E + P$$

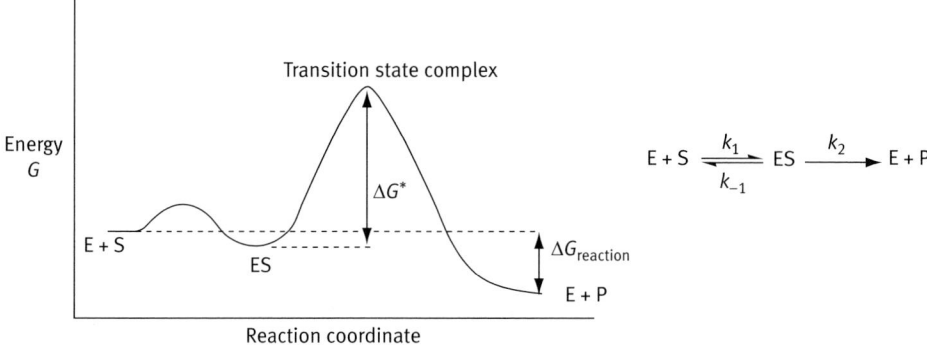

Figure 8.11. Reaction coordinate for a simple enzyme-catalysed reaction.

Enzyme (E) reacts reversibly with substrate (S) to form an enzyme–substrate complex (ES). In a second step, ES is converted to product (P) with release of the enzyme E. The rate constants (k) for the proposed stages are also shown.

Assumptions of the model:

1) There is no appreciable conversion of P back to S.
2) The concentration of ES quickly reaches a steady state, that is [ES] soon remains constant.

A reaction profile shows the energy levels at each step and provides a useful visual representation of the process. In kinetic terms, ΔG^* is the rate-limiting step (Figure 8.11).

In this profile the potential energy of the ES complex is lower than that of E + S. This is reasonable as ES is proposed to form spontaneously and rapidly. The process E + S ⟶ ES will have a negative ΔG. The above figure gives a more complete description than that in Figure 8.10, but it is still far from describing what happens in a real enzyme-catalyzed reaction. Current ideas propose a much more dynamic process, with multiple transition states. Many conformational transitions take place and modification of the enzyme protein quite distant from the catalytic site can influence the conformational changes. Such results suggest that motions of the entire enzyme molecule are essential for catalysis and reinforce the idea that only 'adaptable' macromolecules are capable of producing the observed rates of reaction (Boehr et al. 2006; Benkovic et al. 2008; Hammes 2008).

The Michaelis–Menten equation can be expressed in mathematical form, which allows the experimental determination of the constants used to describe the enzyme activity in quantitative terms. A plot of rate of reaction vs substrate concentration for an enzyme-catalyzed reaction that follows Michaelis–Menten kinetics is hyperbolic in shape (Figure 8.12).

As the substrate concentration [S] is increased, the rate increase levels off, until it reaches the maximum achievable rate (V_{max}) for the conditions. Any further increase in [S] has no effect on the rate, as the enzyme active sites

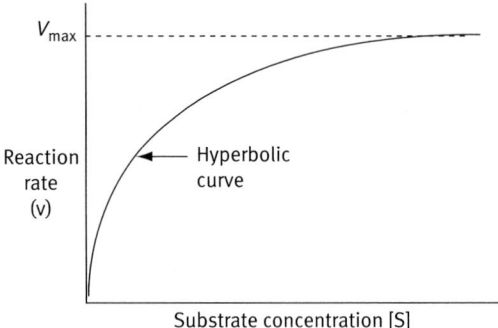

Figure 8.12. A plot of reaction rate vs substrate concentration for a reaction that follows Michaelis–Menten kinetics. The rate increase falls away as the substrate concentration approaches the saturation level of the enzyme. At saturation point of the enzyme with substrate, the maximum rate, V_{max}, is achieved.

are saturated. The V_{max} is an important kinetic characteristic of the enzyme under the given conditions. Enzyme kinetic data are now analyzed quickly using suitable computer programs. The Michaelis–Menten approach is a model, and like all models has its limitations. It doesn't apply to allosteric enzymes (see below). Several criteria and provisos must be considered when evaluating results from this and other models of enzyme kinetics, but I don't intend to pursue them here. If we imagine what is happening during an enzyme-catalyzed reaction, it would be something like this:

1) The reactants, substrates, of the enzyme usually attach quite specifically to the binding site, which positions them close to the active site. The active site is the region on the enzyme where the substrates actually undergo whatever bond-breaking and/or bond-making processes are involved in the reaction. This specific binding positions the substrates in an optimal orientation (the near-activity conformation (NAC)) for the formation of the transition state. The optimal orientation and associated proximity of the substrates substantially increases the probability of forming the required collision complex. NAC formation has been shown to occur from 1 to 7% of the time in active sites compared with 0.0001% for the uncatalyzed reaction (Bruice 2002). This effect contributes significantly to the rate of the reaction.

2) The binding of the substrates to the active site removes their solvation shells and leads to the exclusion of water from the site. Water is often a competing reagent and can alter the desired course of a reaction. This is important for many reactions. Release of 'bound' water molecules increases their mobility, often leading to a favourable increase in overall entropy.

3) As a result of interactions between the substrate and the amino-acid side-chains of the enzyme, there is an actual modification of the shape of the enzyme active site, so that the protein and substrate geometries are more complementary; they fit together better. The enzyme also forces the substrate into a conformation that approaches that of the transition state.

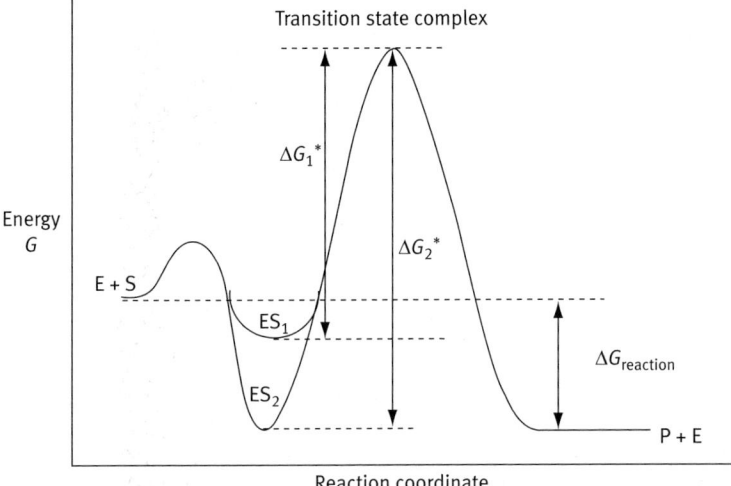

Figure 8.13. The case against tight binding of the substrate. ES_1 = the energy level of the ES complex for a typical enzyme-catalysed reaction. ES_2 = the energy level of the ES complex for a tightly bound substrate. As $\Delta G_2^* \gg \Delta G_1^*$, rate 2 $\ll$ rate 1.

This proposed dynamic adjustment of protein and substrate is part of the induced-fit hypothesis. An important result is stabilization of the transition state. The stabilization results in a great reduction in the activation energy needed to form the transition state, and leads to an increase in the rate of the reaction being catalyzed.

Although it is difficult to estimate quantitatively the relative contributions of the three factors mentioned above, it is generally agreed that the stabilization of the transition state is the most important. It is not the tight binding of the substrate. Very tight substrate binding would lower the energy level of the ES complex, thereby increasing the magnitude of ΔG^* and slowing the reaction (Figure 8.13).

The transition state stabilization idea is supported by the observation that so many enzymes have very high affinity for analogues of the transition state. A transition state analogue is a molecule that is very similar in structure to the transition state that forms with the natural substrate. However, in contrast to the real transition state, which is unstable and decomposes to form the products, the transition state analogue is a stable compound, and if well designed has the effect of blocking the activity of the enzyme. This is called inhibition of the enzyme. Many transition state analogues have been synthesized by chemists for the purpose of studying enzyme mechanisms and their inhibition. A practical application of such analogues is their use as drugs (see below). The study of inhibitors is important in enzymology. Inhibitors can be classified as reversible or irreversible. Reversible inhibitors fall into two main categories: competitive and non-competitive.

Competitive inhibitors involve a molecule similar in shape to the substrate, which binds to the substrate binding site and inhibits the binding of the natural substrate (Figure 8.14).

184 Introducing Biological Energetics

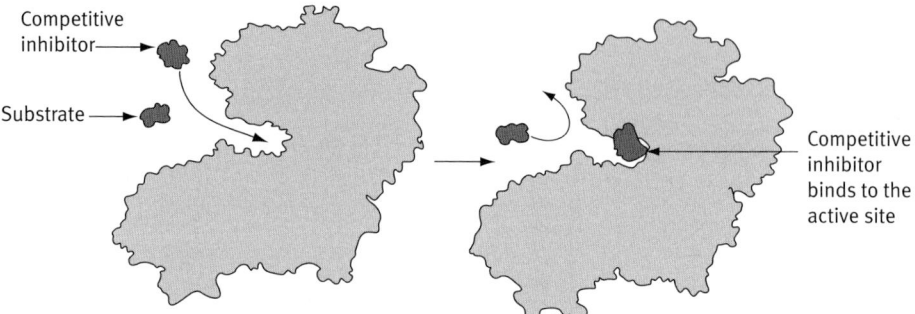

Figure 8.14. Competitive inhibition. A molecule similar in shape to the natural substrate competes with it for the active site. (Courtesy of Professor T. Steitz.)

A characteristic of competitive inhibition is that increasing the concentration of substrate will increase the likelihood of formation of an ES complex and increase the reaction rate in the presence of inhibitor. Allopurinol is a competitive inhibitor that reduces the production of uric acid in those who suffer from gout. It inhibits xanthine oxidase. Note the change in position of the nitrogen N_a in hypoxanthine to N_b in allopurinol (see below). The allopurinol is very similar in shape to hypoxanthine and competes with it for the binding site. The nitrogen atom N_b in allopurinol takes the place of a CH in hypoxanthine, so allopurinol cannot be oxidized to xanthine.

hypoxanthine $\xrightarrow{\text{xanthine oxidase}}$ xanthine $\longrightarrow$ uric acid

allopurinol

Drugs called statins reduce the rate of synthesis of cholesterol; Atorvastatin (Lipitor; Pfizer) is a transition-state analogue that inhibits the enzyme HMG-CoA reductase, a control-point enzyme in the cholesterol biosynthetic pathway.

The influenza drug Zanamivir (Relenza; GlaxoSmithKline) is a transition state analogue of the viral enzyme neuraminidase. Neuraminidase assists the invading virus to attach to host cells, so its inhibition reduces viral infectivity (von Itzstein et al. 1993).

Non-competitive inhibition involves interaction of the inhibitor with both enzyme E and the ES complex. The inhibitor is therefore not binding at the same place as S, so the inhibition cannot be reduced by increasing the concentration of substrate.

Irreversible inhibition occurs when the inhibitor attaches covalently to the enzyme and cannot be removed by dialysis. There is usually a time-dependent decrease in enzyme activity and the kinetic pattern resembles that of non-competitive inhibition. During studies of the group of enzymes called the serine proteases, a diagnostic test for the presence of serine in the active site was used. Serine has the side-chain CH_2OH, which reacts covalently with diisopropylphosphofluoridate (DIPF) and the enzyme is irreversibly inhibited.

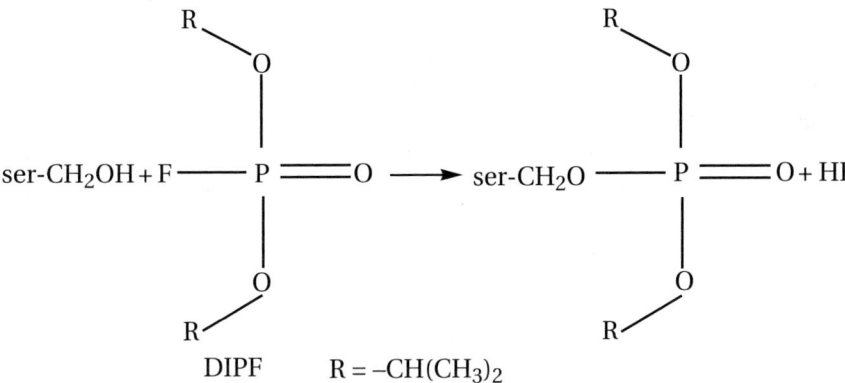

The penicillin antibiotics covalently attach to the serine group in the active site of glycoprotein peptidase, an enzyme that is involved in the synthesis of the bacterial cell wall. When this enzyme is inhibited, the cell wall is not fully crosslinked; it is easily ruptured and bacterial growth is stopped.

With the exception of irreversible inhibition, the attachment of the inhibitor to the enzyme depends on binding energy supplied by secondary forces, such as hydrogen, ionic, and van der Waals bonds, plus hydrophobic effects. The importance to biological activity of many individually weak bonds, precisely located in space, cannot be over-emphasized.

How does an enzyme actually work at the level of a chemical reaction? I shall use chymotrypsin as an example of the mode of action of an enzyme. By mode of action, I mean details of the way in which the enzyme catalyses the reaction. Chymotrypsin is a digestive enzyme that breaks certain peptide linkages in proteins, thus helping to degrade them in the small intestine. Enzymes that act in this way on proteins are known collectively as proteases. Chymotrypsin has been well studied for a number of years and its mode of action is known in detail. Chymotrypsin originates in the pancreas, but not as the active enzyme. If it were active there, it would do damage to proteins in the pancreas, as its 'job' is to degrade proteins. Instead, the pancreas produces an inactive precursor of chymotrypsin, called chymotrypsinogen. Just as allosteric interactions can be used to alter the tertiary structure (and hence activity) of proteins, so the precursor of an enzyme can be cleaved and the tertiary structure altered. This transforms the precursor into an active enzyme. The inactive precursors are collectively known as zymogens.

The chymotrypsin zymogen, chymotrypsinogen, consists of a single polypeptide chain and has five disulphide (−S−S−) bonds which help determine its tertiary structure and keep it inactive as an enzyme. When chymotrypsinogen is released into the small intestine another protease, trypsin, cleaves it quite specifically between arginine-15 and isoleucine-16, forming π-chymotrypsin. The π-chymotrypsin then acts on itself to cleave out two pieces. This leaves the final form of the enzyme, α-chymotrypsin, which because of the cleavages of the original chain now has three shorter polypeptide chains held together by disulphide bonds. The cleavages bring about changes in the tertiary structure, so the final form is an active enzyme.

Chymotrypsin cleaves peptide bonds preferentially at amino acids which have aromatic (non-polar) side-chains, such as phenylalanine, tyrosine, and tryptophan. The selectivity for these side-chains is due to the size and hydrophobic nature of the side-chains lining the binding pocket of chymotrypsin. Precisely how does chymotrypsin achieve this cleavage of bonds, i.e. what is its mode of action? There are many ways in which enzymes catalyze reactions, but as a result of careful, detailed studies of hundreds of mechanisms, general principles applicable to all types have been recognized. In many enzymes the active site is located in a kind of pocket in the surface. A number of amino acids residues line the interior of the active site, and although not all are directly involved in the catalysis, they play a role in positioning the substrate or maintaining a water-free environment, should that be necessary.

Chymotrypsin is one of a group of enzymes called the serine proteases because the amino acid serine is located in the active site and is an essential residue involved in the catalysis. By means of studies making use of various chemical 'probes' it has been shown that serine-195 and histidine-57 are essential for the activity of chymotrypsin. These two residues must therefore be located close together in space. The amino acid sequence (primary structure) certainly doesn't give a clue to this. A look at their numbering will show that they are 138 residues apart! Obviously there has been a lot of folding of the chain to bring them into close proximity in the active site. This emphasizes the importance of the enzyme being folded in exactly the right way into its final conformation. The final conformation is maintained, as mentioned above, mainly by the cooperative energy of a large number of secondary bonds. The above conclusions based on chemical probes have been confirmed by means of X-ray crystallography, a powerful technique used to determine the three-dimensional structure of the entire enzyme. The chymotrypsin 'backbone' is folded mostly in pleated-sheet arrays and positions the essential residues precisely in the active site pocket. The role of all 240-odd amino acid residues is basically to act as a framework to achieve this result. What sublime design!

Chymotrypsin cleaves peptide bonds on the carboxyl side of aromatic amino acid residues. How does the enzyme 'know' how to do this? There are good

Proteins and Enzymes 187

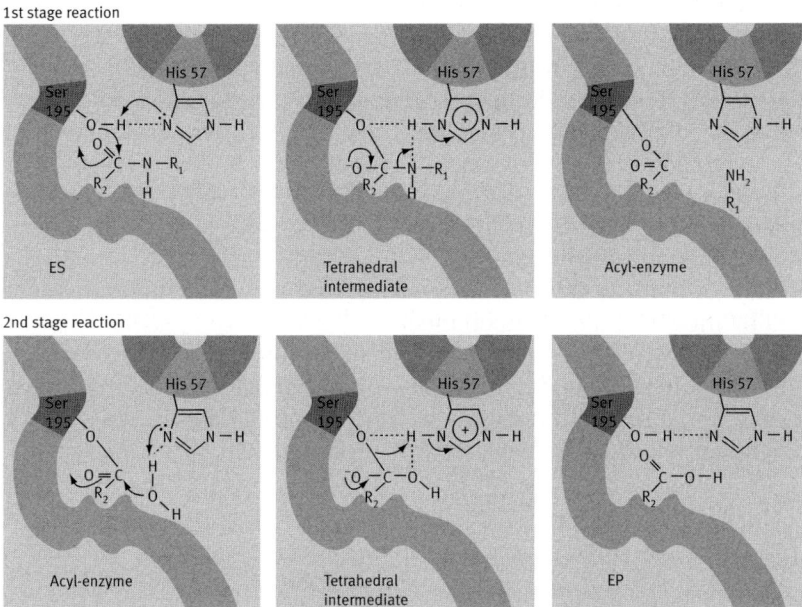

Figure 8.15. Mechanism of chymotrypsin action on a protein chain. (Illustration from Hammes, G.G. *Enzyme Catalysis and Regulation.* © 1982 Academic Press.)

structural reasons, leading to mutual recognition and interaction between enzyme and substrate, such that the peptide bond to be cleaved is in the correct position and orientation for catalysis. This is another example of de Duve's molecular complementarity at work (Figure 8.15).

The hydrolysis of the peptide linkage can be considered as taking place in two stages:

1) In an example of a nucleophilic reaction mechanism, the lone pair of electrons in the −OH group of serine-195 'attack' the partial positive charge on the carbon of the −C=O group in the peptide linkage. This leads, by a rearrangement of bonds as shown, to the acyl enzyme, where one piece of the cleaved substrate is covalently attached to serine-195. This demonstrates why the serine is essential to the activity of chymotrypsin and other serine proteases. If the serine −OH is blocked by a chemical probe specific for −OH groups, all enzyme activity is lost.
2) The acyl enzyme is attacked by the electrons on the oxygen atom of a water molecule, hence the formal term hydrolysis, or cleavage by water, is given to this type of reaction. After more bond rearrangements, as shown in Figure 8.15, both parts of the cleaved peptide linkage are released. One has a −COOH end group and the other an −NH_2 end group. These were the two groups formerly involved in the peptide linkage prior to cleavage. The serine-195 −OH group is restored, leaving the enzyme ready for the next substrate molecule. Note that histidine-57 was involved in both stages of the reaction. The lone pair on the imidazole nitrogen is protonated/deprotonated during the process.

Later research has revealed further details of the chymotrypsin mechanism. Aspartic acid-102 is also involved. Its γ-carboxylate group (γ–COO⁻) is located close to a nitrogen atom in histidine-57 and assists in the protonation/deprotonation steps. Otherwise, the mechanism is the same as shown.

The overall hydrolysis reaction has a negative ΔG, which does not change whether an enzyme is involved or not. The thermodynamics of a reaction are not altered by the presence of a catalyst, but the kinetics most certainly are. The lowering of the activation energy barrier by the enzyme, by a combination of the several effects mentioned earlier in the chapter, has the result of greatly increasing the rate of hydrolysis. Strictly speaking, the enzyme-catalyzed reaction is achieved by means of a new mechanism.

How much faster rates are enzymes able to achieve, relative to an uncatalysed reaction? The rate enhancement varies greatly depending on the enzyme involved, but is commonly thousands of times the uncatalysed rate. Let's look at the decomposition of hydrogen peroxide, with and without catalysis:

$$2H_2O_2 \longrightarrow 2H_2O + O_2$$

Let the uncatalyzed rate = 1:

Conditions	Relative rate	Activation energy (kJ mol^{-1})
Uncatalysed rate	1	75
Platinum catalyst	2.8×10^4	49
Catalase (enzyme)	6.5×10^8	23

Platinum is commonly used in chemistry and catalyses the decomposition of H_2O_2 almost 30 thousand times faster than no catalyst. Catalase does the job a staggering 650 million times faster! This example is exceptional, as catalase has one of the highest turnover rates of any enzyme. Hydrogen peroxide and other highly active oxygen derivatives are intermediates in a variety of human disorders. Catalase is the predominant H_2O_2 – degrading enzyme in human erythrocytes. The importance of the protective role of catalase is reflected in its very fast catalytic rate (Meuller et al. 1997).Catalase is a good example of the enormous effect that lowering the activation energy from 75 to 23 kJ mol^{-1} has on the rate.

This doesn't seem such a great reduction for such an increase in rate, but the reason for the drop in rate becomes clear when the full calculation is made. The relationship between the rate constant of a reaction and the activation energy is given by the Arrhenius equation:

$$k = Ae^{-\Delta G^*/RT}$$

where k is the rate constant and A is a constant for the particular reaction concerned. From this we can see that k is inversely proportional to $e^{\Delta G^*/RT}$, so as ΔG^* increases, k decreases exponentially and the reaction rate will fall dramatically.

The above discussion of protein enzymes brings us to the point when we can answer an interesting and important question. Why are enzymes so large and complex?

Many people, including some chemists, are puzzled and amazed at the complexity of enzymes. Why should all that energy and information go into making a catalyst? I have mentioned before that evolution is parsimonious, especially in the expenditure of energy, so why spend so much? Chemists have devised some very successful catalysts, but such catalysts are incredibly simple and less sophisticated than enzymes. The answer to our question has been provided by research over recent years. To achieve the rate increases observed, all the subtleties of protein structure and movement have been harnessed by aeons of natural selection. Even parts of an enzyme molecule quite distant from the active site or binding site have been recruited to reduce the activation energy. The Arrhenius relationship shows that a linear reduction in activation energy results in an exponential increase in reaction rate, so every little bit helps. Also, some of the most fundamental electron-transfer reactions take place over what chemists consider to be large distances. Examples include the redox reactions of the electron-transport chain and photosynthesis. Electron transfer over distances of the order of 2 nm are involved. These can only occur because special channels exist between the electron donors and electron receivers in the enzymes involved. The dimensions and chemical nature of the channels are crucial, and need to be supported and surrounded by other parts of the protein. Although free water is excluded from the channels, in some cases particular hydrogen bonds are important in the electron-transfer processes. It appears that the parsimony biologists have noted at the organism level is maintained at the molecular level. Attempts by chemists and biochemists to 'clip' pieces from enzymes and reduce them to the bare minimum have usually shown that not much can be removed. A molecular weight of about 10,000 Da has been estimated as the lowest size likely for an active enzyme (Hammes 2008).

8.3 Regulation of enzyme activity

All the different modes of inhibition and activation may be demonstrated and studied quantitatively by various mathematical approaches to enzyme kinetics. Different models need to be applied to enzyme kinetics because it is important to emphasize just that—they are models of reality and simplifying assumptions often need to be applied. It is necessary for scientists to be aware of the assumptions and of the limitations of each model. Models are a vital part of science. An initial model may be proposed, based on the best available evidence at the time. As more results and/or better interpretations arise, the model may be modified, until it becomes as refined as possible. Models are seldom perfect, as they often attempt to deal with quite complex phenomena that are not always amenable to precise mathematical analysis or description.

Enzyme catalysis of the thousands of reactions in a living cell cannot be haphazard. Synthesis vs degradation, energy storage vs energy use, provision of substrates here, removal of products there, all require careful regulation and constant monitoring. Much of the monitoring is done by regulatory enzymes that sense the instant-by-instant needs of the cell and adjust their activity in response. We have spoken of enzymes in isolation, but most enzymes in living organisms are part of a metabolic pathway. Metabolic pathways need various controls or regulatory mechanisms to modulate their activity and maintain their metabolites at optimal levels. Regulatory mechanisms in specific pathways are dealt with in some detail in Chapter 10, so I'll give just an overview of the principles here. There are two rather obvious requirements that depend on factors outside the actual metabolic pathway under discussion. These are availability of precursors and availablity of coenzymes. The former is often the 'responsibility' of other metabolic pathways that feed in to the one under discussion. Not all enzymes require coenzymes, but those which do obviously cannot work without an adequate supply. A good example is that of the coenzyme couple NAD/NADH. At high levels of cell activity, the reduced form, NADH, is channelled into the respiratory chain from the citric acid cycle. High activity depletes the supply of the oxidized form, NAD, and this could be limiting if the respiratory chain for some reason failed to process the NADH back to NAD.

The activity levels of the enzymes of a metabolic pathway are crucial. These activities are controlled at three independent levels.

1) The expression of the enzyme protein at the level of the gene, that is the actual number of enzyme molecules being produced at a given time. The control here is exerted over the amount of messenger RNA coding for the enzyme protein and is termed transcriptional control. Transcriptional control involves hormones or metabolites that act on the gene control mechanisms. Increasing the levels of an enzyme in this way is called induction, whilst decreasing of the enzyme levels is called repression.
2) Enzyme interconversion, where the enzyme is already present, but in an inactive form, as in the case of chymotrypsin. When needed, the active form is produced by an activating enzyme. The activating enzyme often acts in response to a hormone signal, such activation being one of the roles of hormones. Some interconversions involve ATP-dependent phosphorylation of the enzyme involved. Protein kinases are enzymes that catalyze phosphorylation of serine, threonine, or tyrosine hydroxyl groups in proteins. The introduction of a doubly charged phosphoryl group can have large effects on protein functionality, and serves as a means of regulation. Interconversion control is more rapid than transcriptional control, as the enzyme (or its immediate precursor) is already present and ready for rapid modification. Modifications other than phosphorylation include carboxylation, acetylation, and glycosylation (addition of a sugar).
3) Regulation of key enzymes by ligands, which are small molecules that bind to allosteric proteins and change their conformation, as we have seen for haemoglobin. Allosteric enzymes are a major means by which the cell regulates enzyme activity to suit changing requirements. A ligand may be the

immediate product of the reaction or the end product of a metabolic pathway. The process is termed feedback inhibition. An example occurs in the biosynthesis of the amino acid isoleucine. Isoleucine is synthesized from threonine via five enzyme-catalyzed steps:

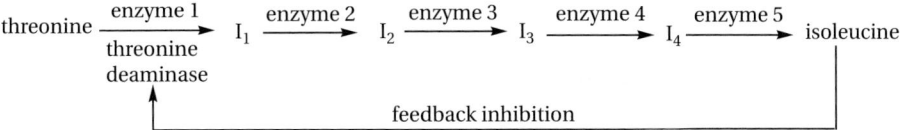

The first step is catalyzed by threonine deaminase. When the pathway is active, isoleucine accumulates and binds to its allosteric site on threonine deaminase, eventually stopping its own synthesis and saving cellular resources.

The feedback ligand might also be a metabolite from an entirely different pathway, but one intimately linked to the pathway under discussion. Depending on the requirements of the cell or tissue, both the levels and the activities of enzymes may be regulated to keep metabolic pathways running optimally.

Allosteric enzymes don't follow Michaelis–Menten kinetics. The rate vs substrate curve is sigmoidal rather than hyperbolic (Figure 8.16).

A sigmoidal curve is characteristic of cooperativity between subunits of the enzyme to bind the substrate and is a feature of allosteric enzyme kinetics. It means that binding of one substrate molecule to the first subunit makes it easier for additional substrate molecules to bind. Allosteric enzymes usually consist of two or more subunits. Each subunit has a substrate binding site and an allosteric ligand-binding site. The regulatory effect may be positive or negative, that is the rate can be increased or decreased relative to the unregulated rate (Figure 8.17).

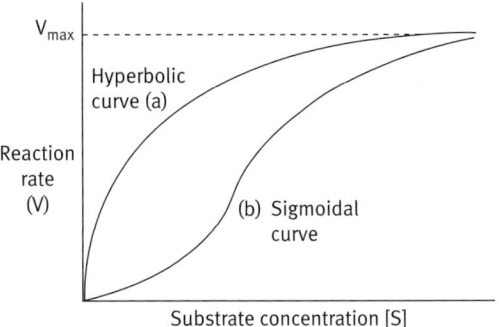

Figure 8.16. Reaction rate vs substrate concentration curves: (a) for an enzyme-catalysed reaction that follows Michaelis–Menten kinetics; (b) for an allosteric enzyme. The hyperbolic curve reflects the cooperativity characteristics of allosteric regulation.

192 Introducing Biological Energetics

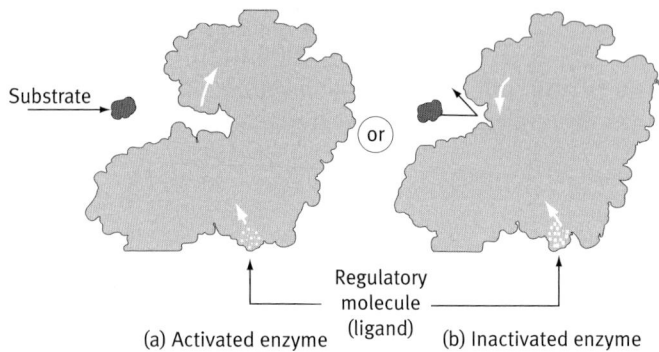

Figure 8.17. Allosteric control of enzyme activity. The regulatory molecule or ligand binds to a site distant from the active site. A regulatory molecule can (a) activate or (b) inactivate the enzyme by changing its conformation, and hence the access of substrate to the active site. Most allosteric enzymes consist of two or more subunits that act cooperatively. (Courtesy of Professor T. Steitz.)

8.4 Coenzymes, vitamins, and enzyme classification

For a substantial number of biochemical reactions coenzymes are required, so a supply of coenzymes must be available to the cell. Why are coenzymes necessary?

While some enzymes carry out their functions relying on their protein structures only, others need the help of non-protein compounds called cofactors. The term cofactor itself is usually reserved for certain metal ions that are essential for enzyme activity. The other cofactor types are organic molecules referred to as coenzymes. Coenzymes are further classified into soluble coenzymes, which bind to the protein portion during a reaction, assist in the reaction, and are then released, and prosthetic groups, which remain tightly bound, sometimes covalently linked, to the protein portion. Coenzymes are so important in biological reactions that they warrant discussion in their own right. It is not possible to cover them all here, but the following examples should give the reader an appreciation of their essential biological roles.

Redox enzymes or oxidoreductases are enzymes that catalyze oxidation/reduction (redox) reactions (see the enzyme classification below). They all require coenzymes and can be of the soluble or prosthetic group type.

1) The pyridine nucleotides NAD^+ and $NADP^+$ are widely distributed and have been mentioned a number of times already, but without detailed discussion. They are both soluble and move about the cell in this form. A number of enzymes use them as coenzymes and the NAD^+/NADH (oxidized/reduced) and $NADP^+$/NADPH ratios are important in cells, as is discussed in Chapter 11. It is instructive to look at the mechanism involving the pyridine nucleotides (Figure 8.18).

The full dinucleotide structures of NAD^+ and $NADP^+$ shown in Figure 8.18(a) are partly replaced by R in Figure 8.18(b) to allow focus on the transfer of H and electrons. When NAD^+ is complexed with an enzyme and the appropriate substrate, there is net removal of two H atoms from the substrate (e.g. a foodstuff)

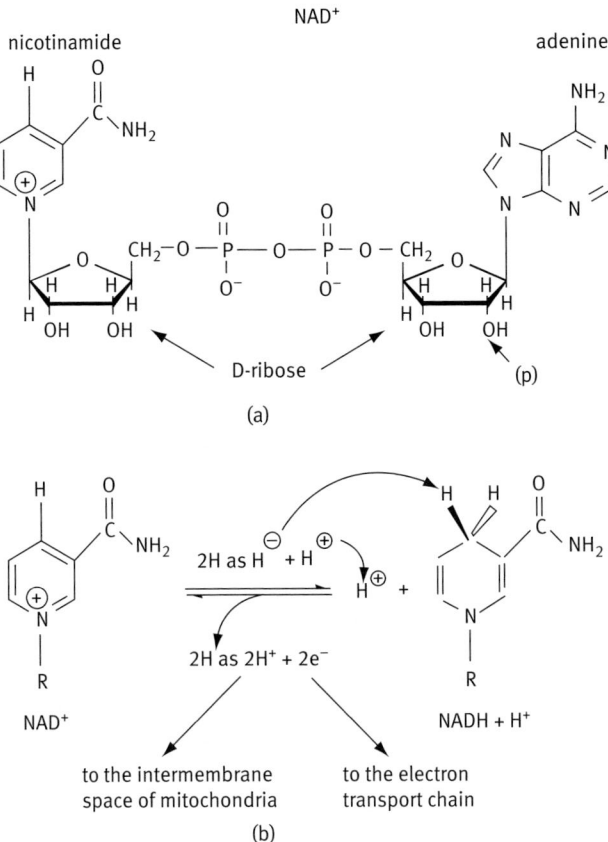

Figure 8.18. (a) The structure of NAD$^+$, nicotinamide adenine dinucleotide. (p) indicates the point of attachment of a phosphate group to form NADP$^+$. (b) The reduction of NAD$^+$ to NAD + H$^+$. Transfer of hydride ion H$^-$ to nicotinamide is stereospecific. It always adds to the same 'face' of the ring. In the reverse direction the release is in the form of protons and electrons, the latter to special receiver molecules.

being oxidized. These H atoms are removed as hydride ion H$^-$ and a proton H$^+$, as shown. In the reoxidation step, there is net removal of two the H atoms as two protons (pumped into the intermembrane space of the mitochondrion) and two electrons that are transferred along the ETC. The reoxidized NAD$^+$ is returned to the coenzyme pool for reuse. In Chapter 10 we discuss the roles of NAD$^+$/NADH, the coenzyme couple involved in many energy-producing (catabolic) oxidations in the cell. The role of NADP$^+$/NADPH is mainly in anabolic pathways (Chapter 11) that utilize energy to build complex molecules.

2) The flavin coenzymes flavin mononucleotide (FMN) and flavin adenine dinucleotide (FAD) have flavin as the redox group (Figure 8.19).

FMN has the same flavin moiety as FAD, but its side-chain R is less complex. In FMN, R is ribitol with one attached phosphate group at position 5 as shown in

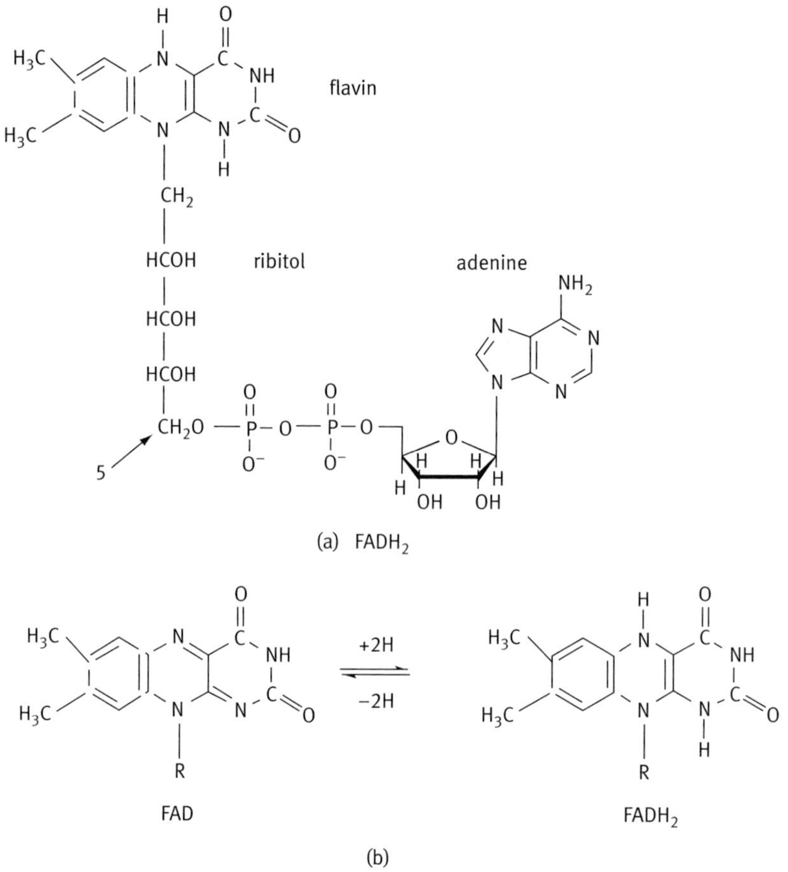

Figure 8.19. (a) The structure of FADH$_2$. (b) The flavin moiety of FAD undergoes reduction by two hydrogen atoms (as H$^+$ and H$^-$) from a substrate, such as succinate from the tricarboxylic acid cycle (see Chapter 11).

Figure 8.19(a). FAD and FMN are covalently linked to proteins as prosthetic groups, and form a group called flavoproteins. As well as transferring two hydrogen atoms, FAD and FMN can transfer one proton and one electron at a time when necessary. During this process, a radical intermediate is formed. Radicals are highly reactive and potentially damaging species, so FAD and FMN remain bound to the enzyme protein and are not allowed to be 'free' to move about the cell. FAD and FMN are found in dehydrogenases, oxidases, and monooxygenases.

Group-transferring coenzymes transfer entire groups of atoms as a single unit.

1) Pyridoxal phosphate is an important coenzyme in amino acid metabolism. Its role in transamination reactions is illustrated in Figure 8.20.

The pyridoxamine phosphate (Figure 8.20(a)) is involved in the transfer of −NH$_2$ groups from one amino acid to an oxoacid to form a new amino acid, catalyzed by a transaminase:

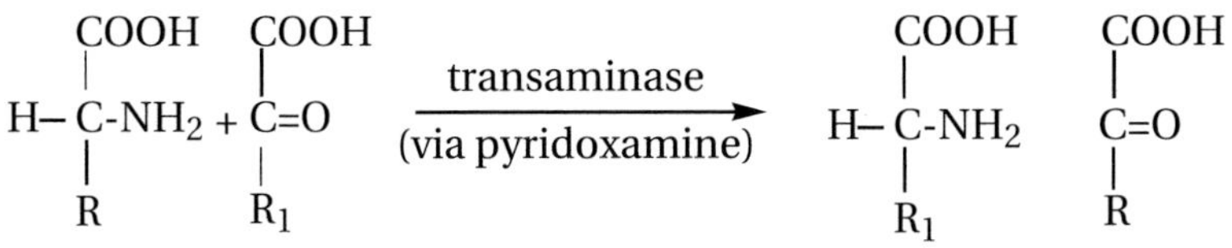

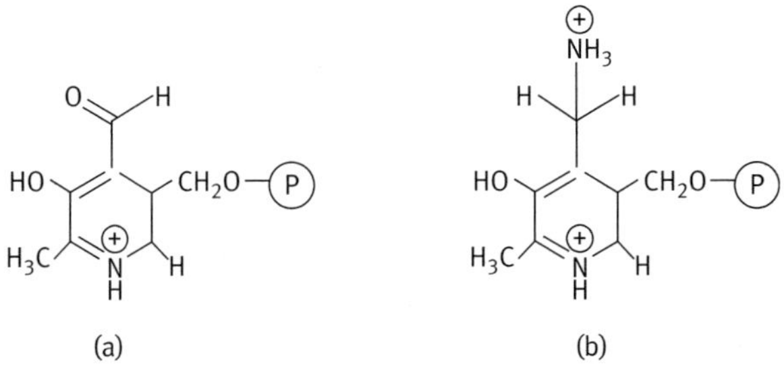

Figure 8.20. (a) Pyridoxal phosphate, aldehyde form. (b) Pyridoxamine phosphate, an intermediate in transamination reactions.

2) Biotin is the coenzyme used by carboxylases. Using ATP, biotin reacts with HCO_3^- to form *N*-carboxybiotin, from which CO_2 is then transferred to other molecules (Figure 8.21).

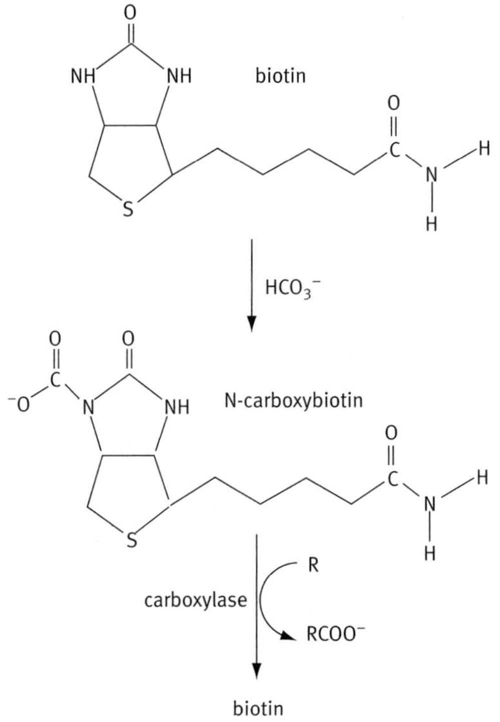

Figure 8.21. Biotin assists in the transfer of carbon dioxide units. Example: R = pyruvate, CH_3COCOO^-; $RCOO^-$ = oxaloacetate, $^-OOCCH_2COCOO^-$.

Many of the compounds we know as vitamins are the precursors of coenzymes. As vitamins are not made in the human body, but are essential for good health, they must be provided in the diet. Traditionally, the vitamins are divided into two groups: water and fat soluble. With the exception of vitamin C, the water-soluble vitamins are components of or precursors of coenzymes. The side-chains of amino acids that make up proteins have only a limited range of chemical activities. The association between the enzyme protein and a coenzyme extends the number of reaction types that can be carried out. The coenzyme is usually modified during the reaction and is later converted back to its original form by other enzymes (e.g. the NAD^+/NADH system). This allows small amounts of these precious substances to be used repeatedly, time after time.

Vitamin deficiencies due to insufficient intake in the diet are the cause of a number of diseases in humans. Considering their key roles in metabolism, this is not surprising. The story of the gradual realization that such compounds existed is long and rather complex, partly because the biochemistry of their action was not known at the time. Early researchers found that certain diseases could be cured by extracts from certain foods, but they had no idea why. The division into water- and fat-soluble arose when researchers tried to purify and identify these mysterious factors. The fat-soluble group could be dissolved in ether, and thus could be separated from the water-soluble group, which could not. Eventually, separation and purification techniques were developed that enabled the individual compounds to be fully identified. This was no easy task, as the chemical structures of the various vitamins vary widely. Table 8.1 summarizes some of the important aspects of the vitamins, the coenzymes, and their associated deficiency diseases.

Table 8.1. Coenzymes, their precursors, reactions catalysed, and deficiency diseases.

Coenzyme	Precursor	Reaction type	Deficiency disease
Biotin	Biotin	Carboxylation	Anorexia, dermatitis
Coenzyme A	Pantothenic acid	Acyl transfer	Dermatitis
FAD, FMN	Riboflavin (B_2)	Redox	Growth retardation
NAD^+, $NADP^+$	Niacin	Redox	Pellagra
Pyridoxal phosphate	Pyridoxine (B_6)	Transamination	Dermatitis
Thiamine pyrophosphate	Thiamine (B_1)	Aldehyde transfer	Beriberi

The fat-soluble vitamins are not directly related to coenzymes; they are involved in vision (vitamin A), maintenance of bone structure (vitamin D), blood coagulation (vitamin K), and antioxidant processes (vitamin E).

Let's look at two further aspects of enzymes that concern us all: deficiencies in the control of their expression and the use of enzymes as markers for disease.

Some diseases are caused by what are known as 'inborn errors of metabolism'. These are genetically transmitted diseases caused by abnormalities in the DNA of genes, resulting in diminished levels of a particular enzyme. An example is phenylketonuria (PKU). The name arises from the fact that a metabolite called a phenyl ketone accumulates and appears in the urine. A deficiency in the level of the enzyme phenylalanine carboxylase causes this accumulation. The problem caused is mental retardation, probably as a result of a build up of the phenyl ketone and related metabolites in the brain. This results in an osmotic imbalance in which water flows into the cells, causing them to expand and crush one another in the developing brain. PKU is easily detected in newborn babies and a test for its presence is mandatory in the USA. The treatment is a modified diet, low in phenylalanine and enhanced in tyrosine (the missing product). Sufferers must avoid the artificial sweetener aspartame, which is a peptide containing phenylalanine. A substitute for aspartame, which contains alanine rather than phenylalanine, is available.

Some enzymes may be used as 'markers' or indicators for disease. Lactate dehydrogenase (LDH) has two isoforms that are slightly different in amino acid composition, so with a suitable test they are distinguishable. One form occurs in heart muscle and the other in skeletal muscle. Any increase in LDH levels in the blood indicates tissue damage of some kind. A heart attack can be diagnosed with a high degree of certainty by detecting increased levels of the heart-muscle LDH in the blood. Similarly, acetylcholinesterase, an enzyme involved in controlling certain types of nerve impulse, is a useful marker. A number of pesticides will interfere with this enzyme, so farm workers suspected of excessive exposure to pesticides may be tested for acetylcholinesterase activity. More than 20 enzymes are regularly used as clinical markers. They are used to examine the function of the liver, heart, brain, prostate gland, pancreas, and red blood cells, to name a few. All this is made possible by the depth of knowledge biochemists now have of so many enzymes and their unique distribution patterns in certain organs and tissues.

There is still a long way to go. Not all proteins believed to be encoded on our genes have been identified, let alone characterized. This identification task is being attempted by practitioners in the field of proteomics. The human genome projects have succeeded in identifying and sequencing essentially all our genes. Functional genomics aims to look at all the genes expressed during particular metabolic process or during embryonic development. It involves identifying the transcriptome, the set of messenger RNAs transcribed from DNA under a defined set of conditions.

Glycomics has been added to the growing 'omics' list recently. A large percentage of proteins are glycosylated, that is have carbohydrate attached. Glycomics seeks to identify the roles of all the carbohydrates that are attached to proteins and cell membranes. Metabolomics is immensely ambitious as it seeks to identify and quantify all the metabolites in a system or organism. Such ambitions have only become realistic with the availability of modern instrumentation and the expansion of computational capabilities. The task ahead is huge.

Just how many different types of chemical reaction can enzymes catalyze? Biochemists have naturally been vitally interested in this. Much work has been carried out on classifying enzymes into the minimum number of groups possible, based on the basic mechanism of the reaction catalyzed. Perhaps surprisingly, it has been possible to fit the large number of enzymes into just six major groups or classes. All enzymes, once fully characterized, are entered in the Enzyme Catalogue (EC) and given a four-digit EC number. The first digit indicates the major class, and the rest the subclass, sub-subclass, and position in the sub-subclass. As an example, let's take alcohol dehydrogenase (ADH), which catalyses the reaction:

$$\text{ethanol} + \text{NAD}^+ \longrightarrow \text{acetaldehyde} + \text{NADH} + \text{H}^+$$

or

$$CH_3CH_2OH + NAD^+ \longrightarrow CH_3CHO + NADH + H^+$$

The EC number for ADH is 1.1.1.1.
The name comprises several levels:
Class 1. Oxidoreductases (catalyze oxidation–reduction reactions)
Subclass 1.1 −CHOH is the electron donor
Sub-subclass 1.1.1 NAD^+ is the electron acceptor
Its place in the sub-subclass is 1, so the whole EC number is 1.1.1.1[Zn^{2+}].
The fact that zinc, as the ion Zn^{2+}, is necessary for activity, is also indicated.
The six enzyme classes are:

1) oxidoreductases: catalyze the transfer of reducing equivalents between redox systems
2) transferases: catalyze the transfer of functional groups from one substrate to another
3) hydrolases: also involve group transfer, but the acceptor is always a water molecule
4) lyases (synthases): catalyze the removal or formation of a double bond
5) isomerases: catalyze the movement of groups from one place to another, within the same molecule
6) ligases (synthetases): catalyze the formation of C–C, C–O, C–N, or C–S bonds. They are energy dependent and are coupled to the hydrolysis of nucleoside triphosphates (e.g. ATP).

Enzymes having the same function are often found in a wide range of organisms. This is particularly the case with enzymes of the major metabolic pathways that are common to most forms of life. The study of the same enzyme across a number of species may be useful in a number of ways. Depending on how long ago two species may have diverged, there will be various changes in the amino acid sequence of an enzyme (or indeed any protein) that is common to both. These changes are due to mutations in the DNA of the genes, whose base sequences determine the protein amino acid sequence, as described in Chapter 9. Comparison of protein sequences in different species is used in studies of the relationship between amino acid sequence and protein structure. If the protein has maintained the same function in different species, despite having some sequence changes caused by mutation, then the mutations must not have altered the structure significantly, that is the protein function must have been 'tolerant' of such mutations. No doubt there could have been other mutations that did seriously affect the function, but the individuals carrying them would have been unlikely to survive. Protein chemists can thus use these protein homology studies to derive general principles of protein sequence/structure relationships.

This assists them in studies on the all-important three-dimensional structure of proteins and enzymes.

The number and types of mutations found in proteins can also provide information on the evolutionary relationship between species, as described for nucleic acids in Chapter 9. Until recently, it was quicker to sequence proteins than nucleic acids. One well-studied example is the pair of oxygen-binding haem proteins haemoglobin and myoglobin of humans, where there is considerable sequence homology, strongly suggesting evolutionary divergence from a common ancestral molecule. Another example, ranging across a number of species, involves the amino acid sequence of cytochrome *c*, a component of the respiratory chain (see Chapter 10). In fact, a phylogenetic tree, a diagram illustrating the evolutionary relationships between a group of organisms, has been constructed using the sequences of cytochrome *c* in another example of molecular evolution. The tree constructed on the basis of differences in primary structure of cytochrome *c* shows close agreement with results derived from more classical methods, such as the fossil record. These results support the idea that closely related proteins share a common evolutionary origin and provide further convincing evidence in favour of the theory of evolution in its modern form.

REFERENCES

Benkovic, S.J., Hammes, G.G., and Hammes-Schiffer, S. (2008) *Biochemistry* **48**, 3317–21.

Boehr, D.D., McElbeny, D., Dyson, H.J., and Wright, P.E. (2006) *Science* **313**, 1638–42.

Bruice, T. (2002) A view at the millennium: the efficiency of enzymatic catalysis. *Accounts of Chemical research* **35**, 139–48.

Garrett, R.H. and Grisham, C.M. (1999) *Biochemistry*, 2nd edn. Saunders College Publishing, Fort Worth, TX, p. 191.

Hammes, G.G. (1982) *Enzyme Catalysis and Regulation*. Academic Press, New York.

Hammes, G.G. (2008) How do enzymes really work? *Journal of Biology and Chemistry* **283**, 22337–46.

Luisi, P.L. (2006) *The Emergence of Life*. Cambridge University Press, Cambridge, pp. 245–46.

Meuller, S., Riedel, H.-D., and Stremmel, W. (1997) Direct evidence for catalase as the predominant H_2O_2 – removing enzyme in human erythrocytes. *Blood* **90**, 4973–8.

Riemer, J., Bulleid, N., and Herrmann, J.M. (2009) Disulfide formation in the ER and mitochondria. Two solutions to a common process. *Science* **324**, 1284–7.

von Itzstein, M., *et al.* (1993) Rational design of potent sialidase-based inhibitors of influenza virus replication. *Nature* **363**, 418–23.

9
Molecular Genetics—the Chemical Basis of Heredity

In keeping with the theme of the book, enough genetics will be introduced to allow us to follow the role of energy in the propagation of individuals and species. We will still be speaking in general terms in this chapter, rather than discussing specific groups of animals or plants.

The earliest forms of life on Earth were single-celled, possessed no discrete nucleus and as such are called prokaryotes. Single-celled organisms which do possess a membrane-enclosed nucleus, the eukaryotes, developed over time from these simpler life forms. Eventually, evolution led to complex organisms such as mammals and flowering plants. These consist of a number of organs, such as the heart, liver, brain, skin, bark, leaves, and roots, which in turn consist of specialized groups of cells called tissues. The individual cells making up the various tissues and organs of these organisms can be considered the basic units, or modules, of complex life. There were advantages to becoming multicellular, but there were also additional problems to be solved. Fortunately, there was built into the primordial life forms the adaptability that was to allow them to evolve into more complex organisms.

Chromosomes were first observed in 1882 by Walther Flemming as tiny threads in the nuclei of dividing salamander larvae cells. It was established that the number of chromosomes in each cell of a particular organism depends on the species to which the organism belongs. Genes, the functional units that contain the code to make the entire organism, are arranged in linear fashion along the chromosomes. We now know that genes consist of molecules of DNA, which typically in eukaryotes are entwined in a framework of proteins that assists them to fold up into extremely compact structures. All of the somatic cells, or body cells, in animals and plants contain the full complement of chromosomes and hence the full complement of genes. In sexually reproducing organisms, the sex cells, or gametes, usually contain half the number of chromosomes found in somatic cells. When sexual reproduction occurs, half the chromosomes of the resulting embryo come from the sperm of the male parent and the other half from the ovum of the female parent, so that the embryo inherits characteristics of both parents and has a complete set of chromosomes.

Genetics is the study of inheritance and of genes. Details of the molecular nature of heredity have been revealed steadily over the last 200 years or so, and most aspects are now known in detail at the level of molecules. The science of genetics was well developed long before genes were actually proven to exist.

The Austrian monk Gregor Mendel (1822–1884) carried out breeding experiments with plants. Mendel was by training a physicist, encouraged by his abbot of the Augustinian monastery in Brunn (now Brno) in Moravia (located in the current Czech Republic) to study methods of agricultural improvement. In his pioneering work with sweet pea plants, he showed that inheritance takes place by means of discrete particles, which Mendel termed 'factors' and were later called genes. Thus, different factors were shown to be responsible for certain physical characteristics of the sweet pea, such as flower colour or seed shape. Previously, it was believed that characteristics in the offspring were inherited by a kind of 'blending' of the characteristics of each parent, by some unknown mechanism. Mendel showed that inheritance involved discrete units, the genes, which themselves remained intact. One of his greatest insights was that genes occur in pairs and that one member of each pair is inherited from each parent.

Mendel used his skills in mathematics to produce a theory of inheritance, first reported in 1865, which explained his quantitative plant-breeding results. It was largely ignored by biologists. At the time, his 'factors' were abstractions, the proofs mathematical, and the journal where he published the results obscure. Mendel's work lay dormant until 1900, when it was 'rediscovered', independently, by deVries, von Seysenegg, and Correns. At the time of Mendel, and indeed until 1953, the actual nature of genes was unknown. The very existence of genes as real chemical entities was debated for many years. For some time it was believed that genes were in fact special proteins. In 1902, Walter Sutton was first to point out the connection between chromosomes and Mendel's factors, but it was not until the 1930s that genes were shown conclusively to reside on the chromosomes in the cell nucleus, after important advances coming from studies in the USA of fruit-fly genetics by Thomas Hunt Morgan and later by Barbara Mc Lintock on maize chromosomes. Studies up to about this time are generally known as classical genetics. It was believed that the structure of DNA was too 'simple' to allow the encoding of the enormous amount of information for a whole organism. This belief proved to be wrong. When James Watson and Francis Crick published the structure of DNA in 1953, the science of molecular genetics became possible. The actual molecules that held all the information for life had been discovered. In a frenzy of scientific activity, it was soon shown that DNA possessed all the requirements for replication (i.e. to reproduce itself) and was also coded to carry all the information to produce an entire organism. The cells of all organisms, from bacteria to humans, carry DNA unique to the species. This fundamental complement of DNA carried by the members of each species is called the genome of the species. The genome of many species is divided into chromosomes, each of which consists of DNA plus various other structural bits and pieces, in mammals these being mainly special proteins called histones. Humans possess 46 chromosomes, in two equivalent sets of 23, in each somatic or body cell. The sperm and ova, that is the sex cells or gametes of humans, each contain only 23 chromosomes, so that when fertilization occurs, the full complement of 46 chromosomes in the form of 23 pairs, half from each parent, becomes part of each cell in the developing embryo.

Each chromosome is a long, continuous DNA molecule. A chromosome contains thousands of special functional regions, the elusive genes, and also

DNA regions of uncertain function. A number of scientists proposed that some of these regions actually have no useful function and called such regions 'junk DNA'. More recent opinion is that this dismissive attitude was probably premature.

Mendel was correct. Each gene has a specific function, or more than one function in some cases, that helps to determine characteristics such as flower colour, eye colour, sex, skin colour, etc. Some characteristics are controlled by more than one gene, so we can't say 'there is *a* gene for skin colour' because there is more than one gene controlling this characteristic. Similarly, it is probably premature to say there is a single defective gene responsible for asthma or for schizophrenia or for alcoholism. The full information is as yet unavailable. It is also likely that many diseases are partly of genetic and partly of other (environmental) origin, so the so-called 'gene therapy' envisaged by some genetic engineers is unlikely to be able to cure all our ailments.

The genome, the total complement of genes working together, is able to determine all the characteristics of the organism that contains it. The human genome contains some 20,000–25,000 genes, while the colon bacillus *Escherichia coli* has about 4000. In general, the more complex the organism, the larger is its number of genes (as one would expect intuitively) because there needs to be more coding to define fully the more complex structures and functions. Paradoxically, however, the total amount of DNA in a cell varies enormously between species. Humans have about 3.3 billion base pairs (see later) of DNA, while the single-celled *Amoeba dubia* has a staggering 670 billion base pairs. As noted above, not all of the DNA of either species is in the form of active genes.

Not all the genes in a particular tissue type are capable of being activated or expressed, to use a genetic term. The particular group of genes that is capable of being expressed in liver cells differs from the particular group of genes that may be expressed in brain tissue cells, even though each type of cell contains the full complement of genes. In fact, the pattern of gene activity in a tissue determines the type of tissue it is, or vice versa depending on how you like to think of it. This is one of the directions in which complex organisms have evolved and we can see why such a development is useful. The evolution of cells capable of different gene expressions, and of course the appropriate mechanisms for controlling such expressions, was essential to the development of specialized tissues and organs. The cells of a developing embryo undergo a process called differentiation, during which the previously undifferentiated cells (essentially, these are cells that are 'uncommitted' at the early stage of development) become specialized into groups, which on further development form into tissues. Eventually, in the fully developed organism, each specialized organ, such as the brain, the liver, or the heart, consists of its own types of tissue, made up of one or more types of differentiated cell.

Just to complicate things further, in any given mature tissue, not all the genes capable of being expressed are actually expressed all the time. The expression of all the genes, all the time, in fact would be disastrous, as not all the chemicals produced by genes are needed all the time. Strict control of gene activity is essential, as energy and many biological molecules are too precious to waste and are usually difficult to store. Genes are usually activated by feedback mechanisms (see Chapter 11) so if the level of a particular gene product (such as an enzyme) should fall below the required level, this will be detected and the gene will be 'switched on', 'activated', or expressed, for example by the action of a hormone or a key metabolite. When the enzyme reaches the required level, the feedback mechanism will ensure that the gene activity will maintain it as long as required (see below). The mechanisms for control of expression of some microbial genes have been studied intensively and the general principles are also known for more complex organisms.

What is the most obvious common feature of all successful species? They and their ancestors have developed effective means of reproduction and survival. New species evolve from pre-existing ones and so on back through time. This constitutes the basis of the theory of common descent. A gene (DNA) from one species can be inserted chemically into the DNA of another species by so-called genetic engineering. Not only that, but if the insertion is done with the appropriate precautions, the gene can be induced to form its specific protein product (s), demonstrating that the chemistry of replication is similar for all life on Earth. Replication refers to the ability of DNA to make exact copies of itself. By implication, it also means that DNA-based replication is flexible, in that it succeeds for all the enormous variety of species which inhabit this planet. It succeeds for organisms which reproduce asexually or sexually, for plants and mammals, fish and fungi, for bacteria and beetles, and has done so for billions of years. As far as we know, DNA and its structurally close partner RNA are the only molecules generally used by living organisms for the process of replication. Evolution on the whole tends to be conservative, maintaining the systems that work well, rather than having to invent new ones.

The replication of DNA is only a part of the process of reproduction, which in a mammal involves fertilization of the ovum by a sperm, implantation in the uterus, development of the foetus, birth, and development to maturity. The fascination scientists have with DNA lies in the fact that, as far as information is concerned, everything is there. Within a single, identifiable structure, wholly accessible to modern science, lies the entire code for each finished organism, and apparently this code can be fully interpreted. This is the biologists' Rosetta stone, characterized by Watson and Crick a mere five decades ago and decoded soon after by others. Not only that, but all living organisms use essentially the same code. No wonder that the Human Genome Project, with its predicted completion date of about 2003 blown away by 2001, continues to cause excitement in both the scientific and the general population.

Modern genetics has provided a unifying principle that links all parts of biology. Previously disparate fields of biology can now be related by the study of DNA and genes. Areas such as physiology, developmental biology, morphology, ecology, and phylogenetic studies have all been advanced by an understanding of the underlying genetics. The term molecular evolution has recently come into existence and this field of study uses gene sequences to document the history of life on Earth.

Before I proceed further, let me emphasize that DNA isn't everything. By this I mean that although DNA carries all the information that eventually will produce an entire organism, this is not the end of the story. The development of an adult mammal from a fertilized egg is an extremely complex process, which is still far from being fully understood. What happens at each of the many steps on the way to adulthood is not necessarily under the exclusive control of the genetic material. There is growing evidence that some of the developmental processes are dictated by what has been called 'necessity'. I sometimes describe evolution as the art of the possible, meaning that evolution by natural selection (or indeed by any other means) could only act, in a constrained way, on what was already there. Here I am saying something similar about events at the molecular level. At any stage of development, what an embryo can 'do' next is limited by its physical and chemical surroundings. The DNA in the genes cannot 'tell' each cell in the embryo *exactly* what to do, even though it may communicate via numerous chemical messengers to various cells. To some extent (as yet we don't know to what extent, or to what extent it may vary between species) the DNA loses exclusive influence, and to some extent the environment takes over. This realization has allowed biologists to look more broadly at development in an effort to solve hitherto intractable problems.

The environment in this case comprises the immediate physical and chemical surroundings of the developing cell or foetus, which of course are continually changing during the developmental process. The surroundings of a cell constantly send chemical messages, some of which directly or indirectly influence particular genes either to become expressed, and begin manufacturing their product, or to 'switch off' and cease to produce their product. This is an example of feedback control, which, as well as affecting genes, is common in most metabolic pathways (see Chapter 11). Maintenance of the correct ensemble of chemicals, at appropriate concentrations, around a developing organism is essential. The composition and concentrations of such compounds vary with the stage of development. In contrast, an imbalance, or the presence of unwanted chemicals in the surroundings of a human foetus, may have drastic effects on its development. A woman unfortunate enough to contract the viral infection rubella (German measles) during the first 8 weeks of pregnancy may have a child who has hearing defects, heart abnormalities, or liver disorders. It has been shown that some infections, rubella being one, some kinds of drug taking, and cigarette smoking during pregnancy can cause defects in the development of the foetus with consequences that may be life-long.

The extent and type of problem which may arise depends on the stage of pregnancy at which the foreign influences occur. Different organs develop at different times during a pregnancy—the two sides of the brain develop at slightly different times, as do other parts of the nervous system, the limbs, etc. Which organ is affected depends on the timing of an infection or on the ingestion of some foreign chemical compound that can penetrate the placenta and reach the foetus. There is medical evidence suggesting that infection in the mother during a period when the development of the two sides of the brain in the foetus is out of phase (i.e. one side is slightly ahead in development compared with the other) can result in an imbalance of receptors for the neurotransmitter dopamine. The left and right brain hemispheres may develop a difference of up to 20% in the number of dopamine receptors. This imbalance, according to some medical opinion, may be a major contributor to the condition of schizophrenia, although this interpretation is by no means universally accepted by researchers.

A point related to what I have been saying about development is that:

The genome is more like a recipe than a blueprint or plan.

A blueprint or plan will show exactly the way the object will look in its final from. All detail is shown in the directions and the diagrams. In the case of a house, so long as standard building procedures are followed, its shape and appearance will be as shown on the plan.

In contrast, the shape and appearance of an animal or plant (its phenotype in biological terminology) is not obvious from its genetic makeup (its genotype) alone. The full sequence of all the DNA (i.e. the entire genome) of some organisms has already been determined. The first free-living organism to have its genome characterized was the bacterium *Haemophilus influenzae*. Another example is the bacterium *E. coli*, which is found in the intestines of us all. All their genes have been mapped, that is their locus positions on the DNA are precisely known and the entire DNA base sequence has been determined.

At the beginning of the year 2000, over 20 prokaryotic and some three eukaryotic genomes had been published, with the sequences of hundreds of more species expected over the next few years (*Science* 287, (2000) 605–6). Until very recently there was by no means a consensus on the number of human genes. Experts differed in their opinions and estimates ranged from as low as about 30,000 to nearly 200,000 genes in the human genome. Celera, a US company, and the multinational study called the Human Genome Project arranged to have a preliminary human genome sequence announced simultaneously in June 2000. In February 2001, the groups simultaneously published in the journals *Nature* and *Science* more details, including the estimate of 32,000 genes for the human genome. This was at the lower end of the previous estimates and one of the implications is that there is probably not a specific gene for every human characteristic. Rather, there are probably more control genes, which fine-tune the timing and extent of expression of other genes, than previously thought. Current reports suggest an even lower number of human genes, perhaps 20,000–25,000.

Another proposal is that the DNA of the same gene may be processed to yield more than one messenger RNA, and hence more than one protein. Despite much detailed sequence knowledge, it is not as yet possible to predict the phenotype (shape) of an organism from its DNA sequence alone. The analogy of a recipe for the genome now becomes clear. A recipe will not tell us the shape of the final cake. This will be determined by the shape of the baking dish (roughly analogous to the species to which the organism belongs). The recipe will hopefully tell us the approximate taste type and even give some idea of the texture of the cake, so long as the cook is competent. If there is an unexpected change in the environment during cooking (analogous to the change in environment during foetal development) then the final product may not turn out as expected in these respects, for example opening the oven door or using too high or too low a temperature can have drastic effects on the final taste and texture of the cake. To push the imagery about as far as possible, we could say that a genetic (DNA) defect in a foetus is the analogy of an inaccurately followed recipe, leading to, say, cystic fibrosis or diabetes. On the other hand, a problem (e.g. rubella) occurring during foetal development resulting in a baby that has physical defects or behaves abnormally is analogous to a cooking error that results in a cake that looks or tastes unusual.

I won't go further into the problems of inherited diseases and developmental problems here. I want to describe the way in which the DNA in cells carries out two essential processes for the continuation and the sustenance of life. The first is replication of the genetic material (the genome) and the second is translation of small parts of the genetic message we have called the genes into something useful. Each type of gene is potentially capable of producing a specific type of protein chain, which may then do its designated job. I say potentially capable because as we have seen, each gene in each cell is not always active, that is it is not always expressed. A little thought will show that a careful balance between the thousands of possible gene products is essential. The maintenance of this balance is managed, in part, by controlling the 'on' and 'off' messages to the genes. In, say, a mammal, how does the body control this switching of genes? It is usually achieved by some kind of feedback mechanism, for example when the concentration of insulin (a small protein involved in glucose uptake by cells) in the blood becomes too low, this is detected by cellular sensors, which send a 'message' or 'signal', usually a chemical, to the special cells in the pancreas that produce insulin. The signal informs the cells that insulin is low and soon the cell activates (or, more accurately, stops the repression of) genes that control insulin synthesis and insulin is produced for transport to the bloodstream. Once the insulin in the bloodstream reaches normal levels, this too is sensed and the message goes all the way back to the pancreas cell genes, which are switched off till required again. You can imagine that this feedback loop is part of an incredibly important balancing mechanism that must be precisely controlled. There are many hundreds of other essential processes that need to be controlled by similar means. The internal balance of a complex organism such as a mammal must be monitored and maintained at all times. The interplay of molecules and energy is involved in all these processes. As we have seen previously, cells require a constant input of energy to maintain their physical integrity, to carry out metabolic processes, and thereby to remain alive. Energy is involved in the recognition processes taking place between the molecules involved in information transfer. The situation in single-celled organisms, which lack specialized organs and tissues, is less complex, but a high degree of control is necessary nevertheless.

How does all this work? What is so special about DNA? The full story is a fascinating one, which would take much space to describe, so I shall cut technicalities to the bone. DNA is a large molecule. Rather than referring to the size of DNA in Daltons, molecular biologists usually refer to the number of base pairs. The human genome has some 3.3 billion of these base pairs, so sequencing this enormous number correctly is a mammoth task. Each member of a base pair occurs on a different strand of DNA, the two strands being wound into the famous double helix. There are four bases in DNA: adenine (A), guanine (G), thymine (T), and cytosine (C).

An abbreviated way to represent the sequence of the bases on a single strand of DNA is

—AGTCAATGCTCAGGCTA—, etc.

We shall see later a similar way to represent double stranded DNA.

The sequence or order of bases in DNA in a particular gene is of extreme importance, as it determines the structure of the protein that the gene codes for. The base sequence is referred to as the primary structure of DNA. The maintenance of the correct sequence of bases is so important that each cell invests a considerable amount of its precious energy in DNA repair. Should an inadvertent change in the sequence in DNA occur, the structure of its protein 'offspring' could be changed, perhaps drastically. This is what happens in some inherited diseases such as cystic fibrosis, and in sickle-cell anaemia, which infects hundreds of thousands of people in parts of East Africa. Such inherited problems are often referred to as molecular diseases, and the precise causes of many are now known. The causes of many more molecular diseases will become known as the benefits of the Human Genome Project begin to flow through to medical science.

At the molecular level DNA consists of two intertwined double helices, with specific base-pairing holding the two strands together. What holds them in this highly organized, and apparently unlikely, conformation? The answer is hydrogen bonding between the base pairs. Figure 9.1 shows that the base pairs bridge across the two strands of DNA rather like steps on a spiral staircase. The double helix is referred to as the secondary structure of DNA (Figure 9.1).

The steps, however, have gaps in the middle of the tread that are occupied by hydrogen bonds; these are essential stabilizers of the double helix. Each base-pair step contributes the energy of either two (A-T) or three (G-C) hydrogen bonds to stabilize the double helix. We saw in Chapter 5 that many weak hydrogen bonds (about 20 kJ mol^{-1} in water) working simultaneously are able to stabilize a particular molecular conformation quite successfully and that is what is happening here. One of the great insights of Watson and Crick was to make their models of the four bases A, T, G, and C in the correct form, such that the hydrogen-bonding possibilities became obvious. The insight came, after much unsuccessful and frustrating model building, after one of them had spoken to a chemist who understood that some molecules are capable of existing in two forms, called tautomers. It so happened that for most of their model building Watson and Crick were using the wrong tautomeric form for the bases. As soon as the correct forms were used, the solution essentially 'fell out'. They saw that A and T could form two hydrogen bonds, while G and C readily formed three (Figure 9.1). However, A didn't readily form such bonds with G or C and neither could T.

The specific pairing of A-T and G-C also explained an experimental finding that had puzzled biochemists for many years. It was well known that in DNA, the molar amount of A always equalled that of T, and similarly the molar amount of G equalled that of C. The specific pairing noted above shows why. The Watson/Crick double helix in fact explained a number of things, and several of these involved specific hydrogen bonding.

Firstly, it explains replication, which is the process during which DNA makes copies of itself and which makes reproduction of the whole organism possible.

208 Introducing Biological Energetics

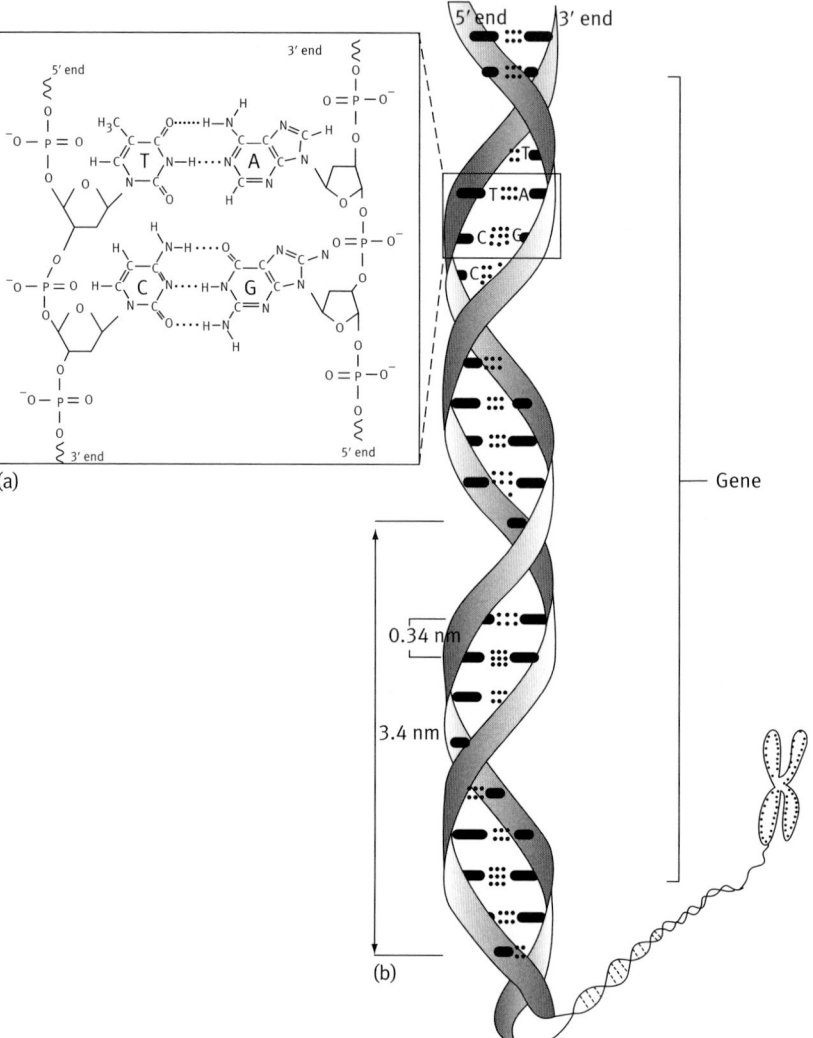

Figure 9.1. (a) Base pairing in DNA. (b) From double helix to metaphase chromosome.

Offspring need their own copy of the genome and this comes about by means of DNA replication. The process involves base pairing directed by hydrogen bonding, illustrating the highly specific role of energy at the most fundamental level of life.

The average length of DNA in a human chromosome is about 50 mm. The entire genome is the equivalent of about a metre of DNA. Each somatic cell in your body (a typical cell being about 200 microns (1 micron = 0.000001 or 10^{-6} m) across) contains the full genome. All the DNA must fit into the cell nucleus, which at most is only about 5 microns (0.000005 m) in diameter. This is a most amazing packaging feat. In the broadest sense we can readily believe that the orderly packing of DNA into the nucleus is a process that involves a decrease in entropy, and this overall will be energy consuming. The supercoiling of the DNA double helix is known as the tertiary structure of DNA.

Replication of DNA is shown in Figure 9.2. There are many details to this process, all of which are known essentially in full, but we will concentrate mainly on the energetics of the major steps. DNA replication involves separation of the two strands and, using these as templates, the production of two new strands. As half of the original DNA molecule is conserved in each replicated molecule, this process is called semi-conservative replication, which takes place

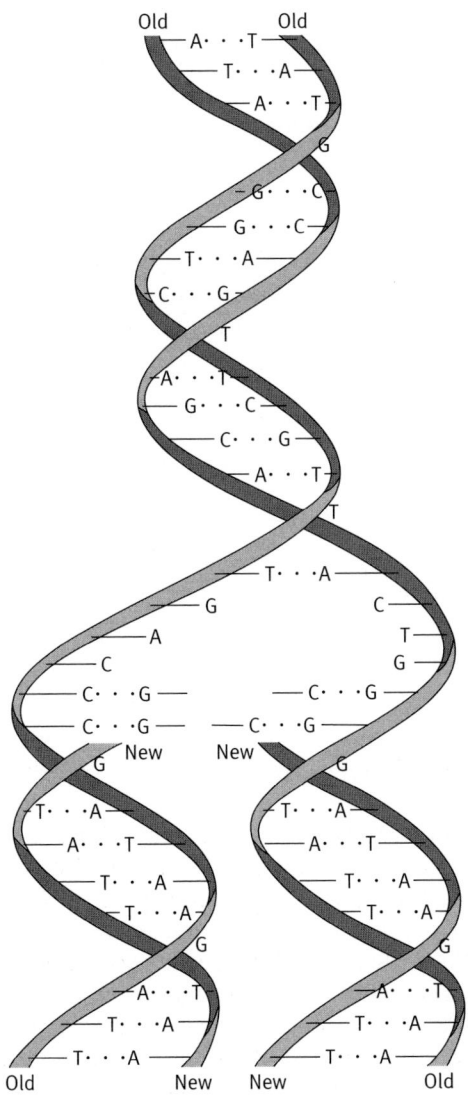

Figure 9.2. Semi-conservative replication of DNA. (Illustration from Voet, D., Voet, J.G., and Pratt, C.W. *Fundamentals of Biochemistry*, upgrade edn, p. 54. © 2002 John Wiley & Sons Inc.)

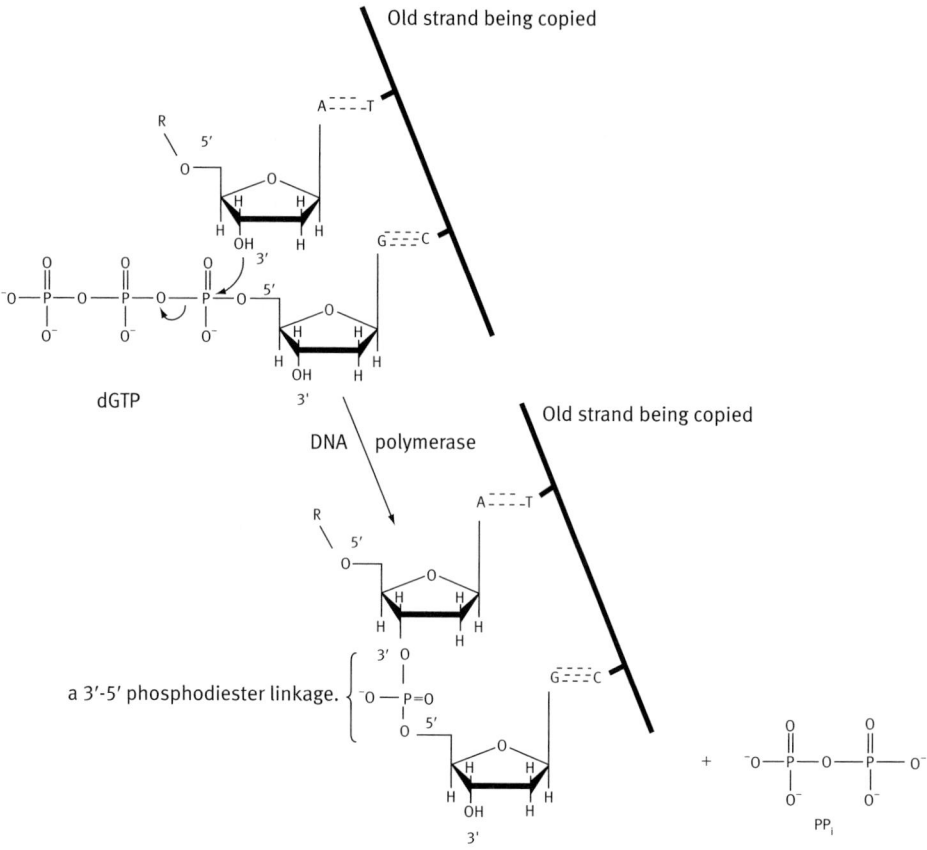

Figure 9.3. Replication of DNA. Extension of the new DNA strand by formation of a 3'–5' phosphodiester linkage. The reaction is driven by the ΔG of hydrolysis of dGTP, with the release of pyrophosphate (PPi).

in both prokaryotes and eukaryotes. The description is a general one, illustrating the principal steps common to most organisms.

The enzyme performing the actual replication is called a DNA polymerase, of which there are up to five types in prokaryotes, for example *E. coli,* and five (slightly different) types in eukaryotes. Overall there are several other enzyme types involved in replication. A DNA gyrase introduces a swivel point ahead of the replicating strands and a helix-destabling protein (a helicase) binds and promotes unwinding. The single-stranded regions of the unwound DNA are stabilized by a single-stranded DNA-binding protein, which also protects the single strand from hydrolysis. The helicases possess ATPase activity, which harnesses the energy of ATP to break the hydrogen bonding between base pairs and separate the DNA strands. Two molecules of ATP are consumed for each base pair separated, so full replication requires substantial amounts of energy.

Further energy is now involved in the formation of 3'–5' phosphodiester linkages to extend the new strand. One of the four trinucleosides dATP, dTTP, dGTP, or dCTP will be added. For example, the G of dGTP hydrolysis is harnessed via a polymerase enzyme to couple it to the growing new strand. (Fig. 9.3). This process is continued until he entire piece of DNA undergoing replication has been completed.

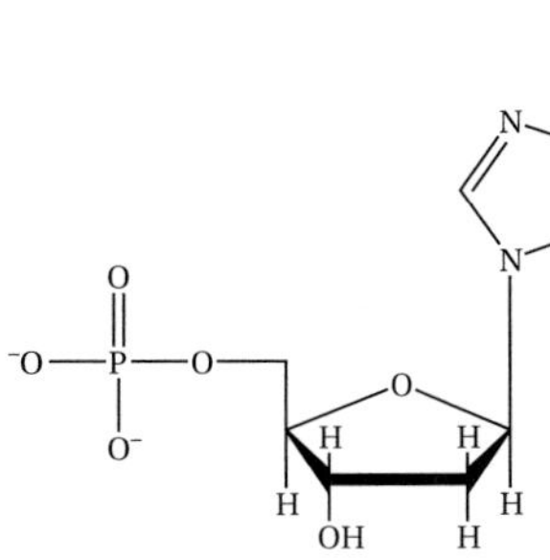

dAMP

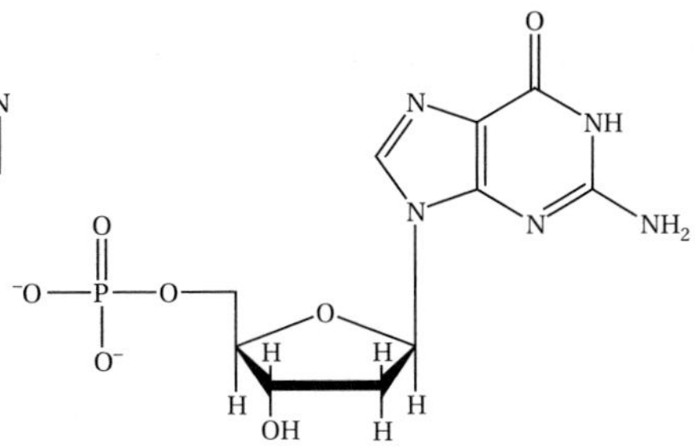

dGMP

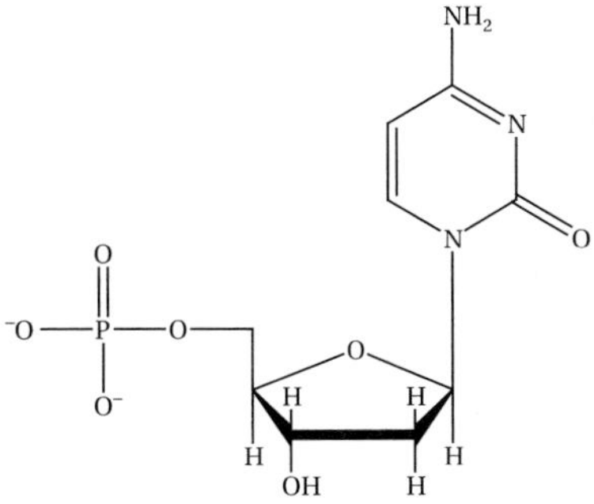

dCMP

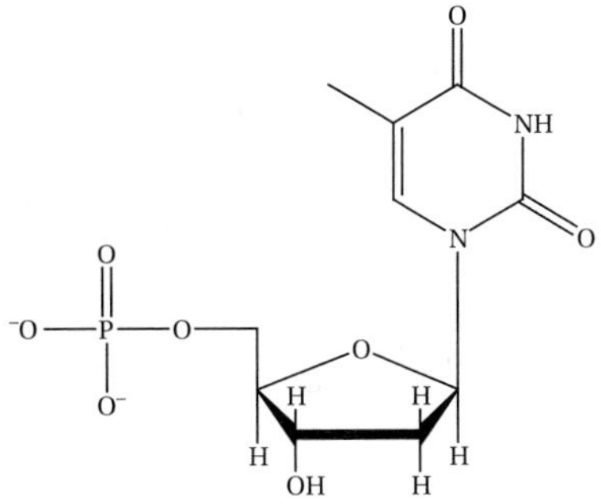

dTMP

The prefix 'd' stands for 'deoxy', to distinguish them from the ribonucleoside phosphates, such as AMP, UMP, GMP, and CMP:

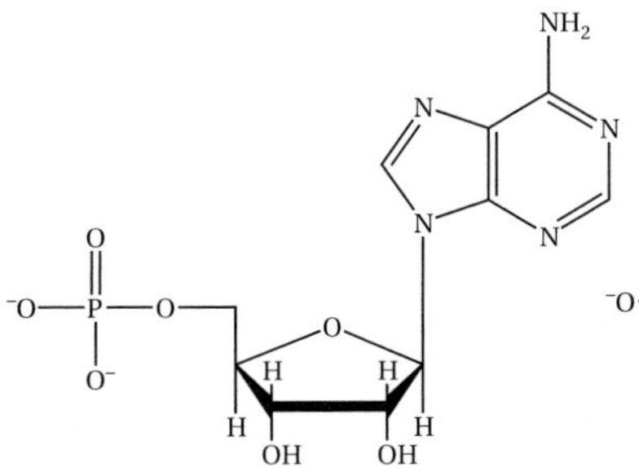

AMP

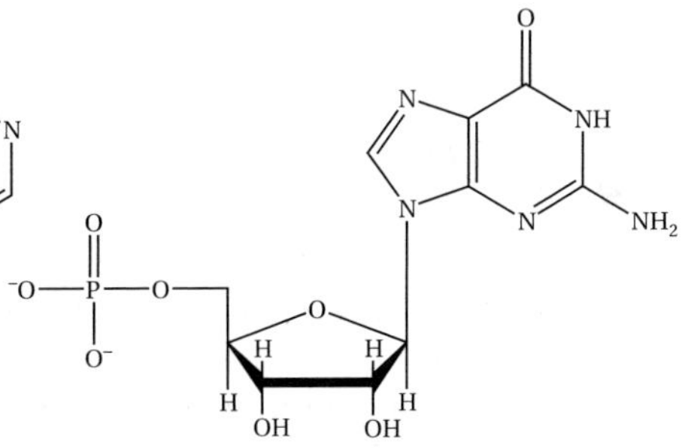

GMP

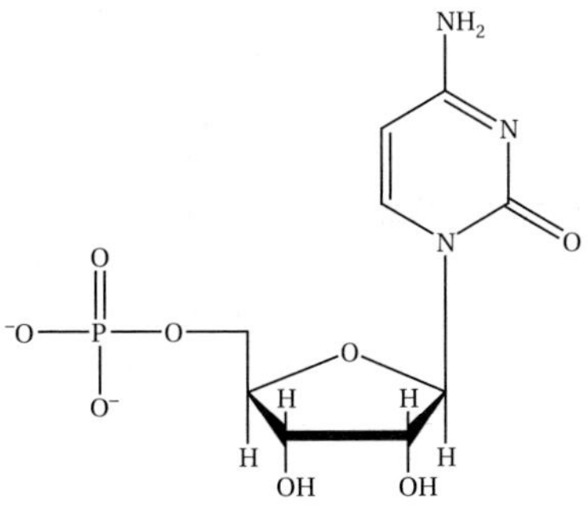

CMP

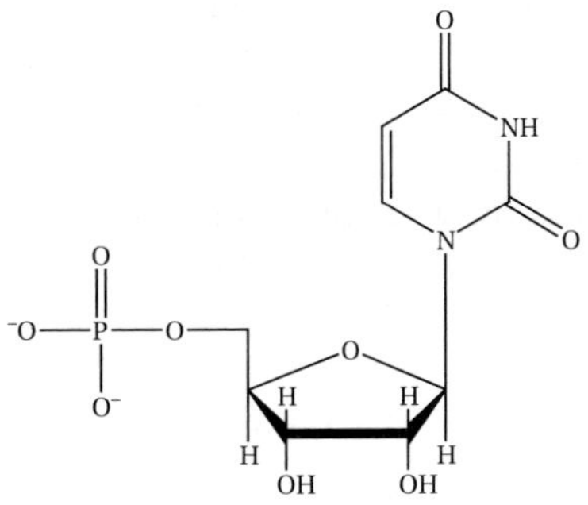

UMP

(The ribonucleosides, the precursors of the other major type of nucleic acid, RNA, differ from their deoxy equivalents only in the type of sugar they contain: ribose instead of the very similar, but importantly different, 2-deoxyribose. Like DNA, RNA contains four bases, the familiar A, G, and C, but instead of T it has U (uracil). Apart from its sugar group, RNA is structurally like a single-stranded DNA). All the deoxyribo- and ribonucleoside triphosphates are 'high-energy' compounds like ATP, that is they have a large negative standard Gibbs energy of hydrolysis, ΔG^0.

Although replication of DNA is essential for propagation of the species, for an individual member of that species it is essential to access the DNA that has all the information needed for its life, but this coded information must be decoded and processed into whatever molecules are required and precisely when they are required. When a gene is expressed, the part that codes for the product (usually a protein) must be copied to form a corresponding RNA. This RNA, called messenger RNA (mRNA) is then moved outside the nucleus to special structures called ribosomes, where it directs the actual protein synthesis. For the DNA of a gene to be copied to form mRNA, other molecules must be able to have access to it. As was the case for replication, the double helix must be unwound at least temporarily so that the machinery for its transcription into mRNA can act. The 'message' is transcribed from one strand of the DNA only, the so-called template strand. The enzyme that unwinds the DNA double helix, catalyses the polymerization of new bases into mRNA, and rewinds the DNA is called RNA polymerase, of which there are three types in eukaryotes and one type in prokaryotes.

The mRNA acts as a messenger molecule, but other types of RNA have different roles, such as major structural elements of the ribosomes. We will encounter some other roles of RNA later.

During transcription, U on the growing RNA strand pairs by specific hydrogen bonding, with A on the template DNA strand, and G pairs with C. As the RNA is produced we can represent it in our familiar abbreviated form as AUGCCAUG-CAU, etc., as shown in Figure 9.6.

There are special signals built into the transcription process to start and stop the synthesis of mRNA so that each molecule is exactly of the length and base sequence dictated by the DNA of its gene. The completed mRNA is transported out of the nucleus to structures in the cytoplasm of the cell, the ribosomes, which are the sites of protein synthesis. In the ribosomes, the mRNA itself is used as a template, directing amino acids (in an activated form, so there is energy required here also) to be joined in the precise sequence that was coded by the DNA back in the nucleus. The DNA has thereby 'masterminded' the synthesis of a protein with the precise sequence that is most suitable for its function in the life of the cell or organism. No wonder DNA must be protected from mistakes such as replication errors and/or mutations, which alter the base sequence. An altered base sequence means that there is the possibility (but not the certainty) that one or more different amino acids could be coded for and inserted into the protein, which could alter the function of the protein. Examples where such mutations have occurred are the genetic diseases cystic fibrosis and sickle-cell anaemia, as mentioned above.

We have seen that expression of a gene leads via transcription from DNA to the corresponding mRNA, but precisely *how* does the mRNA become translated into proteins in the ribosomes?

This is the role of the now-famous genetic code. The genetic code is what is known as a triplet code, meaning that three bases on mRNA form the code for each amino acid.

There are 20 amino acids used to form proteins. Can four bases, A, U, G, and C, taken three at a time, form enough unique triplet codes for 20 different amino acids? Easily. The first base of a triplet can be any one of the four. The same is true for the second and third places of the triplet, leading to $4 \times 4 \times 4 = 64$ possible triplets, more than enough. What about the extra possible combinations that are 'not needed?' All amino acids have more than one triplet code. Some have up to six, and because of this occurrence of several triplets for the same amino acid, the genetic code is said to be degenerate. The coding triplets in the mRNA are called codons.

In many cases the degenerate codons for an amino acid differ in the third base, but this is not always so. Of the triplets, 61 code for amino acids. The one coding for methionine, AUG, also forms part of the 'start' or initiating signal for the first amino acid in the new protein chain. There are also three special triplets that code to 'stop' the protein chain (UAA, UAG, and UGA; Figure 9.4).

Note that while there are several triplets for each amino acid, there is not any triplet which codes for more than one amino acid. The reason is obvious. If this were the case, the ribosome would be 'in confusion' as to which of the two amino acids to add. Put another way, as the sequence of amino acids in a protein strongly affects its function, the chance that two amino acids could yield the same functionality is low. Just how low is shown by the fact that such a situation simply does not occur in living organisms. Other points to note are that the code is non-overlapping (no bases are shared between consecutive codons) and is essentially universal, that is it is the same for nearly all organisms, underlining the common origins of life on Earth. (For a long time the code was believed to be truly universal, but recently it has been shown that the mitochondrial genomes of some species use slightly different codes. Note that mitochondria have some of their own genes, that is these genes do not occur in the cell nucleus, but in the mitochondria themselves. The reasons for this will be discussed in Chapter 12.)

When mRNA binds to a ribosome, it is arranged in such a way as to have its coding triplets accessible to another type of RNA called transfer RNA, or tRNA.

In each cell there exists a 'pool' of tRNA molecules available for protein synthesis. There is one specific type of tRNA available for each amino acid. Thus, the tRNA for phenylalanine will become linked only to phenylalanine by a specific synthetase enzyme, and become linked in such a way that phenylalanine is in an

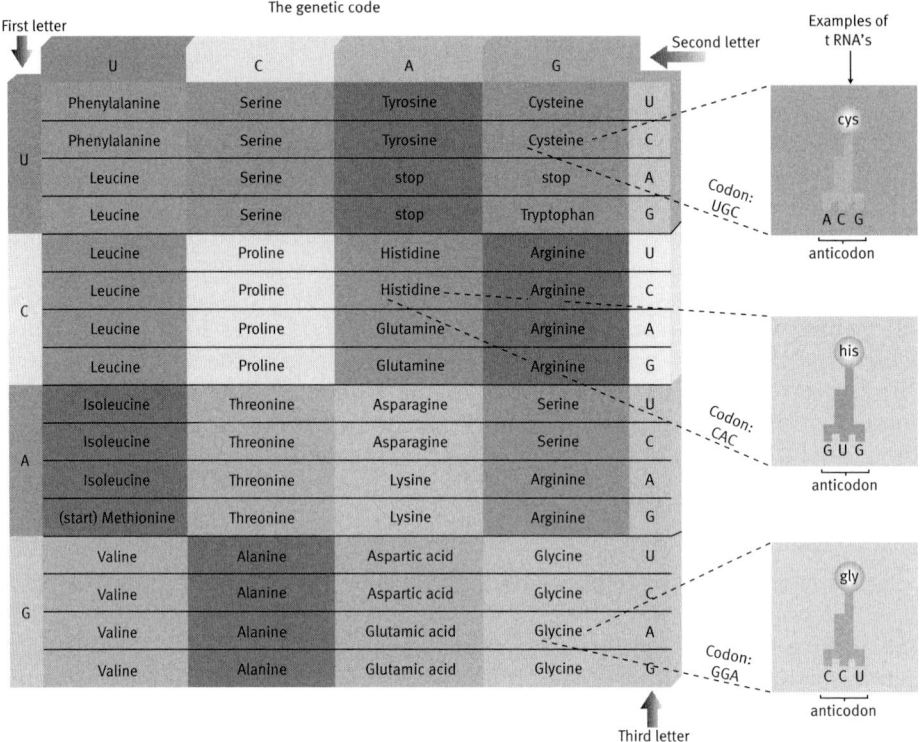

Figure 9.4. The genetic code. (Illustration from Trefil, J. and Hazen, R.M. (2001) *The Sciences—an Integrated Approach*. p. 520. © 2001 John Wiley & Sons, Inc.)

'activated' form, ready to form a covalent bond with the previous amino acid in the growing protein chain. This aminoacylation process consumes one mole of ATP per mole of amino acid (Figure 9.5).

The overall aminoacylation reaction is:

$$(tRNA^{Phe}) + phenylalanine + ATP \rightarrow phenylalanine\text{-}(tRNA^{Phe}) + AMP + PP_i$$

The details of the reaction are more complex than indicated. Note that the ATP forms AMP and PP_i (inorganic pyrophosphate) rather than the familiar ADP + P_i. Cells sometimes use this method of group transfer and the ΔG is similar.

How does the phenylalanine-tRNA 'know' its correct position on the mRNA waiting on the ribosome? This is where the genetic code performs one of its most important functions. On a region of the phenylalanine tRNA is a triplet sequence (AAG) that is complementary to the codon for phenylalanine on the mRNA (UUC). This triplet sequence is called an anticodon.

As a result of the specific pattern of hydrogen bonding between the UUC and AAG, the phenylalanine tRNA binds to the mRNA in the ribosome in exactly the correct position, the phenylalanine molecule is brought into close contact with the preceding amino acid in the growing protein chain (which is already bound

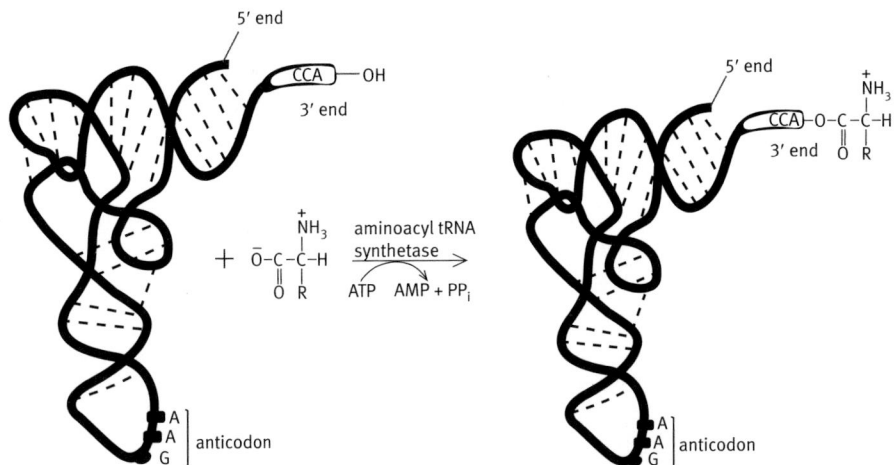

Figure 9.5. Structure of phenylalanine tRNA (tRNA^{Phe}) from *E. coli*, showing the tertiary folding and major hydrogen bonds (dashes). The anticodon for phenylalanine (AAG) is bottom left. Phenylalanine attaches to an adenosine at the 3′ end (top right) to form an aminoacyl tRNA, in readiness to add phenylalanine to the growing peptide chain being synthesized on the ribosome.

to the ribosome), and a covalent bond is formed, using the energy in the phenylalanine-(tRNA^{Phe}) acyl bond to drive the process via a coupling mechanism. The ribosome then moves along the mRNA, which becomes ready for the next amino acid tRNA to bind, and the process is repeated. When the final amino acid has been added, the completed protein chain is removed from the ribosome and is ready for further processing. This further processing may involve cleavage of parts of the chain, as in enzyme precursors, the addition of special groups, such as haem groups in haemoglobin, folding into a specific three-dimensional shape as in many enzymes, or the addition of sugar chains (glycosylation), as in most of the blood proteins. The processing stages also require the input of energy via ATP.

A very simplified scheme of eukaryotic protein synthesis is shown in Figure 9.6. Note the direction of each DNA or RNA chain. The directionality, 5′ to 3′ or 3′ to 5′, is important. The expanded region shows mRNA having codons for phenylalanine, threonine, and serine. These are translated on the ribosome and are shown as part of the completed product of gene 2. Processing of the pre-mRNA to remove introns has not been included.

We have now seen the fundamentals of the way in which DNA is replicated, a process essential to the reproduction of organisms, and the way it is transcribed and finally translated into protein. It is worth reiterating the importance of the recognition processes between a coding strand and mRNA, and between codon and its anticodon. It is largely the energy of the hydrogen bonds between the appropriate base pairs that allows this to happen with such precision. The functions of DNA are essential, so it is protected from change or degradation as much as possible. Each organism spends considerable amounts of energy on

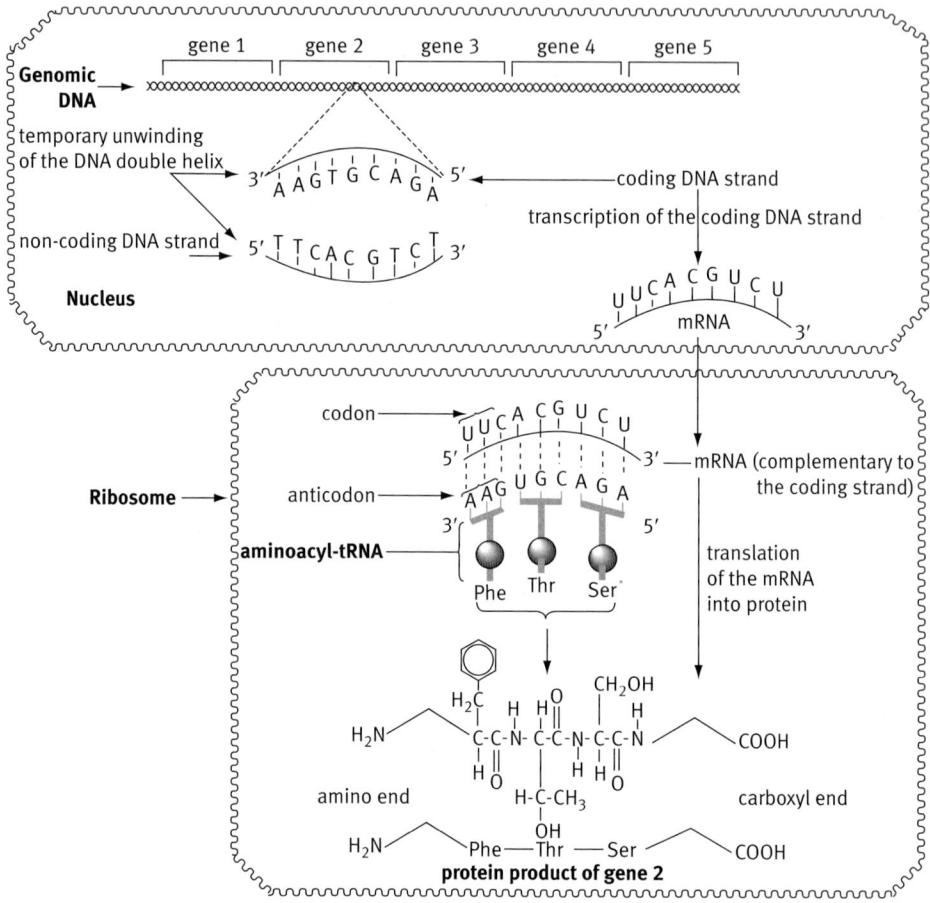

Figure 9.6. Protein synthesis in a eukaryotic cell. The nucleus provides a separate compartment for transcription. In eukaryotic cells, the original RNA transcript, called pre-mRNA, is processed in various ways before leaving the nucleus as mRNA. This is not shown. Translation occurs after the mRNA is transferred from the nucleus to the ribosomes.

DNA repair and on ensuring that the base sequence of its DNA is not altered. Such alterations lead to errors in replication, which of course affect the product of the gene to which the altered DNA belongs. Errors in hydrogen bonding that lead to the incorporation of the wrong base occur about once in every 10^4 to 10^5 base pairs and are called copying errors. In addition to copying errors, changes in the base sequence called mutations can occur. These may be caused by exposure to mutagens, agents which lead to mutations. Examples of mutagens include ultraviolet light, radioactivity (ionizing radiation), and certain chemicals. Other problems leading to genetic defects may be inherited. The copy of a 'defective' gene may be inherited from one or other parent, generation after generation. Well-documented examples include those of several European royal families, whose members suffered from an obvious disease such as haemophilia. Ultimately, however, such defective genes must have arisen from mutation(s) in an ancestral 'normal' or 'wild-type' gene.

Fortunately, proofreading and repair mechanisms have developed over evolutionary time and come into action so that replication errors overall are kept to an absolute minimum, normally occurring about once in 10^9 to 10^{10} base pairs. Despite these very efficient protective mechanisms, changes in DNA have always occurred and will always occur. Defects in DNA repair processes can be disastrous. The disease xeroderma pigmentosum causes the development of skin cancers at an early age because the sufferers do not have the repair mechanism to correct damage to skin-cell DNA caused by ultraviolet light. An enzyme that cuts out the damaged portion of DNA is believed to be missing. The cancer eventually spreads through the body and usually causes death.

Mutations which end up in the gametes of one generation will, as long as they are not so serious as to cause immediate death or infertility in the offspring, appear in the next generation of a population. As a result, mutations are the basis of much genetic variation (diversity) in populations, which may or may not be advantageous. Genetic variation may act to protect populations from extinction. There are many examples of this, although not all are favourable in the eyes of humans. Bacteria and other microbial organisms develop resistance to drugs, insects eventually resist pesticides such as DDT, and unwanted plants avoid extinction by developing resistance to herbicides. This is not good news, unless you happen to hold shares in a drug or agrochemical company.

Crops that are too homogeneous genetically are highly susceptible to pathogens, such as potatoes in the nineteenth-century Irish famine and sugar cane strain Q124, which was widely used in the Mackay region of Queensland, Australia during the year 2000 growing season. Both crops were devastated. More genetic variation would have allowed increased numbers of pathogen-resistant plants to survive and might perhaps have averted the disastrous results. Mutations are also responsible for some cancers, such as skin cancers, which can be caused by high-energy ultraviolet radiation penetrating the nuclei of surface skin cells and altering the precious sequence of their DNA. Thus, as with many things in biology, there are several facets to genetics, at least when looked at from our anthropocentric viewpoint. Evolution is driven by mutations. As mentioned in a previous chapter, with no mutations there would be essentially no evolution of new species. The living world as we know it would not exist and, probably, neither would we.

Over evolutionary time, mutations have been recorded in the genomes of all species. One spinoff of the recent upsurge in the studies of genomics is its application to so-called molecular evolution. The work of Cambridge biologist George Nuttall and others in the early twentieth century eventually led to acceptance of the important principle that the degree of similarity between genes reflects the closeness of the evolutionary relationship between them. That is, if a gene sequence for one type of organism is very close to that of the same gene in another organism, then the two organisms are likely to be closely related. Comparative studies of gene sequences have been used to assist the more classical means of classification of organisms. Such studies, now termed molecular systematics, have led to the restructuring of the phylogenetic tree, which until the 1970s divided cellular organisms into prokaryotes, which possess no discrete nucleus, and eukaryotes, which do. Karl Woese and his colleagues showed that there were in fact two distinct groups of prokaryotes, the Eubacteria (now called the Bacteria) and the Archaebacteria (now known as the Archaea). This work led to the recognition of three domains of cellular life—the Bacteria, the Archaea, and the Eukarya—and was based on sequence analysis of the 16S ribosomal RNA (rRNA). The three domains will be discussed further in Chapter 12.

The 16S rRNA, being essential to the synthesis of proteins in all organisms, is highly conserved. Being highly conserved means that it is so important a molecule that organisms with mutations in the 16S rRNA are unlikely to survive, with the result that very few mutations survive over evolutionary time in the gene that codes for this rRNA. This helped the researchers in that they had a simpler task in sorting out the implications of the relatively few differences in RNA sequence and led to the revelation of the two prokaryote groups quite early in the development of nucleic acid sequencing. The rate of DNA sequencing has developed greatly since the 1970s, with the complete genomes of several organisms from all three domains having been sequenced since 1995.

In contrast to that for the highly conserved 16S rRNA, other genes mutate at much higher rates. An example is that of the human immunodeficiency virus (HIV), which is the causal agent of AIDS. HIV evolves about a million times faster than the average human gene, which is one reason why the development of AIDS drugs and vaccines is such a problem.

A major feature of HIV evolution are the many and rapid changes in the carbohydrate/protein (glycoprotein) 'coat' of the virus, and these changes help HIV to stay ahead of developments designed to destroy it.

The long sequences of non-coding DNA that occur in many genomes (the badly named junk DNA) can 'tolerate' mutations to a much greater extent than can coding DNA. As the mutations cause no apparent harm, there will be little or no selective pressure to remove them and so such mutations accumulate in genomes, giving rise to a high degree of diversity in the junk regions. This diversity in the non-coding regions of the human genome is used in DNA fingerprinting or profiling to identify individuals, as the likelihood of two individuals having the same profile is extremely remote. The tests are often used for determination of paternity or for forensic purposes. As well as convicting a number of true criminals, DNA profiling has also led to the release of wrongly convicted persons, for whom the genetic triplet code and hydrogen bonding have come to the rescue. Having said that, recent legal challenges to DNA-based evidence have shown that the technique is not infallible.

Another area that is receiving benefits from gene sequencing was actually developed before DNA sequencing was technically feasible. This is based on protein sequencing. Historically, it happened that the techniques for the full sequencing of proteins were developed before those for full sequencing of nucleic acids. Incidentally, the same man was involved in both sequencing breakthroughs: Frederick Sanger from Cambridge University, who received two Nobel prizes, one for the development of each technique. As the amino acid sequence of a protein largely determines its three-dimensional structure and thus its function, it is therefore important to conserve the correct sequence.

The amino acid sequence of a protein is in turn determined by the messenger RNA that directed its synthesis, and as the mRNA sequence is determined by the DNA sequence of its gene, the protein sequence is maintained by the corresponding gene, as we have seen. Changes (mutations) in the gene DNA may be reflected in the amino acid sequence of the protein. As an example of all this, let's take the inherited disease sickle-cell anaemia. As we have seen earlier, haemoglobin is a protein in mammalian red blood cells that transports oxygen around the body. The way in which haemoglobin achieves this is known in great detail. Red blood cells are shaped rather like doughnuts, except that there is no central hole, and this shape is just perfect for the uptake and release of oxygen and several other simple, but important, chemicals, such as carbon dioxide and bicarbonate. Up to 20% of people in certain parts of East Africa have sickle-shaped red blood cells. These unfortunate individuals suffer from anaemia, as the misshapen red cells cause blocking of small blood vessels, thereby cutting off the circulation and causing tissue damage. The sickled cells are also more fragile and rupture easily, causing anaemia. What causes sickle-cell anaemia?

The answer lies in the amino acid sequence of sickle-cell haemoglobin (Hbs). It was shown experimentally that in position six of the two β-chains of haemoglobin (which is a tetramer of two α- and two β-chains) the normal glutamic acid residue had been replaced by a valine residue in the Hbs. Glutamic acid has a polar side-chain with acid properties, whilst valine has a non-polar hydrocarbon side-chain, which is quite hydrophobic in its properties. As a result, the Hbs crystallizes into long aggregates and in so doing pulls the red blood cell out of shape, causing the problem. This is another example of energy influencing molecular properties. Why did the Hbs crystallize? In general, and given the appropriate conditions, molecules may crystallize when their arrangement into crystals generates a lower energy state than the non-crystalline form. In this case, the deoxygenated form of the Hbs has hydrophobic pockets into which the valine side-chains of adjacent Hbs molecules pack nicely. The small interaction energy between the valine and the hydrophobic pockets lowers the overall energy level just enough to cause aggregation of a large number of Hbs molecules. The packing of Hbs molecules to form elongated crystals inside the red blood cell, and thus deform the cell wall, is an energetically favourable process, but it is unfortunate for the sufferers of sickle-cell anaemia. This is an example of the exquisite molecular detail in which it is now possible to study certain diseases.

What happened to change the glutamic acid to a valine in Hbs? If you look at the genetic code, you will see that the codons for glutamic acid are GAA and GAG, while two of those for valine are GUA and GUG. The substitution of U instead of A will change both the codons of glutamic acid to the two codons for valine, and apparently this is the mutation that occurred many years ago in the African population now afflicted with sickle-cell anaemia.

Molecular medicine is an apt term for studies on diseases such as these, and no doubt many more examples will emerge as a result of further work. At present more than 400 mutants of haemoglobin are known, most of which are single amino acid substitutions. Not all of these are harmful, and the availability to scientists of such a variety of structures has been useful in structure/function studies of proteins. The haemoglobin molecule, being so widespread in living organisms, has also been extremely useful in molecular evolution studies.

Before leaving sickle-cell anaemia, let's consider a rather puzzling aspect. If sickle-cell anaemia is so dangerous to health, how is it that the sickle-cell gene persists in those east-African populations? On the basis of natural selection, one might expect the defective gene to be eliminated, as the survival rate of the parts of the population with the Hbs gene would be less than the survival rate of others in the general population without it. The reason proposed is that resistance to a virulent form of malaria appears to be closely linked to the Hbs gene. Resistance to malaria has been a survival advantage in Africa, so presumably its genetic component has persisted, dragging the Hbs gene along with it. On balance, the two features have allowed the survival, at least until now, of the deleterious gene.

I will use this example to illustrate some of the possible complications of the emerging gene therapy. Even though we will soon have the complete sequence of all the DNA in the human genome, plus the exact locations of all the genes, there will still be an enormous amount to learn. We will need to discover more about the control of gene expression and about the ways in which genes and gene products interact. Who would have anticipated the protective effect of the malaria-resistance gene on the Hbs gene, if indeed this is the full explanation? Scientists are now aware of this and other examples, and as the data comes in new solutions will be found for the problems that are bound to emerge. There is an exciting future in store for those involved in gene therapy.

As a final illustration of the role of energy in molecular genetics, I want to link one of the most ancient examples of life on Earth to one of our most recent scientific developments.

As part of the processes of genetic engineering and genetic fingerprinting, it is often necessary to increase the quantity of a particular section of DNA. A small quantity of the DNA may be obtained from some natural source or from the scene of a crime. The quantity may not be enough to enable all the necessary experiments to be carried out.

How can more of the identical DNA be obtained? In principle, there are three possibilities. Firstly, there is chemical synthesis. If the exact base sequence of the section of DNA required is known, there are automated techniques available to carry the synthesis out. They are relatively slow, complex, and expensive. Secondly, there is cloning, where the required piece of DNA is inserted into the genome of a suitable organism such as a yeast or a bacterium, the organism is grown in culture, and the DNA is isolated and purified. This is also a complex and expensive process.

In the 1980s Kary B. Mullis developed what is known as the polymerase chain reaction (PCR), which greatly improved on the first two procedures for the amplification of sections of DNA (Figure 9.7). For this he was awarded the Nobel Prize for Chemistry in 1993, together with Michael Smith, who developed the technique of site-directed mutagenesis.

In PCR a segment of double-stranded DNA containing the base sequence of interest is heated to about 95°C to separate the two strands. Remember that the

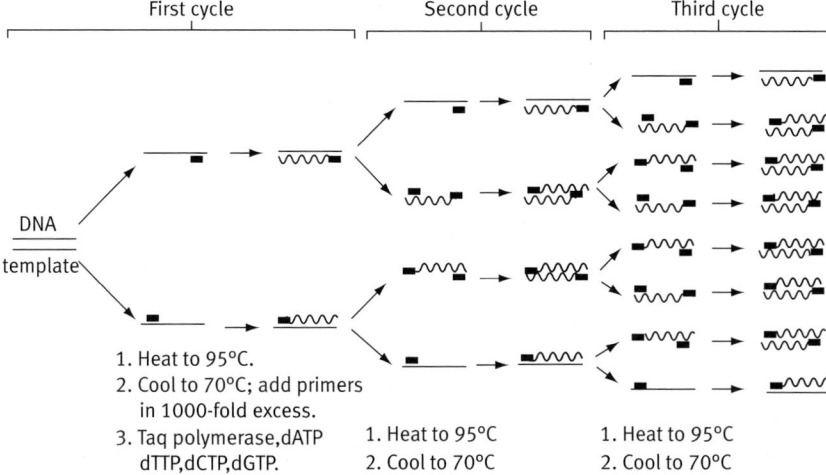

Figure 9.7. The polymerase chain reaction (PCR). Each cycle consists of three steps, as shown, though DNA, *Taq* polymerase, dATP, dTTP, dCTP, and dGTP are added only once. The small black rectangles represent DNA primer oligonucleotides.

two strands are held together by cooperative hydrogen bonds between A and T (two bonds) and G and C (three bonds). Heating to 95°C is sufficient to dissociate the hydrogen bonds without affecting the covalent bonds that maintain the primary structure of the DNA chain. The separation of the two strands is necessary because during PCR each strand is synthesized separately, in a kind of semi-conservative replication process similar to natural replication, except that both DNA strands are replicated simultaneously in PCR. A large excess (1000-fold) of short oligonucleotides, called primers, which have been chosen on the basis of their base-pair complementarity to the 3' end of the DNA sequence chosen for amplification, is then added and allowed to cool with the DNA, a process called annealing. During annealing, the short oligonucleotide primers line up, their As to the DNA's Ts and their Gs to the DNA's Cs, etc. and so in their specific places adhere by hydrogen-bonding to each separate strand of the DNA. After annealing, therefore, we have each strand of the section of DNA destined for amplification bounded by these short primers, which in effect act as markers for DNA polymerase to start and stop DNA synthesis. DNA polymerases require primers to add to. They cannot synthesize DNA from single nucleotides such as ATP, etc. Our DNA template system for the synthesis of new DNA is now ready.

Also present is a suitable DNA polymerase, an enzyme capable of synthesizing DNA in the presence of a DNA template containing suitable primers, and a supply of nucleotides (our A, T, G, and C bases). The DNA polymerase and the nucleotides, directed by the primed DNA template, go to work and after one cycle will have synthesized an extra copy of the double-stranded DNA of our choice, that is we have doubled the amount of our DNA in one step. The process of unwinding the two lots of double strands by heating, then annealing, and copying the four separated strands is repeated, so that after two cycles of PCR, we have four times our original amount of DNA. There is no need to add more primer as it is present in large excess, but what about the DNA polymerase?

We know that proteins in general, and enzymes in particular, are sensitive to heat and often become inactive after such treatment.

The DNA polymerase that was initially used in PCR was isolated from *Thermus aquaticus* (*Taq* DNA polymerase) and it remains active under the heating conditions necessary for PCR. *Thermus aquaticus* is one of the Archaea that live in hot springs, and the biotechnology industry is constantly searching for similarly useful organisms as they sometimes produce such enzymes as the *Taq* DNA polymerase. The inspired idea of Kary Mullis was to recognize that such enzymes would be able to survive the temperatures that are required to separate the DNA double helices formed in one cycle into separate strands, ready for the next DNA polymerase cycle.

The heating/cooling cycles described above for the amplification of DNA have been automated in equipment called thermal cyclers, which are available to molecular biologists. After about 25 thermal cycles, the amplification of our DNA will be about 1 million in practice, usually enough for most purposes, and will have taken just a few hours. Considering the amount of such PCR work now being undertaken, few would deny Kary Mullis his Nobel Prize.

The uses for PCR are several, and the number is increasing steadily. PCR can amplify minute amounts of DNA. By using appropriate primers, small amounts of bacteria and viruses can be detected in tissue samples. This has proven useful in diagnosis. Forensic scientists use PCR routinely, but their collection, handling, and PCR protocols must be extremely robust to avoid contaminating DNA, which can sometimes be a real problem, in the legal sense. Screening for human genetic diseases and amplifying archaeological samples are further useful applications.

The reason for my choice of the PCR to illustrate the application of ancient organisms, via thermal energy cycling, to modern science should by now have become apparent.

GENERAL REFERENCES

Campbell, N.A., Reece, J.B., Urry, L.A., Cain, M.L., Wasserman, S.A., Minorsky, P.V., *et al.* (2008) *Biology*, 8th edn. Pearson Benjamin Cummings, San Francisco.

Garrett, R.H. and Grisham, C.M. (2010) *Biochemistry*, 4th ednBrooks/Cole, Cengage Learning, Boston.

Voet, D., Voet, J.G., and Pratt, C.W. (2002) *Biochemistry*, upgrade edn. Wiley & Sons Inc., New York.

10

Electron Gymnastics: Energy Revisited

Here we begin to explore some of the fundamental processes of life: mechanisms of energy flow and transfer in cells. Current scientific research on the mechanisms of electron transfer and energetics is opening up possible alternatives to photosynthesis for energy production. In the near future, finding alternatives to fossil fuels will be essential if the world is to continue to increase energy consumption at current rates. Don't be surprised to see a reaction such as

$$2C_nH_{2n+2} + (3n+1)O_2 \longrightarrow (2n+2)H_2O + 2nCO_2$$

Replaced by

$$2H_2 + O_2 \longrightarrow 2H_2O$$

as a major supplier of transportation energy.

So far we have established that thermodynamics, the science of energy flows or fluxes, underlies and directs all real processes, including those concerned with living matter.

I stated earlier that to approach too close to equilibrium was to approach death. This is true not only for living organisms, but also for essentially any complex dynamic system, so long as we change 'death' to something like 'destruction' or 'oblivion'. If energy flow through the system ceases, the structure of the system will head towards equilibrium and oblivion. Consider some naturally occurring, non-biological phenomena that depend for their existence on energy fluxes: a cyclone (or hurricane, depending on where you come from), a large swell at sea, even a fire can be considered to be structures that will decay once the energy source keeping them in existence is dissipated.

The familiar, often destructive, atmospheric features are temporary structures of moderate complexity that have arisen spontaneously because of energy gradients set up due mainly to differences in atmospheric pressure. Under favourable circumstances, a hurricane will form (with a decrease in the local entropy) and its 'purpose' is to even out those energy gradients. For a while hurricanes have a very definite structure, but only so long as some energy gradients persist. After that, they die. The local atmosphere will go back to 'normal', to something approaching dynamic energy balance (but only crudely equivalent to equilibrium, as thermodynamically aware people know), with a commensurate increase in entropy. An ocean swell will become lower when strong winds across a long stretch of ocean abate; a fire will run out of fuel eventually.

Life on Earth can also be considered as a group of interacting complex structures, having as its 'purpose' the maintaining of its energy gradients in a manner advantageous to its survival. The point I am making is that energy has the capacity to flow through, and indeed to maintain, structures that are more complex than their surroundings, and to maintain the integrity of such structures. This capacity extends to life itself. This is no mere assertion on my behalf; the evidence is there in the form of yourself and every other living organism on Earth. More contentiously, some scientists believe that such capabilities of energy flow, given the right circumstances and the presence of some basic molecules, are sufficient to explain the origin of life itself (Morowitz and Smith 2007). Some even believe that life was an inevitable result of the properties and composition of the early Earth, although nothing has been proven to the satisfaction of all. Considering the staggering improbability involved, perhaps nothing ever will be.

Nature, via thermodynamics, works to reduce gradients of all types—gradients of temperature, pressure, concentration, gravitational potential, electrical potential, and chemical potential. Gradients are improbable, and improbable entities if left alone will spontaneously tend towards equilibrium. A characteristic of life is that living organisms must maintain many gradients for survival and must work (in both the scientific and the physical senses) to maintain them. Once an organism is no longer able to maintain its essential gradients, death is inevitable and often rapid, although equilibrium may still be some time off. Bodies take time to decompose, even with the help of a few of our microbial cousins, but eventually we alive today will go the way of all flesh.

One mechanism of energy transformation we have yet to discuss is that of electron transfer, which must be dealt with before we can discuss the ways in which organisms, over evolutionary time, have 'solved' the problems imposed by 'demands' for increased complexity. There seems to be no predetermined drive for complexity, but if it confers a survival advantage, increased complexity is likely to occur provided there is a viable mechanism.

Electron transfer is an extension of what we have said about chemical reactions. Although not explicitly shown or acknowledged, in many reactions electron transfer is the underlying process taking place. Most of us are familiar with the concept of a flow of electrons in everyday electrical appliances. Electrons will flow from higher energy intensities (higher voltage, or potential) to lower ones, and are capable of doing work in the process. The maximum potential difference that can be supplied by different chemical systems varies depending on the chemicals involved. Electron transfer reactions must involve an electron donor and an electron acceptor, and are known to chemists as redox reactions, or redox systems. The term redox comes from *red*uction and *ox*idation. The substances involved in biological redox systems are capable of switching readily between their reduced and oxidized forms. The chemical species that loses (donates) electrons during the process is said to be oxidized and the species that receives (accepts) electrons is said to be reduced. These two processes must always occur together, linked by the transfer of one or more electrons between donor and acceptor. Written to show that when the electron-transfer process occurs spontaneously, energy is produced:

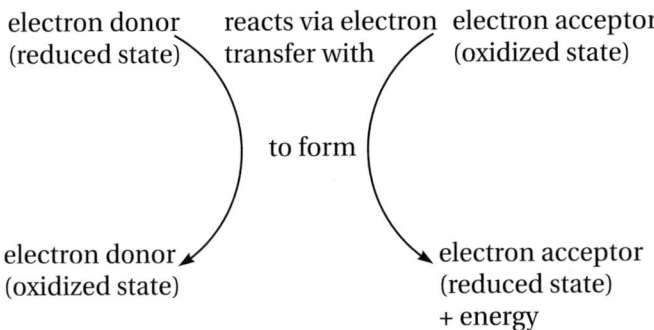

The above conventional way of depicting a redox system is convenient, especially when multiple redox systems are coupled, as in some of the biological processes discussed later in this chapter.

Box 10.1 Standard reduction potentials

As with electrical flow in a wire, we speak of different redox systems in terms of the difference in electrical potential between the oxidized and reduced states. This is the redox potential of a particular redox pair, which is also measured in volts. As with all chemical energetics, there is a set of standard conditions specified, leading to a series of standard reduction potentials. Standard reduction potentials, designated E^0, are measured relative to the redox potential of hydrogen, which is taken to be zero volts (V).

The half-reaction for the reduction of hydrogen is:

$$2H^+ + 2e^- \rightleftharpoons H_2 \quad E^0 = 0.0 \text{ V} \tag{10.B1}$$

Thus redox potentials are a measure of the affinity of a substance for electrons relative to that of hydrogen. A positive redox potential means that a substance has a greater affinity for electrons than hydrogen does, that is it will accept electrons from H_2 (which will end up as H^+). A negative redox potential, on the other hand, implies that the substance will donate electrons to $2H^+$, to form H_2. By convention, tables of standard reduction potentials record the reduction potential for the half-reactions that accept electrons. For a redox reaction to occur, both a reductive half-reaction and an oxidative half-reaction must be coupled in some way. A simple experiment to demonstrate a full redox reaction visually is to dip a strip of zinc metal into an aqueous solution of copper sulphate. The zinc soon becomes covered with a pink layer of deposited

Box 10.1 (continued)

copper and eventually can be completely eroded away. The overall equation is:

$$Zn(s) + Cu^{2+}(aq) \longrightarrow Zn^{2+}(aq) + Cu(s) \quad (10.B2)$$

The two half-reactions show the electrons explicitly. E^0 values below were obtained from tables.

oxidation step: $Zn \longrightarrow Zn^{2+} + 2e^- \quad E^0 = 0.76\,V \quad (10.B3)$

reduction step: $Cu^{2+} + 2e^- \longrightarrow Cu \quad E^0 = 0.16\,V \quad (10.B4)$

Add (10.3) and (10.4): $Zn + Cu^{2+} \longrightarrow Zn^{2+} + Cu$

$$E^0 = 0.92\,V \quad (10.B2)$$

Reaction (10.B2) is spontaneous with a standard reaction potential, E^0, of 0.92 V. As the reaction is spontaneous, ΔG will be negative. The reaction can do work. The work of redox reactions, just as that of group transfer reactions, can be harnessed by biological systems to do biological work. To calculate the total voltage for a redox reaction we use the expression:

$$\Delta E^0 = E^0_{(electron\ acceptor)} - E^0_{(electron\ donor)} \quad (10.B5)$$

This is exactly the process we carried out with copper and zinc above. The advantage in using expression (10.B5) is that we don't need to worry about changing the sign of the redox potential for the electron donor. Recall that to write the equation for a reaction that proceeds spontaneously as written, the overall voltage must be positive. Applying expression (10.B5) to the Cu/Zn example:

$$\Delta E^0 = E^0_{Cu} - E^0_{Zn}$$
$$= 0.16\,V - (-0.76)\,V$$
$$= 0.92\,V$$

Another advantage in using expression (10.B5) is that it will indicate whether or not the redox reaction as written will actually proceed. If expression (10.B5) is used correctly, and the resulting reaction is calculated to have a positive voltage, the reaction will proceed spontaneously. If the voltage you calculate turns out to be negative, the reaction will proceed in the opposite direction.

Table 10.B1. Some standard reduction potentials.

	$E^{0\prime}$ (volts)
$\frac{1}{2}O_2 + 2H^+ + 2e^- \rightleftharpoons H_2O$	0.815
Cytochrome a_3 (Fe^{3+}) + $e^- \rightleftharpoons$ cytochrome a_3 (Fe^{2+})	0.385
$O_2(g) + 2H^+ + 2e^- \rightleftharpoons H_2O_2$	0.295
Cytochrome a (Fe^{3+}) + $e^- \rightleftharpoons$ cytochrome a (Fe^{2+})	0.29

Box 10.1 (continued)

Table 10.B1. (continued)

	$E^{0\prime}$ (volts)
Cytochrome c (Fe^{3+}) + e^- ⇌ cytochrome c (Fe^{2+})	0.235
Cytochrome c_1 (Fe^{3+}) + e^- ⇌ cytochrome c_1 (Fe^{2+})	0.22
Cytochrome b (Fe^{3+}) + e^- ⇌ cytochrome b (Fe^{2+})	0.077
Ubiquinone + 2 H^+ + 2 e^- ⇌ ubiquinol	0.045
Fumarate$^-$ + 2 H^- + 2 e^- ⇌ succinate$^-$	0.031
FAD + 2 H^+ + 2 e^- ⇌ $FADH_2$ (*flavoproteins*)	~0.
Oxaloacetate$^-$ + 2 H^+ + 2 e^- ⇌ malate$^-$	−0.166
Pyruvate$^-$ + 2 H^+ + 2 e^- ⇌ lactate	−0.185
Acetaldehyde + 2 H^+ + 2 e^- ⇌ ethanol	−0.197
NAD^+ + H^+ + 2 e^- ⇌ NADH	−0.315
$NADP^+$ + H^+ + 2 e^- ⇌ NADPH	−0.320
H^+ + e^- ⇌ $\frac{1}{2}H_2$	−0.421
Acetate$^-$ + 3 H^+ + 2 e^- ⇌ acetaldehyde + H_2O	−0.581

Data from Loach, P.A., in Emerson, G.D. (ed.) (1976) *Handbook of Biochemistry and Molecular Biology*, 3rd edn, Physical and Chemical Data, Vol. 1, pp. 123–131. CRC Press.

With the exceptions of Cu and Zn, these reduction potentials are $E^{0\prime}$ values. This means that they have been adjusted from the standard E^0 values of thermodynamics tables in a similar way to that in which ΔG^0 values were adjusted to $\Delta G^{0\prime}$ to make them compatible with biological conditions, namely pH = 7, 1 M concentration, 25°C.

A redox reaction will proceed spontaneously as written when its redox potential is positive, although by convention the ΔG is negative for a spontaneous process. The anomaly arises from the convention that defines redox potentials. Such anomalies require one to remember many special cases and are the bane of scientist and student alike. The main point is that, whether expressed as a redox potential or in the form of a ΔG, the amount of energy involved is the same. The conversion factor between standard redox potentials and $\Delta G^{0\prime}$ values is

$$\Delta G^{0\prime} = -nF\Delta E^{0\prime} \tag{10.B6}$$

where n is the number of moles of electrons transferred, F is Faraday's constant (96.485 kJ V^{-1} mol^{-1}) and $\Delta E^{0\prime}$ is the total voltage for the two half-reactions. To calculate the total voltage for a biological redox reaction, we use the expression:

$$\Delta E^{0\prime} = E^{0\prime}_{(electron\ acceptor)} - E^{0\prime}_{(electron\ donor)} \tag{10.B7}$$

A coupled system consisting of electron carriers 1, 2, 3, and 4 is shown below. Note that as the coupled redox processes occur, each successive redox state is lower in energy than its predecessor's state. Each redox process has its own ΔG. If suitably coupled, some of the energy liberated at each step in such a redox system can be utilized to do useful work.

228 Introducing Biological Energetics

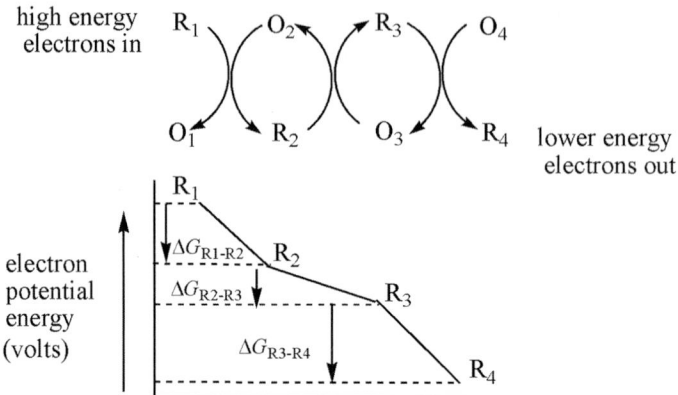

In order for such a system to continue working, each oxidized species must be re-reduced by supplying it with further electrons. Thus, O_1 must receive electrons to form R_1, which passes these on to reduce O_2 to R_2 and so on down the chain. In living systems such a process is run by channelling some of the high-energy electrons produced during metabolism into the redox system, often via the electron donor NADH. In a biological system, the electrons from R_4 would be passed on to another electron acceptor, as discussed below.

Let's look at a biological example. We saw in Chapter 8 the involvement of the coenzyme NAD^+ and its reduced form $NADH + H^+$ in the oxidation of lactate to pyruvate. The redox pair $NAD^+/NADH + H^+$ constitutes a half-reaction system that is involved in many biological reactions. Showing the electrons explicitly indicates clearly that a redox reaction is involved. For the half-reaction:

$$NADH + H^+ \rightleftharpoons NAD^+ + 2H^+ + 2e^- \quad E^{0\prime} = 0.32 \text{ V} \quad (10.1)$$

In half-reaction (10.1), the change in redox potential going from left to right ($\Delta F^{0\prime}$) is positive, so $\Delta G^{0\prime}$ will be negative. Using expression (10.B6) to convert from volts to our more familiar kilojoules:

$$\Delta G^{0\prime} = -(2)(96.485)(0.32)$$
$$= -61.75 \text{ kJmol}^{-1}$$

Thus, suitably coupled, half-reaction (10.1) is capable of driving an endothermic reaction, as we will see below.

Let us consider synthesis of one of the most important 'high-energy' compounds in biology, ATP. When ATP is used to provide energy in a living cell, it usually ends up as ADP and inorganic phosphate, designated P_i. Thus:

$$ATP + H_2O \longrightarrow ADP + P_i + H^+ \quad \Delta G^{0\prime} = -30.5 \text{ kJ mol}^{-1} \quad (10.2)$$

As discussed previously, coupling to an endergonic reaction is carried out, rather than the simple hydrolysis of ATP implied by eqn (10.2). The source of energy being exploited is the group transfer potential of the phosphate group. As ATP is the principal source of immediately available energy in living cells, it needs to be continually replenished. There are several ways by which this can be achieved. ATP synthesis from ADP is the reverse of eqn (10.2):

$$ADP + P_i + H^+ \longrightarrow ATP + H_2O \qquad \Delta G^{0\prime} = +30.5 \text{ kJ mol}^{-1} \qquad (10.3)$$

This is clearly endothermic and will not proceed spontaneously under biological conditions. If, however, the synthesis of ATP could be coupled to the oxidation of NADH by oxygen, for example:

$$NADH + H^+ + \tfrac{1}{2}O_2 \rightleftharpoons NAD^+ + H_2O \qquad \Delta G^{0\prime} = -220 \text{ kJ mol}^{-1} \qquad (10.4)$$

$$\Delta E^{0\prime} = +1.14 \text{ V}$$

There would be more than enough Gibbs energy to drive the synthesis of ATP against the 30.5 kJ required. This is in fact what happens overall in mitochondria. Biologically, the coupling of these two systems does not occur in a single step. Electrons are not transferred directly from NADH to oxygen, as you might think just by looking at eqn (10.4). They are transferred to oxygen along a series of electron carriers called the electron transport chain (ETC) or respiratory chain. Let us look at this process.

During the flow of electrons along the ETC some of their energy is 'extracted' to drive the synthesis of ATP from ADP, as outlined above. As the electrons flow along the ETC and carry out this work, their energy level (potential) decreases progressively. The various steps are now known in considerable detail. Because phosphate is added to ADP and oxygen is involved as the ultimate electron acceptor, formation of ATP by this pathway is called oxidative phosphorylation. It is easily the major source of ATP in the cell. Electron transport is usually tightly coupled to ATP synthesis, that is electrons do not flow along the ETC unless ADP is simultaneously phosphorylated to ATP. As ADP concentrations fall, electron transport slows, in a process known as respiratory control. This ensures that electron flow occurs when ATP is required.

Where does all this happen? In eukaryotic cells the site of oxidative phosphorylation is usually the inner membrane of the mitochondrion, while in prokaryotes the components are located in the plasma membrane (Figure 10.1).

How does it all happen? The processes I shall describe are inherently complex, and there is a limited amount I can do to simplify the description without losing something vital.

The ETC consists of four large protein complexes and two small electron carriers that link them, all embedded in the inner mitochondrial membrane.

The large complexes, NADH dehydrogenase, succinate dehydrogenase, cytochrome bc_1 complex, and cytochrome oxidase, each contain several electron carriers. The two small electron carriers are ubiquinone (or coenzyme Q) and cytochrome c.

Figure 10.2 shows electrons flowing from NADH to oxygen through the ETC. Electrons from the conversion of succinate to fumarate in the TCA cycle enter via complex II. Where do these electrons come from? The NADH arises from metabolic processes, especially from glycolysis, the TCA cycle and fatty acid oxidation. These important systems will be dealt with in Chapter 11.

230 Introducing Biological Energetics

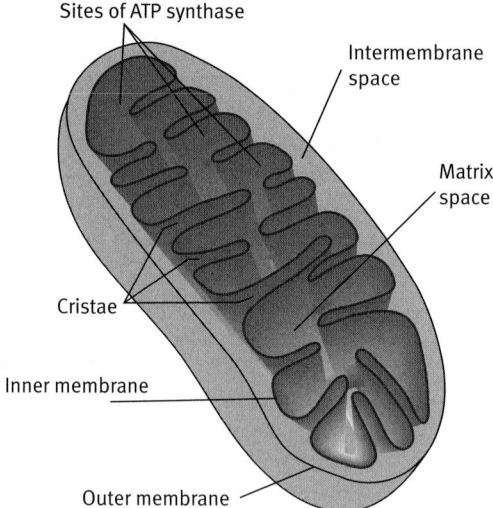

Figure 10.1. Cutaway diagram of the mitochondrion.

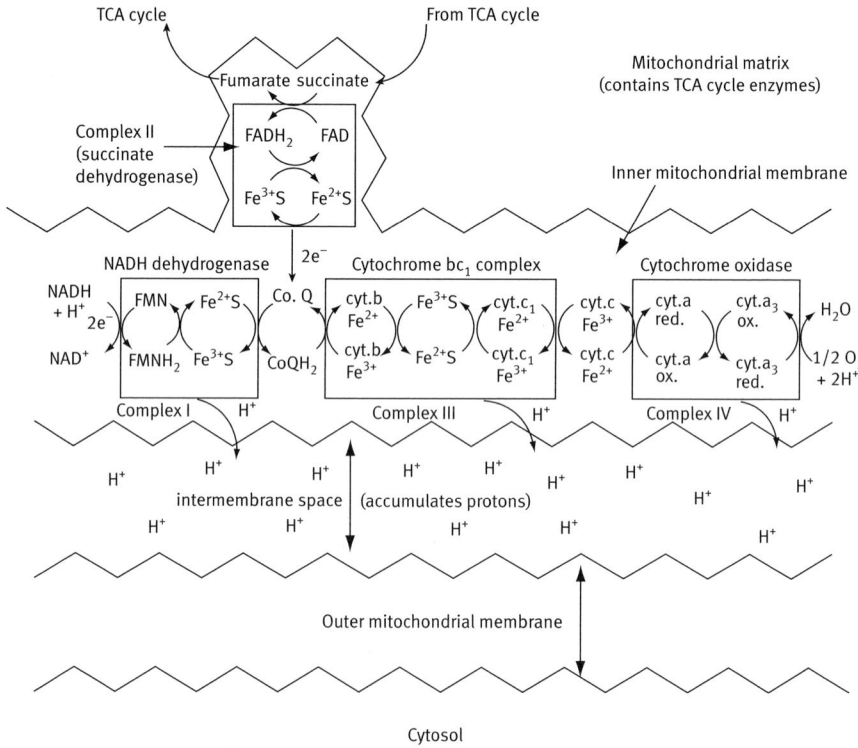

Figure 10.2. The electron transport chain (ETC) of a mitochondrion. Note the four complexes. Each consists of a number of electron carriers—complex proteins plus their prosthetic groups. The prosthetic groups alternate between their oxidized and reduced forms as electrons move along the ETC towards the final electron acceptor, oxygen. Protons are pumped into the intermembrane space at the complexes I, III, and IV as shown.

It is instructive to consider the detail of what is happening when NADH (or FADH$_2$) feeds electrons into the ETC.

$$\text{NADH} + \text{H}^+ \rightleftarrows \text{NAD}^+ + 2\text{H}^+ + 2\text{e}^- \quad \Delta E^{0\prime} = 0.32 \text{ V} \quad (10.5)$$

Each NADH donates two electrons to the ETC, and two protons are evolved. Each electron carrier in the ETC interacts with its neighbours according to its redox potential. At each electron carrier pair, the accepting carrier has a higher affinity for electrons than the donating carrier, and as a result there is a net flow of electrons from NADH (low electron affinity) to oxygen (high electron affinity). Each molecule of oxygen actually needs to accept four electrons to be reduced to water, H$_2$O. Each NADH or FADH$_2$ donates only two, and the electron carriers with metal prosthetic groups can cope with only one electron at a time. To manage the different electron-handling capacities, the ETC components are arranged so all work in harmony. The complete process is not a simple linear sequence as shown in Figure 10.2. Information on exactly how the entire process works is incomplete. Many individual steps are known from studies of the isolated complexes I–IV. Other important components of the ETC include the flavoproteins, which possess the tightly bound prosthetic groups FAD and FMN. These groups are involved in one- or two-electron transfer. Coenzyme Q (CoQ) is also known as ubiquinone. CoQ, which also can transfer one or two electrons as required, is mobile within the membrane and transfers electrons from complex I or complex II to complex III. Complexes I and III contain iron–sulphur proteins (Fe^{2+}S/Fe^{3+}S in Figure 10.2), which carry out one-electron transfers. Cytochromes, abbreviated cyt., are proteins that contain haem groups in which electron transfer involves the Fe^{3+}/Fe^{2+} redox pair. Cytochrome oxidase also involves a Cu^{2+}/Cu$^+$ redox pair. Cytochromes are involved in one-electron transfers only. Cytochrome *c* is not membrane bound. Being water soluble, it is mobile in the intermembrane space and transfers electrons between complex III and complex IV.

I won't describe in detail all of the electron-transfer steps involved, but as an example let us take the final step in the ETC where electrons are transferred from reduced cytochrome *c* (cyt.c Fe^{2+} in Figure 10.2) via the cytochrome oxidase complex to oxygen to form water. Overall:

$$4(\text{cyt.c Fe}^{2+}) + 4\text{H}^+ + \text{O}_2 \xrightarrow{\text{cytochrome oxidase}} 4(\text{cyt.c Fe}^{3+}) + 2\text{H}_2\text{O} \quad (10.6)$$

The electrons being transferred are not shown explicitly. Omitting the cytochrome *c* as the source of electrons but showing them explicitly:

$$4\text{e}^- + 4\text{H}^+ + \text{O}_2 \rightleftarrows 2\text{H}_2\text{O} \quad (10.7)$$

There is a change in redox potential as we progress along the ETC and this is an exact measure of the Gibbs energy changes occurring at each stage (Figure 10.3). All the redox potential changes have now been determined for the mitochondria in a number of organisms. Consensus values for a number of mammalian mitochondria are sometimes used in texts as examples (Garrett and Grisham 2010 p. 597; Voet *et al.* 2002 p. 499).

232 Introducing Biological Energetics

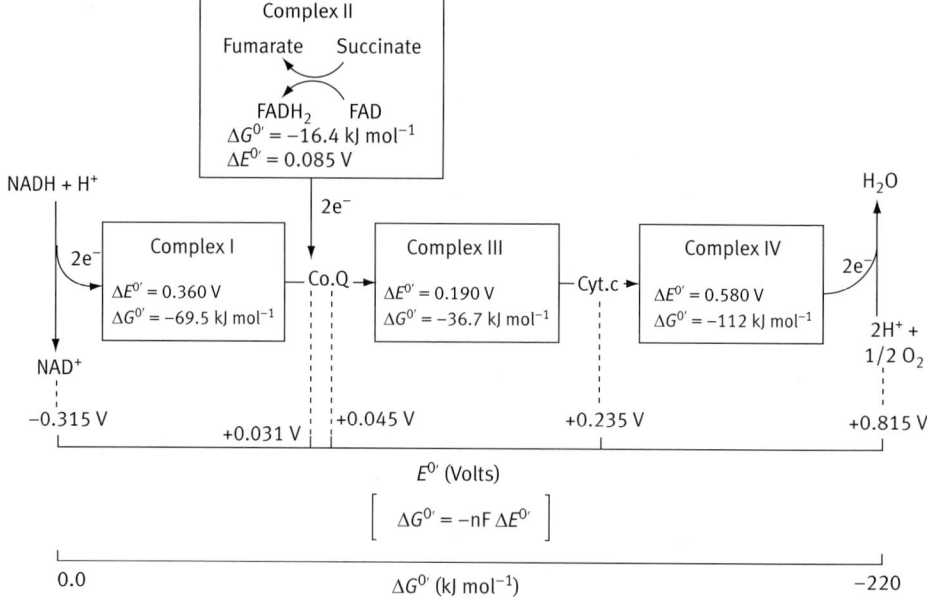

Overall reaction for the NADH reduction of oxygen to form water:

$$NADH + H^+ + 1/2\, O_2 \rightleftharpoons NAD^+ + H_2O \qquad \Delta G^{0\prime} = -220 \text{ kJ mol}^{-1}$$
$$\Delta E^{0\prime} = +0.815 - (-0.315) = 1.130 \text{ V}$$

Figure 10.3. The energetics of the ETC in mitochondria. Note the energy fall, indicated by $\Delta E^{0\prime}$ and $\Delta G^{0\prime}$, across each complex. Much of this energy is preserved in setting up the proton gradient that is used to drive ATP synthesis via the ATP synthase. Complex II does not pump protons (Figure 10.2) as its $\Delta G^{0\prime}$ is only -16.4 kJ mol^{-1}. Complex II functions to inject electrons from FADH$_2$ into the ETC via CoQ.

The potential falls as electrons pass along the chain, the largest falls occurring in the steps involving the large complexes, reflecting loss of energy. As the electrons pass through each large complex, some energy is captured and used to pump protons (H$^+$) from the inside of the mitochondrion across the inner mitochondrial membrane into the space between the inner and outer membranes. This creates a large proton gradient across the inner membrane. Thermodynamics acts to remove gradients, and as we shall see, the potential energy generated by the formation of the proton gradient is used to drive the synthesis of ATP during oxidative phosphorylation. The free energy change required to drive the proton pumps is great enough only across the three large complexes, as indicated in Figure 10.2.

Figure 10.2 also shows that electrons may enter the ETC via FADH$_2$, which feeds them in via CoQ, leaving the oxidized form FAD. There are several enzymes which use the FADH$_2$/FAD redox system as a coenzyme in a manner similar to the NADH/NAD$^+$ system. One important source of FADH$_2$ for supply of electrons to the ETC is succinate dehydrogenase, an enzyme involved in the TCA cycle, a major metabolic pathway in the oxidation of glucose which will be mentioned again later. We can thus see another way in which the oxidation of glucose, the most ubiquitous of the simple sugars, is linked to the production of ATP.

So far we have used high-potential electrons and oxygen to produce water and a proton gradient, but no ATP. What next?

Bound to the inner mitochondrial membrane is what has been described as the smallest rotary engine known. This is the enzyme complex ATP synthase. The protons pumped from inside to the outside of the inner membrane create an electrical potential, or electrochemical gradient, across the membrane. The free energy change generated by transferring an electrically charged proton across a membrane is determined by its concentration plus its charge. Thus, the total free energy available from such a system derives from (a) the chemical gradient of H^+ ions (a concentration effect) plus (b) the membrane potential (the electrical charge potential across the membrane) and is called the protonmotive force. This can be expressed quantitatively.

The free energy required to pump protons across a membrane from a lower (inside) to higher (outside) proton concentration is given by:

$$\Delta G = RT \ln \frac{[H^+_{outside}]}{[H^+_{inside}]} + ZF\Delta\Psi \qquad (10.8)$$

or

$$\Delta G = 2.303\, RT\, [\text{pH(outside)} - \text{pH(inside)}] + ZF\Delta\psi \qquad (10.9)$$

where Z is the charge on the proton, F is the Faraday constant, and $\Delta\psi$ is the membrane potential. Values reported in the literature for the difference in pH and for $\Delta\psi$ vary between about 0.75 and 1.0 pH units, and 0.168 and 0.18 volts, respectively (Garrett and Grisham 2010; Voet et al. 2002). Using the two extremes of these values, eqn (10.7) can be used to show that about 22–24 kJ mol^{-1} are required to form the electrochemical gradient. This means that the gradient can supply 22–24 kJ mol^{-1} of protons when they flow back through the ATP synthase, but one mole of protons cannot provide the 50 kJ required to synthesize 1 mole of ATP. About three protons must flow to produce one ATP molecule.

10.1 How many ATP molecules are produced when electrons traverse the entire ETC?

Recent measurements indicate that about 2.5 moles of ATP are produced per mole of NADH. Starting from FADH$_2$ the electrons only pass through two of the three large complexes, and produce about 1.5 moles of ATP by this route (Garrett and Grisham 2010; Hames and Hooper 2000). The actual rate of oxidative phosphorylation is determined by the availability of ADP. If ADP levels are low, ATP levels in the cell will be high, so oxidative phosphorylation is not so necessary, and the ETC is inhibited. If the ETP is inhibited, NADH, FADH$_2$,

234 Introducing Biological Energetics

and citrate concentrations build up. This slows down the citric acid cycle and another important metabolic pathway, glycolysis. The feedback controls involved in regulating the TCA and glycolysis are known in detail and are discussed in Chapter 11.

Direct regulation of the ETC is proposed to occur at cytochrome oxidase, the last step. It is an irreversible step. Its $\Delta G^{0\prime}$ is large and negative (-112 kJ mol^{-1} (Figure 10.3)). Reduced cytochrome c is coupled to and essentially in equilibrium with the rest of the oxidative phosphorylation process, so the concentration of reduced cytochrome c depends on the concentration ratios of [NADH]/[NAD$^+$] and [ATP]/[ADP][P$_i$] inside the mitochondrion. If [NADH]/[NAD$^+$] is high and [ATP]/[ADP][P$_i$] is low, the concentration of reduced cytochrome c will be high, and cytochrome oxidase plus the ETC will be active. This is the case in exercising muscle tissue. When muscle is at rest, the ratios are reversed and ATP synthesis slows dramatically.

Box 10.2 How much ATP do humans require daily?

Suppose a 70 kg adult male consumes 12,000 kJ per day. Assuming a conversion to ATP of 50% and a yield of 50 kJ mol^{-1} for ATP under cellular conditions (Garrett and Grisham 2010) the body will utilize $6,000/50 = 120$ moles of ATP. Taking the molecular weight of ATP as 550, this converts to $550 \times 120 = 65,000$ g or 65 kg of ATP per day. In reality, a 70-kg human possesses about 50 g of ATP, so a great deal of ATP recycling must occur and in a controlled manner as the demand is not constant throughout the day. Overall, mitochondria may range in number from one to thousands per cell (Campbell et al. 2008). A typical eukaryotic cell contains 1000–2000 mitochondria, which occupy some 20% of the cell volume (Voet et al. 2002).

10.2 What about the rotary engine?

The ATP synthase is visible under the microscope as almost spherical projections from the inner mitochondrial membrane (Figure 10.4).

During ATP synthesis, one subunit of ATP synthase rotates relative to another, thus constituting a small rotary engine. Two large subunits can be isolated. The F_1 subunit is roughly spherical and is the one seen protruding into the mitochondrial matrix. The F_0 subunit sits within the membrane structure. It is made up of a number of proteins arranged into the c ring. The F_1 and F_0 subunits are joined by a stem consisting of one γ, one δ, and one ε subunit. The F_1 subunit is held in place by the b and OSCP subunits. When the F_0 subunit rotates, powered by the flow of protons through the c ring via the a subunit, the γ-stem rotates with it. The γ-stem protrudes into the F_1 subunit and rotates within it. This rotation causes conformational changes within the α- and β-proteins of the F_1 unit. The conformational changes initiate the synthesis of ATP. ADP and P$_i$ must be brought into position in a non-aqueous environment that is favourable to the

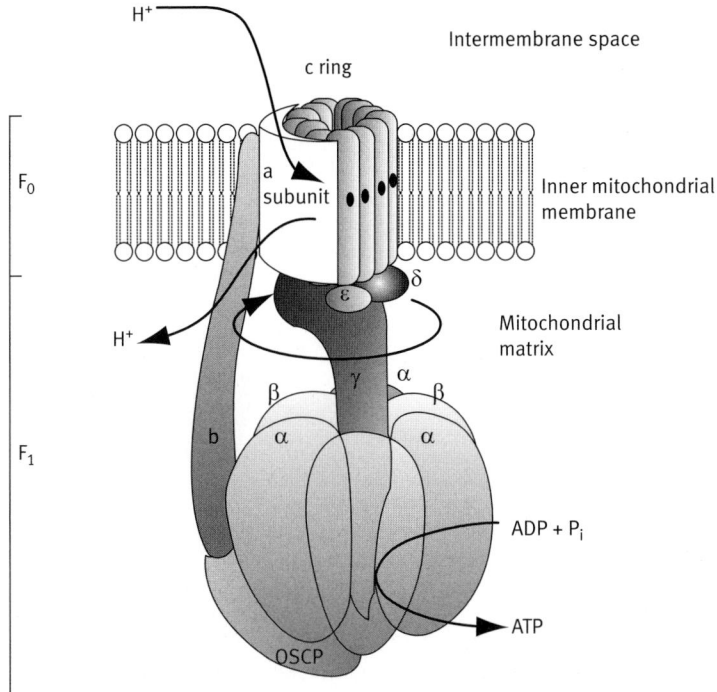

Figure 10.4. The mitochondrial ATP synthase, which appears under high magnification as a spherical shape protruding from the inner mitochondrial membrane into the matrix. Detail of the molecular motor structure is shown. (Illustration from Nicholls, D.G. and Ferguson, S.J. (2000) *Bioenergetics 3*. Academic Press, London, p. 200.

formation of ATP and water. Subsequently, the ATP and water must be released into the environment. There are three pairs of alternating α- and β-proteins in the F_1 unit that are involved in ATP synthesis. These ATP synthesis steps are shown schematically in Figure 10.5 in what is called the three-site alternating binding site mechanism (Nicholls and Ferguson 2002).

ATP synthase is thus a triple-headed enzyme, and the active sites presumably employ the usual enzymic principles: exclusion of water, alignment of the substrates, and stabilization of the transition state. In a sense, the rotating γ unit is analogous to an allosteric modulator that changes the conformation and activity of the active sites.

Some further properties of mitochondria are worthy of mention. Most of us are aware that cyanide (CN^-), carbon monoxide (CO), and azide (N_3^-) are poisons that can cause rapid death. Azide is used widely by biochemists to inhibit microbial growth in protein solutions. These chemicals strongly inhibit cytochrome oxidase in complex IV, thus preventing the running of the ETC, and the victim essentially suffocates. Barbiturates such as Amytal (amobarbital) and the painkiller Demerol (meperidine) inhibit complex I. These and other inhibitors were helpful in early studies on the sequence of events in the ETC. Some chemicals, such as 2,4-dinitrophenol and dicoumarol, uncouple electron transport

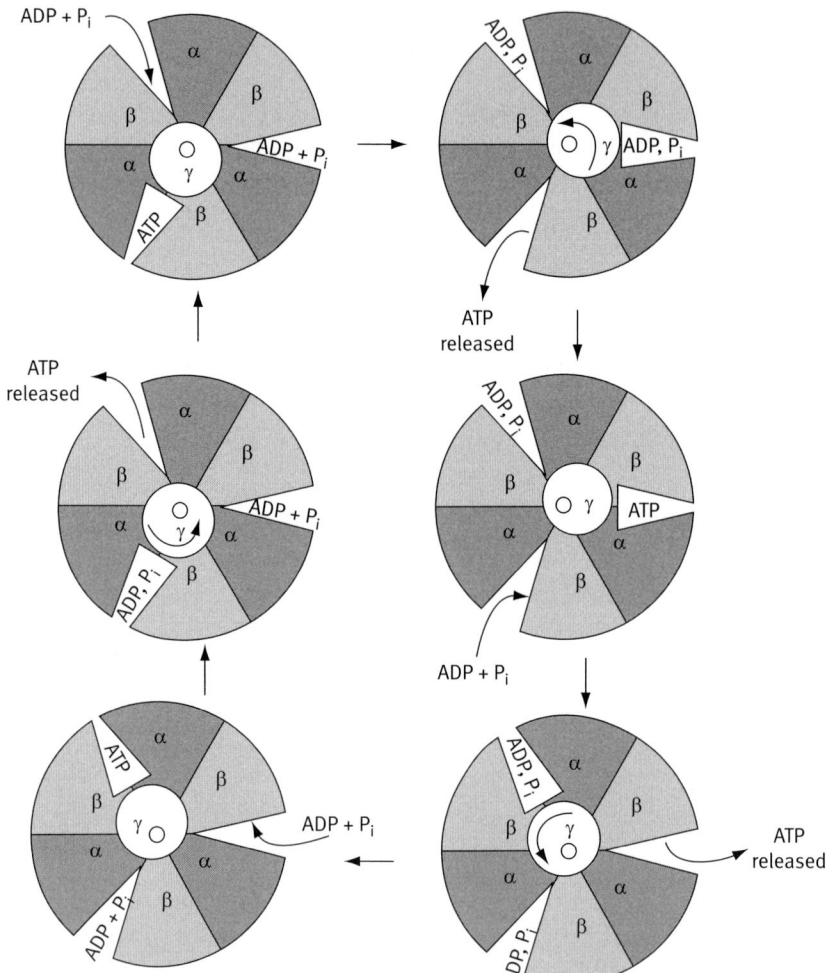

Figure 10.5. The three-site alternating binding site mechanism of ATP synthase. Looking from below F_1 towards the inner membrane, the central γ-stem rotates anticlockwise. The α- and β-chains are held stationary by subunit b and OSCP (Figure 10.4). From top left: ADP and P_i bind to an empty site, with ADP and P_i already bound to a second site, and ATP is tightly bound to the third site. Rotation of the γ-stem as shown causes a conformational change such that ATP is released and its empty site changes conformation so that it can now bind fresh ADP and P_i. Simultaneously, the original ADP and P_i become bound, and the bound ADP and P_i form ATP. Thus in the middle diagram on the right-hand side, the structure has returned to the original (top left) conformation, except that the three conformations have effectively 'migrated' around the ring of α- and β- units. Repeating these steps twice releases two more ATP molecules and returns the whole structure to the original top-left position. (Illustration from Nicholls, D.G. and Ferguson, S.J. (2002) *Bioenergetics 3*. Academic Press, London, p. 209.)

from ATP synthesis. Both compounds are hydrophobic (membrane compatible) and can donate/accept a proton, so can carry protons across the inner membrane, bypassing the ATP synthase. Normally, the ETC is active only when ATP is being synthesized. Uncoupling refers to the ETC being active, although ATP is not being formed. Energy derived from uncoupled electron transport is instead released as heat. An example of uncoupling occurs naturally in some tissues. It is caused by the special protein thermogenin and takes place in brown adipose tissue. Brown adipose tissue is coloured red-brown because it contains many mitochondria with cytochromes of this colour. The production of heat by brown adipose tissue is used by some animals to keep warm during hibernation and by others, including humans, as a protection from cold in the newborn.

There are many other details known about the workings of mitochondria that I won't mention at this point. It is interesting to contemplate the evolution of these cellular powerhouses, the mitochondria. How did such specialized organelles come to be in cells ranging from early eukaryotes to plants and modern humans? To be so widely distributed and conserved over millions of years of evolutionary time, they must be a very effective means of producing chemical energy. We shall return to this later, and you can be assured of a few surprises.

During evolution, another important pathway has emerged. It is older than the electron transport-coupled oxidative phosphorylation discussed above and may occur under both aerobic and anaerobic conditions. This mechanism is based on the coupling of ATP synthesis to the hydrolysis of other 'energy-rich' compounds. One such is creatine phosphate, in which the ΔG of hydrolysis of the phosphate group is coupled to the phosphorylation of ADP to form ATP.

$$\text{ADP} + \text{creatine phosphate} \xrightarrow{\text{creatine kinase}} \text{ATP} + \text{creatine}$$

Why should such a reaction proceed or, in other words, what is the thermodynamics of the reaction?

As with coupled reactions we have encountered previously, the above equation may be considered as the sum of two reactions:

1) The hydrolysis of creatine phosphate to creatine, liberating phosphate. This is strongly exothermic; ΔG for the reaction is strongly negative ($\Delta G^{0\prime} = -43.1$ kJ mol^{-1}).
2) The coupling of phosphate to ADP to form ATP. This is strongly endothermic; ΔG for the reaction is strongly positive ($\Delta G^{0\prime} = +30.5$ kJ mol^{-1}). By coupling the two reactions with the assistance of a suitable enzyme, creatine kinase, the transfer of the phosphate from creatine to ADP becomes thermodynamically favourable ($\Delta G^{0\prime} = -12.6$ kJ mol^{-1}).

Creatine phosphate is an energy-reserve compound found, for example, in muscle tissue, although it is soon exhausted with exercise.

There are only a few metabolic compounds capable of coupling ADP to inorganic phosphate to form ATP. Another is phosphoenolpyruvate ($\Delta G^{0\prime}_{\text{hydrolysis}} = -55$ kJ mol^{-1}), which we have encountered as an example in Chapter 6.

The general name substrate-level phosphorylation is given to such processes, to distinguish them from oxidative phosphorylation, which involves the whole ETC.

Finally, it is interesting to calculate the efficiency of eukatyotic cells in converting a mole of glucose, via glycolysis, the citric acid cycle, electron transport, and oxidative phosphorylation, to usable energy in the form of ATP. Formally, the equation for the overall process of biological oxidation of glucose is

$$C_6H_{12}O_6 + 6O_2 \longrightarrow 6H_2O + 6CO_2$$

It can be shown that a maximum of about 32 moles of ATP can be formed via the above pathways per mole of glucose oxidized (Campbell and Farrell 2006 pp. 562–3; Garrett and Grisham 2010, pp. 622–3). The free energy of hydrolysis of ATP under cellular conditions is about -50 kJ mol^{-1}. This means that a maximum of $32 \times 50 = 1600$ kJ of energy are available per mole of glucose. The chemical oxidation (burning) of a mole of glucose to CO_2 and H_2O yields 2937 kJ mol^{-1}). This is the maximum amount of energy available from a mole of glucose conversion to water and carbon dioxide, whatever the course or mechanism involved. Remember that thermodynamics involving state functions are independent of the pathway taken.

Thus efficiency for glucose oxidation $= 1600/2937 \times 100 = 54\%$ (Garrett and Grisham 2010 pp. 622–3). A modern motor vehicle is perhaps 30% efficient.

Such is the result of millions of years of biological evolution. Where do you thermodynamically aware readers think the rest of the energy from the biological oxidation of glucose has gone?

Each day the Earth is bathed by some 1.5×10^{22} kJ of solar energy, only about 1% of which is absorbed and transformed into chemical energy by photosynthetic organisms. Two-thirds of the rest is absorbed by the land and the oceans, warming them, and one-third is reflected back into space. The conversion of solar energy into chemical bond energy involves what is termed the fixation of carbon dioxide from the atmosphere.

$$6H_2O + 6CO_2 \xrightarrow{\text{light energy } h\nu} C_6H_{12}O_6 + 6O_2 \qquad (10.10)$$

The overall process summarized by the above equation is called photosynthesis (synthesis by energy supplied by light) and this is the other prime biological example of redox reactions and electron transfer. Photosynthetic organisms produce the major energy foods, carbohydrates, from carbon dioxide and water, for use by the Earth's biological consumers.

Note that written simply as above, the equation is formally the reverse of the oxidative phosphorylation of glucose to form H_2O and CO_2, where energy is liberated. Photosynthesis provides a primary source of chemical bond energy in the form of glucose, while oxidative phosphorylation does the reverse, to produce ATP, an immediately available source of bond energy. Together, photosynthesis and oxidative phosphorylation form an energy cycle involving water, carbon dioxide, oxygen, and carbohydrate, which drives life on Earth (at least on the surface, see later). For an organism to stay alive it must consume energy, constantly. Not to do so is to approach overall equilibrium and death.

The above photosynthetic reaction really summarizes two rather complex processes. One, the oxidation of water to form oxygen, requires light energy from the Sun. The second process, which uses sunlight indirectly, is the 'fixation' of carbon dioxide to form carbohydrates. Water is a poor reducing agent, and cannot, as implied in the above simple equation, be persuaded to reduce carbon dioxide to carbohydrates by a direct reaction. However, photosynthesis, using a rather roundabout, coupled series of energetically favourable steps, can achieve the formation of carbohydrates from water and CO_2. Both processes are discussed in more depth below.

A carbohydrate produced very early in photosynthesis is glucose, a prime energy-producing foodstuff. For humans the dietary source of glucose is overwhelmingly the plant storage carbohydrate starch, found in high amounts in grains such as rice, wheat, and corn (maize). Many green plants store the glucose produced by photosynthesis largely as starch. The major storage carbohydrate of some other plants is sucrose. The principal sources of sucrose are sugar cane and sugar beet. Sucrose consists of two six-carbon sugars, glucose and fructose, linked by covalent bonds, whilst starch consists of millions of glucose molecules also linked by covalent bonds. Starch and sucrose are used as energy storage. Direct storage of energy from sunlight cannot be achieved by living organisms, which instead have evolved a storage system involving carbohydrates. During photosynthesis, a molecule of oxygen is produced for each molecule of water and carbon dioxide converted to carbohydrate. This turns out to be a crucial factor in the evolution of modern life. As mentioned in Chapter 2, when life emerged on Earth, the atmosphere was essentially devoid of free oxygen. Life thrived in the absence of air as we know it, but as photosynthesis gradually led to the build up of oxygen, organisms were 'forced' to adapt. Some anaerobic life disappeared, some survived in ecological niches where no oxygen collected, and many still survive today. Largely, however, humans have experience of and rely on aerobic organisms.

How does photosynthesis work? This topic could fill several large books.

In summary, photosynthesis in eukaryotes occurs in one of the three sets of membranes in chloroplasts, green-coloured organelles found in green plants and algae. In photosynthetic prokaryotes such as cyanobacteria, the photosynthetic membranes fill the interior of the cell and are bound to the plasma membrane. In each chloroplast there are stacks of discs called thylakoids (Figure 10.6).

The membrane inside each thylakoid disc faces into the thylakoid space or lumen. In the membrane are structures where the energy of sunlight is captured by the electrons of chlorophyll molecules. Sunlight, and indeed all energy, comes in 'packets' or quanta called photons (see Appendix A). One photon is the smallest amount of radiant energy possible. Photons are absorbed by chlorophyll molecules in such a manner as to promote certain of their electrons to higher energy levels. These 'excited' electrons are passed along an ETC (analogous to the ETC described for oxidative phosphorylation), passing on some of their gained energy as they traverse the various stages. As a result of their passing this energy, a molecule of water is split and oxygen is released into the atmosphere; protons are pumped out of the thylakoid membrane to form a proton gradient (as in mitochondria), which in turn provides the energy to drive an ATP synthase to form ATP; excited electrons also reduce $NADP^+$ to NADPH by removing hydrogen atoms from water, liberating oxygen. $NADP^+$ is a coenzyme with the same

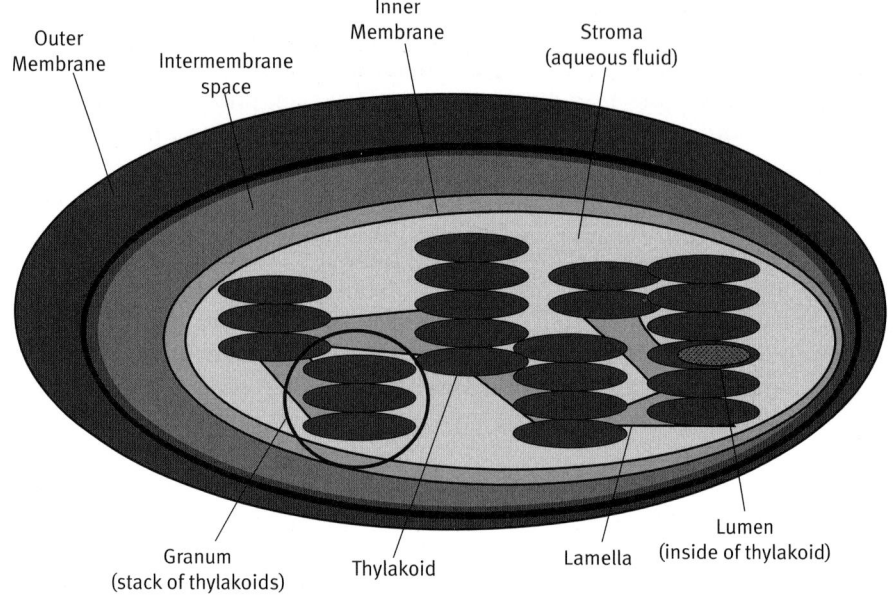

Figure 10.6. The chloroplast – general structure.
Source: http://en.wikivisual.com/index.php/Image:Chloroplast-new.jpg.

structure as NAD^+ but with an extra phosphate group. It is preferred by enzymes in many biosynthetic (anabolic) processes, whereas NAD^+ is mainly used in catabolic, energy-releasing reactions. The reactions described so far are called the 'light reactions'. The NADPH and ATP are used for the synthesis of glucose by the reduction of CO_2. The synthesis of glucose takes place in the region outside the thylakoid membrane called the stroma. This synthesis can occur in the absence of light and the reactions involved are commonly known as the 'dark reactions', although 'light independent' would be a more accurate description.

Let's look a little more deeply into the workings of photosynthesis. Firstly, a reminder of the relationship between wavelength and energy of light. The electromagnetic radiation we know as visible light is the only part of the total electromagnetic spectrum to which our eyes are sensitive. Visible light to which our eyes are sensitive ranges from red (longest wavelength, lowest energy) to violet (shortest wavelength, highest energy). On either side of the visible spectrum are infrared (IR; longer wavelength and lower energy than red), commonly known as heat waves or thermal radiation, and ultraviolet (UV; shorter wavelength and higher energy than violet). We have seen earlier that IR radiation heats the surface of the Earth, sometimes to uncomfortable, even dangerous, levels, and that due to its high-energy properties, UV radiation can be harmful to living organisms. Even shorter wavelength radiation, such as X-rays and gamma rays are generally destructive to life, so plants use some of the visible spectrum to harness energy. Such wavelengths are relatively safe. The visible spectrum ranges from about 400 nm (violet) to 700 nm (red) (1 nm = 10^{-9} m).

The primary event in photosynthesis is the absorption of light energy by pigments whose major molecules are chlorophylls, of which there are two main types, chlorophyll a and chlorophyll b. Eukaryotes such as green plants and green algae contain both types, while prokaryotes such as cyanobacteria contain only chlorophyll a. The two chlorophylls differ slightly in chemical structure and hence in the wavelength of the light they absorb. Both chlorophylls contain a magnesium porphyrin group, which aborbs the light photons, and have a long non-polar (hydrophobic) side chain called a phytol group, which anchors the chlorophyll to the thylakoid membrane by hydrophobic interactions. Certain accessory pigments such as carotenoids absorb at other wavelengths and help to maximize light absorption.

The light reactions of photosynthesis occur in two parts, each part being carried out by a different photosystem. A photosystem consists of an antenna complex and a photosynthetic reaction centre linked to an ETC. The antenna complex is a structure containing a group of several hundred chlorophyll molecules and accessory pigments clustered together in the thylakoid membrane. Of the hundreds of chlorophyll molecules in the antenna complex, only a few in the reaction centres can pass on absorbed light energy. Photosystem II carries out the oxidation of water to produce oxygen, while photosystem I reduces $NADP^+$ to NADPH. As described below, the NADPH and its accessory molecules can reduce CO_2 to glucose.

Figure 10.7 shows that the two photosystems are linked by an ETC that produces a proton gradient across the membrane, which in turn is used to drive an ATP synthase (not shown) similar to the one in mitochondria. In photosynthesis, however, the ATP produced is used, along with NADPH, to drive the formation of glucose from CO_2. The two photosystems absorb light of different wavelengths. The reaction centre of photosystem II contains a special chlorophyll (P680), which has a maximum absorbance at 680 nm. The corresponding chlorophyll in photosystem I, P700, absorbs maximally at 700 nm.

What steps are involved in photosynthesis up to the formation of NADPH?

When a chlorophyll molecule in the antenna complex of a photosystem is struck by a photon of sunlight, one of its electrons is excited into a higher-energy orbital. This captured energy can be transferred in one of two ways (Figure 10.8).

The excited chlorophyll molecule can pass on the energy to a chlorophyll in the reaction centre of the photosystem by means of exciton transfer (also known as resonance energy transfer) and return to the lower-energy state ready for excitation by another incoming photon. Exciton transfer involves interaction between the molecular orbitals of one chlorophyll molecule and a neighbouring one with similar electronic properties. For efficient energy transfer, the chlorophyll molecules must have optimal spacings and orientations (Figure 10.9).

Alternatively, the high-energy electron itself may be passed on. For example in photosystem II, chlorophylls and accessory pigments in the antenna complex absorb visible light photons and via exciton transfer channel the energy to the photosynthetic reaction centre, where chlorophyll P680 becomes excited to P680* and passes it on as high-energy electrons to the electron-transport chain. As an electron is lost, P680 becomes $P680^+$, an electron-deficient cation. This is a

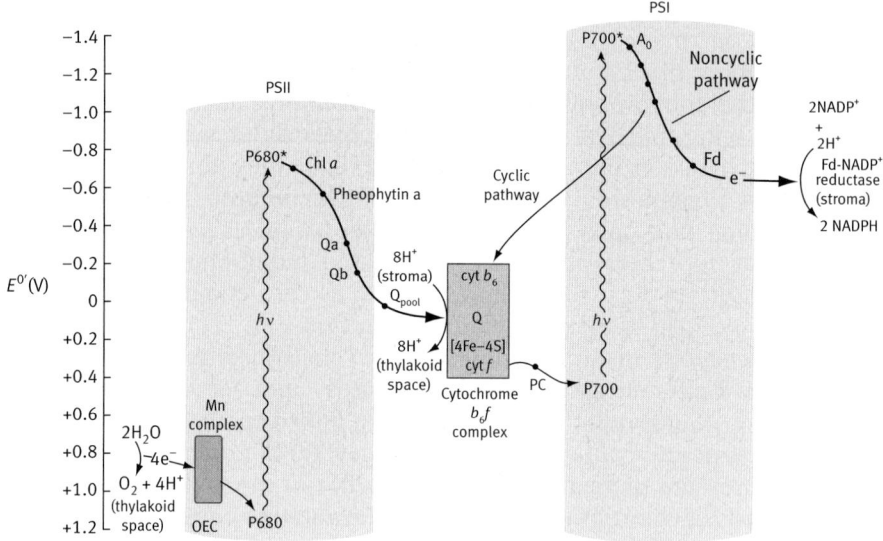

Figure 10.7. The energetics (in volts) of photosystems and electron flow during photosynthesis in green plants. Electrons in chlorophyll P680 are ejected when a photon (hv) is absorbed. These are replaced by electrons extracted from water by an Mn-containing complex called the oxygen-emitting centre (OEC). Each successive electron passes through a chain of electron carriers, ultimately to reduce $NADP^+$ to NADPH (far right). At the cytochrome b_6f complex site (centre) protons are transferred from the stroma to the thylakoid space, forming a proton gradient that provides a protonmotive force to drive ATP synthesis (not shown). (Adapted from Voet, D., Voet, J.G., and Pratt, C.W. *Fundamentals of Biochemistry*, upgrade edn, p. 542. © 2002 John Wiley & Sons, Inc.)

highly reactive species and is one of the most powerful biological oxidants known. The ΔG for the following reaction can drive the 'extraction' of electrons from water:

$$2H_2O \xrightarrow{\text{[four photons / P680]}} 4e^- + 4H^+ + O_2 \qquad (10.11)$$

Oxygen is liberated, as we know occurs during photosynthesis.

In case you missed it, we have just described one of the singular events in energy transformation on Earth—conversion from electromagnetic to chemical energy.

Note that to obtain a molecule of oxygen, two molecules of water are decomposed and the energy from four light photons needs to reach the reaction centre of photosystem II. The electrons then traverse the ETC via a pool of mobile plastoquinone (Q) molecules (similar in structure and function to CoQ in mitochondria) and the protons are pumped, at the site of the cytochrome b_6f complex (cytochrome b_6 plus Fe-S proteins), into the thylakoid space. These protons help provide the proton gradient that drives the formation of ATP in a process called

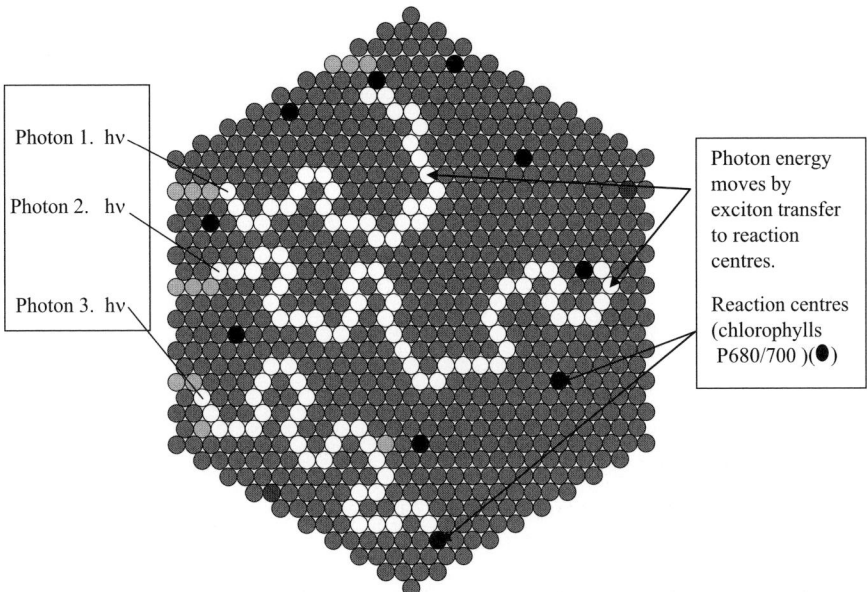

Figure 10.8. Antenna complex of a plant photosystem. A snapshot of the random pathways taken by three absorbed photons that move by exciton transfer between chlorophyll molecules (○) until they reach a reaction centre chlorophyll P680 or P700 (●). (Illustration adapted from Voet, D., Voet, J.G., and Pratt, C.W. *Fundamentals of Biochemistry*, upgrade edn, p. 533. © 2002 John Wiley & Sons, Inc.)

non-cyclic photophosphorylation. As the electrons use energy to pump the protons, by the time they reach the plastocyanine (PC, a blue copper-containing protein) redox system, they require a boost. This is provided by photosystem I, where the chlorophyll at the reaction centre (P700) is excited by photons at this longer wavelength. The resulting high-energy electrons are passed on to the next part of the ETC, leaving P700* (electron deficient). The 'waiting' low-energy electrons at the PC system are then passed on, returning P700 to the unexcited state ready for the next round of excitation by light. When this has occurred, the new high-energy electrons from P700* pass along the rest of the ETC, via the ferredoxin (Fd) redox system to NADP reductase, an enzyme that catalyses reaction (10.12):

$$NADP^+ + 2e^- + H^+ \xrightarrow{\text{NADP reductase}} NADPH \qquad (10.12)$$

The entire sequence of this electron transport scheme can be written simply as the net result:

$$2H_2O + 2NADP^+ \longrightarrow 2NADPH + 2H^+ + O_2 \qquad (10.13)$$

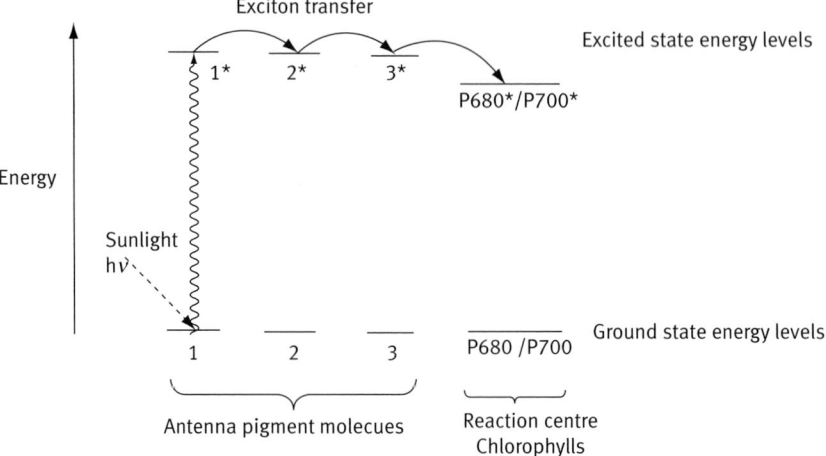

Figure 10.9. Exciton transfer. Energy (hν) from captured photons is transferred between pigments in the antenna complex via exciton transfer to a reaction centre chlorophyll, either P680 or P700, to form P680* or P700*.

As one would expect, this overall reaction is endothermic; $2 \times \Delta G^{0\prime} = +440$ kJ mol^{-1}. The light photons absorbed by photosystems I and II provide the energy to drive the reaction. Figure 10.7 shows that electrons flow from water to NADP$^+$, reducing it to NADPH and releasing oxygen in the process. As we will see below, NADPH together with other components is a strong enough reducing agent to reduce CO_2 (albeit in a combined form) to carbohydrate during the dark reactions of photosynthesis. Sometimes conditions in the cell occur such that the NADPH/NADP$^+$ ratio is high, meaning that there is too little NADP$^+$ to accept all the electrons generated by the excitation of P700. Instead, P700 passes the excited electrons to the cytochrome $b_6 f$ complex instead of to NADP$^+$. They then pass to PC and back to P700 as low-energy electrons, leaving P700 ready for re-excitation by light, thus completing the cycle. As only photosystem I is active, water is not involved so no oxygen is produced. Neither is NADPH formed, but ATP is generated as a result of the proton gradient set up as before at the cytochrome $b_6 f$ stage (Figure 10.7). This second role of photosystem I is called cyclic photophosphorylation.

Let us consider the situation so far, in comparison to that in mitochondria. The details of ATP production have not been discussed for photosynthesis. I won't do so, as they are the same in principle, and largely the same in detail, as that already described for mitochondria—formation of a proton concentration gradient by the pumping of protons at the site of redox complexes to an intermembrane space, and use of the resulting protonmotive force to drive the synthesis of ATP via an ATP synthase. In mitochondria and the cell it inhabits, the ATP is used for a number of metabolic processes. The NADH via the ETC has O_2 as its ultimate electron acceptor to form H_2O.

Overall, the mitochondrion converts chemical energy from food into ATP for general cellular use, plus water, and CO_2. The ultimate electron donors are, broadly speaking, CH bonds in foodstuffs.

So far in our discussion of photosynthesis, light energy has provided (i) a source of reducing power in the form of NADPH and (ii) chemical potential energy in the form of ATP. In photosynthesis, the objective is the fixation of CO_2 to form carbohydrate, and this is where most of the ATP and NADPH will be directed, via the so-called 'dark reactions'. Overall, photosynthesis converts light energy and CO_2 to chemical energy in the form of carbohydrate. The ultimate electron donor is water. Equation (10.14) summarizes photosynthesis to the point we have reached.

$$2H_2O + 2NADP^+ + xATP + XP_i \xrightarrow{\text{light }(h\nu)} O_2 + 2NADPH + 2H^+ + xATP + xH_2O$$

(10.14)

For the reaction as written, we must use $2 \times \Delta G^{0\prime} = 440$ kJ as the free energy change. This means that the net light energy input for reaction (10.14) must exceed 440 kJ, or 220 kJ per mole of $NADP^+$ reduced. The yield of ATP depends on the details of the of photophosphorylation route being utilized at any time, and on the ATP/H^+ ratio, hence the use of $xATP$ in reaction (10.13).

The dark reactions are really carbon-fixation reactions. The term comes from the notion of 'fixing' the mobile gas CO_2 into a more tangible form, that of carbohydrate. The dark reactions use the ATP and NADPH produced by the light reactions to convert CO_2 to carbohydrate. This is not done directly by patching together six molecules of CO_2 with the aid of NADPH. Rather, CO_2 is fixed during the operation of a biochemical cycle called the Calvin cycle (or C_3 pathway) after US biochemist Melvin Calvin, who received the 1961 Nobel Prize for Chemistry for elucidating this path of carbon in photosynthesis.

The first carbon fixation reaction in the dark reaction series is catalysed by a large enzyme located in the stroma, ribulose bisphosphate carboxylase/oxygenase, which, being such a mouthful, is usually referred to as rubisco. Note that it has two functionalities, acting as a carboxylase and also as an oxygenase. The significance of this dual behaviour will be seen later. The single enzyme rubisco makes up about 50% of the protein in a typical chloroplast, and considering the world-wide abundance of chloroplasts, it is also a leading contender for the most abundant protein on Earth.

In the following description, keep track of the number of carbon atoms at each stage. To make this clearer, I have dispensed with the full chemical structures, details of which can be accessed in Appendix D.

Rubisco condenses a single CO_2 molecule with the five-carbon molecule ribulose 1,5-bisphosphate (RuBP) to form a six-carbon intermediate, which rapidly hydrolyses to two three-carbon molecules of 3-phosphoglycerate (3-PG):

$$CO_2(1C) + \text{ribulose1,5-bisphosphate(5C)} \xrightarrow{\text{rubisco}} 2(\text{3-phosphoglycerate}) \ (2 \times 3C)$$

(10.15)

or

$$CO_2 + (RuBP) \xrightarrow{\text{rubisco}} 2(3PG)$$

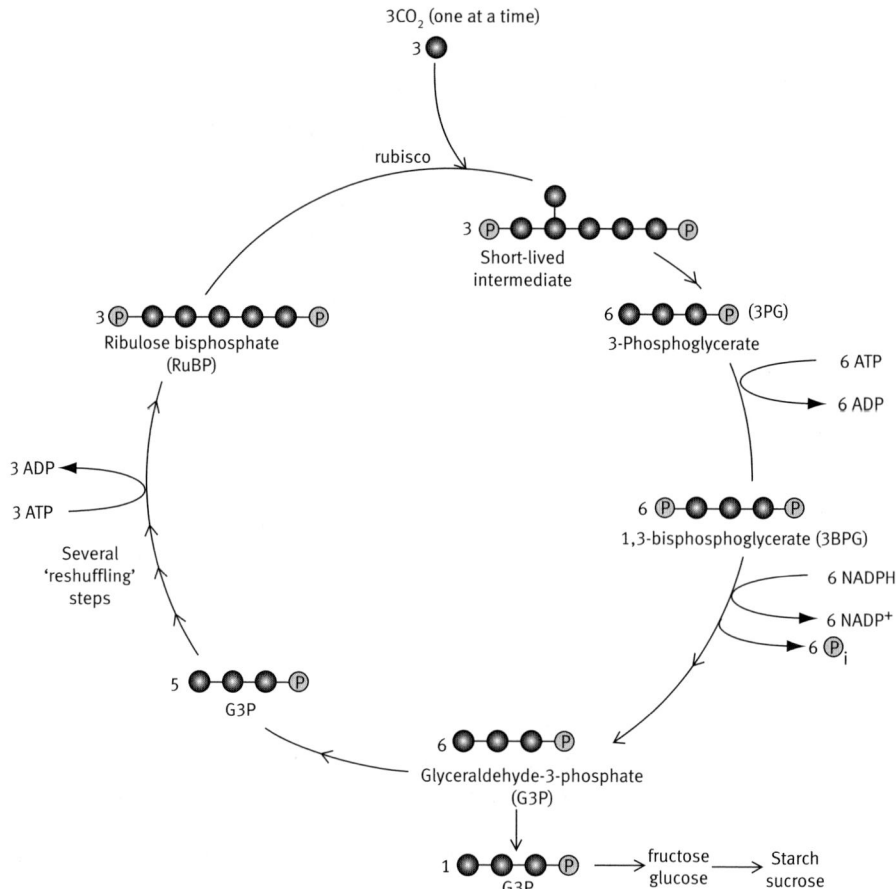

Figure 10.10. The C_3 pathway or Calvin cycle for the fixation of CO_2 during photosynthesis in green plants. The numeral in front of the compound at each step shows the number of molecules reacting in that step for a single turn of the cycle. In one such turn, three CO_2 molecules are fixed to form one net glyceraldehyde-3-phosphate (G3P), which is bled off the cycle to form the six-carbon sugars fructose and glucose. The five remaining G3P (5 × 3 carbon) molecules undergo 'carbon reshuffling' to form three RuBP (3 × 5 carbon) molecules that complete the cycle, ready to fix a further three CO_2 molecules.

This reaction forms the start of the Calvin cycle which, as shown below, regenerates RuBP ready to fix further molecules of CO_2 (Figure 10.10).

The net product of one turn of the cycle is glyceraldehyde-3-phosphate (G3P), a three-carbon (hence the name C_3 pathway) molecule that requires the fixation of three molecules of CO_2. The G3P formed at each turn of the cycle is bled off and used, via formation of the six-carbon sugars fructose 6-phosphate, and glucose-6-phosphate, for metabolism, the synthesis of the storage carbohydrates starch or sucrose, and the synthesis of cellulose for plant growth. The remaining five G3P molecules undergo a complex series of enzyme-catalyzed reactions that convert five three-carbon molecules (5 × G3P) to three five-carbon molecules of RuBP (3 × RuBP). The three RuBP molecules are then ready to fix, via catalysis by rubisco, three more CO_2 molecules in the next round of the cycle. The rubisco-catalysed reaction produces an unstable, short lived, six-carbon intermediate that splits into two 3-phosphoglycerate (3PG) molecules. It is highly exothermic ($\Delta G^{0\prime} = -35$ kJ mol^{-1}). Such strongly exothermic reactions are essentially irreversible and are often the sites of regulation in metabolic pathways. Such is the case with rubisco, a main regulatory enzyme.

The enzymes of glycolysis are present in the stroma of the chloroplast as well as those of the Calvin cycle, so plants require a control mechanism to prevent the Calvin cycle using up their stores of NADPH and ATP produced by glycolysis. The control mechanism needs to be light sensitive so that the Calvin cycle is inactivated at night, but active during the day when photosynthesis can drive it. It happens that rubisco activity is under the control of at least three light-dependent influences.

1) When light drives the Calvin cycle, the pH of the stroma increases from about 7.0 to about 8 as protons are pumped from the stroma into the lumen. A pH of 8 is required to carry out an essential step in the conversion of rubisco from an inactive form to its active form.
2) When protons are transported, Mg^{2+} ions are transferred in the opposite direction—from lumen to stroma. These Mg^{2+} ions bind to rubisco and complete its conversion, commenced in 1, to the active form.
3) The inactive form of rubisco binds its substrate, RuBP, more tightly than the active form, making RuBP an inhibitor. To release RuBP, and allow rubisco to be activated, the plant employs a regulatory protein, rubisco activase. Rubisco activase triggers the release of RuBP in an ATP-dependent reaction. In turn, rubisco activase is activated by light.

In addition to the three effects on rubisco, the pH of 8 activates three other enzymes in the Calvin cycle, so overall regulation of photosynthesis is light dependent.

There are 11 enzymes involved in the Calvin cycle, 8 of which have equivalents in animal tissues. This is a reminder that the reductive, redox processes of photosynthesis have similarities to the reductive, redox processes of anabolism in animals (see Chapter 11). In another comparison, the steps in the conversion of 2 × G3P from the Calvin cycle to the hexoses fructose and glucose are essentially the reverse of glycolysis. Glycolysis (Chapter 11) is a metabolic pathway that can, with a little help, oxidize hexoses such as fructose and glucose to CO_2 and water, generating ATP. As we have seen previously, this is formally the reverse of what happens in photosynthesis. Such reversal of a metabolic pathway provides one example of how metabolic control works. By juggling concentration ratios and continually removing substrates, reactions, and by extension metabolic pathways, can be made to run in reverse (Chapter 6).

As indicated in Figure 10.10, for each CO_2 converted to G3P, three ATP and two NADPH are required. Thus, the overall reaction for the synthesis of one molecule of G3P is:

$$3CO_2 + 6NADPH + 9ATP \longrightarrow G3P + 6NADP^+ + 9ADP + 8P_i \quad (10.16)$$

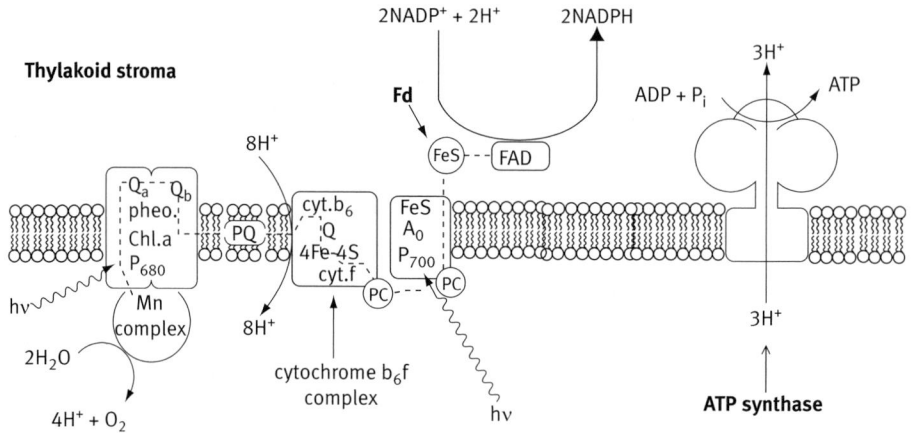

Figure 10.11. Spatial relationships between the components of the photosynthetic system. Refer to Figure 10.7 for the functions and identities of the components. The dotted line shows the path of electrons.

Reaction (10.15) needs to be doubled to make one glucose molecule (six carbons, equivalent to 2 × G3P). Figure 10.11 shows the relationships between the components of the photosynthetic system we have discussed.

Photosynthesis is clearly a large consumer of energy, which is provided by the solar-powered photosynthetic system. Further energy is required to convert the early products of photosynthesis to sucrose or starch, and several enzyme systems are involved in the pathways. The G3P is exported to the cell cytosol for the synthesis of sucrose. Starch is initially produced in the stroma of chloroplasts and is stored there in the form of granules. Many plants move this so-called assimilation starch to other regions for storage. These regions include tubers (potatoes), cereal grains such as corn, wheat, and rice, and the underground stems (rhizomes) of plants such as the iris and the grasses.

When the plant needs to use the stored energy of starch for metabolic processes, hydrolytic enzymes are activated and glucose is progressively released. This glucose usually undergoes the catabolic (stepwise degradation) process called glycolysis. Glycolysis was the first metabolic pathway to be elucidated, by Warburg, Embden, and Meyerhof in the first-half of the twentieth century. The principal steps of this Embden–Meyerhof or glycolytic pathway are anaerobic, and it played a vital role in anaerobic metabolic processes for the first 2 billion years of biological evolution. More about this pathway is explained in Chapter 11.

Some plants, at least 100 species, possess what is called the C_4 pathway (or Hatch–Slack pathway after its Australian discoverers) for the fixation of CO_2. Normally, rubisco fixes CO_2. When CO_2 concentrations are low, as during bright days when photosynthesis has been active, rubisco adds O_2 instead (its oxygenase activity kicks in) to form 3PG plus 2-phosphoglycolate (2PGc). The metabolism of PGc releases CO_2 and utilizes some ATP and NADPH, the net result being a waste of metabolic energy. This process is called photorespiration, and puts a limit on CO_2 fixation. As the oxygenase activity of rubisco increases with a rise in temperature faster than its carboxylase activity, this is a serious problem for hot-climate plants, for example sugarcane and corn. We won't go into details, but these plants have evolved a Calvin cycle that is located in special bundle-sheath cells that are protected from air. The CO_2 is transported from the air (and away from oxygen) to the bundle-sheath cells by combining it with a three-carbon molecule (phosphoenol pyruvate (PEP)) to form a four-carbon molecule (oxaloacetate). The enzyme catalyzing the reaction, PEP carboxylase, has a much higher affinity for CO_2 than does rubisco and, importantly, no affinity for oxygen. In the bundle-sheath cells, oxaloacetate is transformed into CO_2 and pyruvate. The net result is concentration of the CO_2 for processing by the Calvin cycle, with no competition from oxygen. The route is known as the C_4 pathway and the plants using it are referred to as C_4 plants. Corn and sugarcane produce about twice the average growth rate (kilogram of dry weight produced per square metre per week) of typical C_3 plants such as hay or spinach. Although other factors contribute to the productivity of a crop, the possibility of breeding the favourable characteristics of C_4 plants into less efficient but economically important C_3 species has been suggested.

Finally, consider the recurring themes in the energy transformations involving electron and proton transfer in bacteria, mitochondria, and chloroplasts. All involve membranes with embedded functional molecules, ETCs and proton concentration gradients combining to provide energy for ATP and sugar synthesis, and the roundabout but successful coupling of a series of energetically favourable reactions to bring about reactions that are too energetically unfavourable to achieve directly. These undeniably similar properties reinforce the concept of a common ancestry for living organisms, and are underpinned by the ubiquitous but often disregarded influence of thermodynamics on all life processes.

REFERENCES

Campbell, M.K. and Farrell, S.O. (2006) *Biochemistry*, 5th edn. Thomson Brooks/Cole, Belmont, CA, pp. 562–563.

Campbell, N.A., Reece, J.B., Urry, L.A., Cain, M.L., Wasserman, S.A., Minorsky, P.V., and Jackson, R.B. (2008) *Biology*, 8th edn. Pearson Benjamin Cummings, San Francisco, p. 110.

Garrett, R.H. and Grisham, C.M. (2010) *Biochemistry*, 4th edn. Brooks/Cole, Cengage Learning, Boston, pp. 66, 597, 620.

Hames, B.D. and Hooper, N.M. (2000) *Biochemistry*, 2nd edn. Bios Scientific Publishers Ltd, Oxford, p. 255.

Morowitz, H.J. and Smith, E. (2007) Energy flow and the organization of life. *Complexity* **13**, 51–9.

Nicholls, D.G. and Ferguson, S.J. (2002) *Bioenergetics 3*. Academic Press, London, pp. 200, 209.

Voet, D., Voet, J.G., and Pratt, C.W. (2002) *Biochemistry*, upgrade edn. Wiley & Sons Inc., New York, pp. 494, 499.

11
Cells and Metabolism: Putting it all Together

11.1 General aspects of metabolism

Much of what has been discussed so far has concentrated on the how and why of chemical reactions and their energetics. We have seen that some biochemical reactions will not proceed spontaneously in the desired direction, but they can be made to do so indirectly by means of coupled reactions, appropriate concentration ratios, and removal of substrates. Often the energy source to drive such pathways is ATP, the major readily available source of metabolic energy in the cell. Primary generation of ATP occurs by oxidative phosphorylation, which involves a series of redox reactions coupled in an electron transport chain (ETC) to generate a proton gradient. The energy available as a result of the proton gradient, the protonmotive force, drives the synthesis of ATP. The primary energy source driving oxidative phosphorylation in cells is glucose, which originates in photosynthetic cells by photosynthesis, another process involving ETCs and redox reactions.

Having dealt with the principles of chemical thermodynamics and with the major energy-producing processes in living organisms, we are now in a position for an overall look at the way these are integrated in individual cells as the processes of metabolism. The word metabolism derives from the Greek for 'change'. As mentioned in an earlier chapter, metabolism refers to the totality of chemical changes that convert the raw materials (foods and various nutrients) into energy plus the many chemical compounds required to nourish the cell.

Some of the metabolic pathway 'maps' found in biochemistry texts are mind boggling in their complexity; they may show hundreds of reactions and contain many interlocking pathways. I shall not deal with metabolism in such detail, as this is not a biochemistry text. My major objectives in this section are to provide an overview of metabolism within the framework that has already been set, and to show that essentially all organisms have the same basic set of metabolic pathways, underlining their common origins and the continuity of life on Earth from the earliest times to the present.

Metabolism in a cell consists of hundreds of reactions organized into metabolic pathways. Each pathway runs in a series of discrete steps, most of which are catalysed by a specific enzyme. By means of these pathways, chemical substrates are converted into various end products via a series of intermediate chemical types. The term intermediary metabolism is often used to reflect this. Each pathway needs to be regulated, so that the overall balance of chemical metabolites is maintained. The 'correct' or 'desirable' balance of metabolites for each pathway will vary depending on the demands of the cell. When ATP is required, pathways that generate ATP will be activated until the demand is met. Feedback mechanisms that slow down the rate of ATP production come into operation. These feedback mechanisms often involve the allosteric inhibition of a key enzyme or enzymes early in the pathway by a metabolite further along the pathway. This makes sense, as a high concentration of a compound in the pathway close to the end product means that the end product concentration is likely to be high and the pathway can probably be closed down safely.

In Chapter 1 I discussed the concept of producers (autotrophs, or 'self-feeders') and consumers (heterotrophs, or 'feeders on others') in the biological world. This is a simplified classification and we are now in a position to extend it. If we arrange organisms in metabolic pathway terms to include the origins of their carbon-based chemicals, energy sources, and requirements for oxygen, a broader, more useful classification system emerges.

Note the electron donor requirement for all types of organism. Life on Earth depends on electron transfer as a major means of energy transduction. We have discussed redox reactions as a source of energy in the form of electron flow and typical examples of electron donors in such systems are included in Table 11.1.

Phototrophs are photosynthetic organisms, gaining their energy from light. Chemotrophs use organic compounds such as glucose (chemoheterotrophs) or certain simple inorganic compounds such as those listed in Table 11.1 (chemoautotrophs).

Table 11.1. Classification according to carbon, oxygen, and energy requirements.

Metabolic classification	Carbon source*	Energy source	Electron donor
Aerobic photoautotrophs Green plants, algae, cyanobacteria, protists	CO_2	Light/aerobic respiration	H_2O
Anaerobic photoautotrophs Photosynthetic bacteria	CO_2	Light/glycolysis	H_2S, S, and other inorganic compounds
Photoheterotrophs Non-sulphur purple bacteria	Organic compounds	Light	Organic compounds
Chemoautotrophs Nitrifying bacteria; sulphur, hydrogen, and iron bacteria	CO_2	Redox reactions	Inorganic compounds [H_2, H_2S, Fe^{2+}, NH_4^+, Mn^{2+}]
Aerobic heterotrophs All animals; protists, fungi; many microbes	Organic compounds	Aerobic respiration	Organic compounds, e.g. glucose
Anaerobic heterotrophs Fermenting bacteria	Organic compounds	Glycolysis	Organic compounds, e.g. glucose

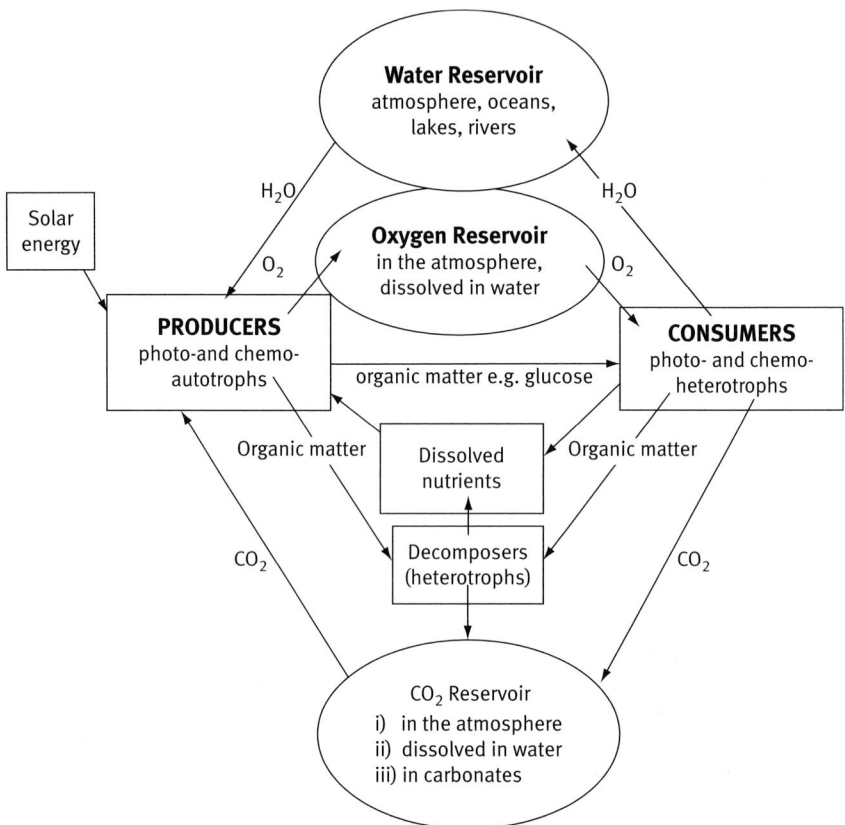

Figure 11.1. Summary of biogeochemical cycles.

Before looking more deeply into metabolic pathways, let's first look very broadly at the flow of energy in Earth's biosphere through some important biogeochemical cycles. This is important to have in mind as a lead-in to metabolic pathways because it reminds us of the interrelated nature of organisms and the biosphere (Figure 11.1).

The scheme in Figure 11.1 outlines the important carbon, oxygen, and water cycles that illustrate the purely chemical interrelationships between living organisms and their environment. Other important cycles are the nitrogen cycle, which refers to the movement of nitrogen through the food chain, the sulphur cycle, and the phosphorus cycle. All these cycles trace the flow of the respective essential nutrients and ingredients in the biosphere. For a healthy biosphere, the cycles should be kept in a state such that the chemical ingredients are present in sufficient amounts and in an accessible form, in other words in 'dynamic balance' or 'steady-state' conditions, terms we have met previously. Taking an idealized example, when a new generation of organisms comes into being, an old generation dies and begins to decompose. The essential nutrients in the decaying old generation enter the cycles and so replace those that have been removed by the new generation.

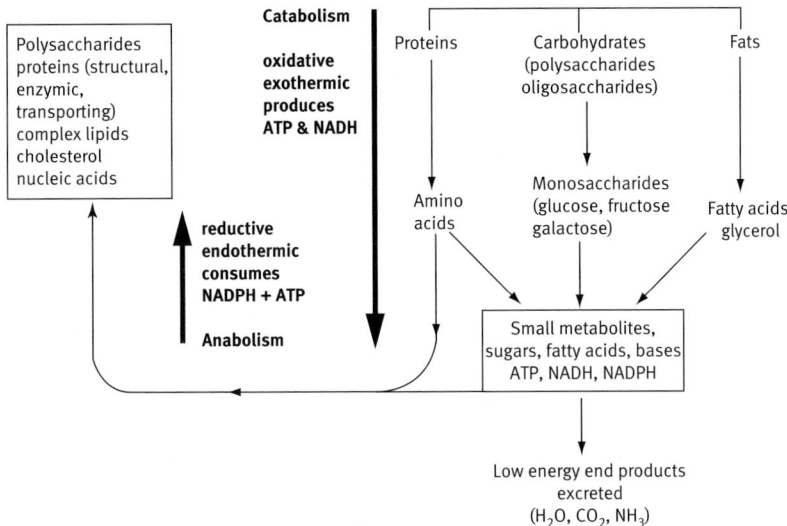

Figure 11.2. General features of anabolism and catabolism for a heterotrophic, aerobic cell.

What the balance may be will vary with the region in question. Knowledge of the state of the cycles can assist biologists and earth scientists to understand what is occurring in the broadest sense and allow planned intervention in cases where environmental deficiencies (or excesses) in nutrients, etc. may occur. If dissolved nutrients from over-irrigation accumulate, combined with a rising water-table, the resulting salination of groundwater can be devastating to crops and a serious problem for human consumers. This has occurred in parts of the Murrumbidgee irrigation area in eastern Australia, and even the billions of dollars which need to be spent might not cure the problems of the region. Too high a level of phosphate from fertilizers in a water system may result in a 'bloom' of toxic algae. Coral polyps on the Great Barrier Reef are dying from pollutants in runoff water from cattle-grazing and sugar-cane operations in Queensland. Recent legislation provides for fines of up to A$30,000 for non-compliance with the pesticide and fertilizer use laws. We are all familiar with one or more of these instances of imbalance.

In a similar fashion, within the autotrophs and heterotrophs embedded in the above cycles, there are metabolic pathways and cycles whose 'balance' and 'health' must be maintained. They are kept in control by various feedback mechanisms, but in addition they must have the correct ingredients at the input end. For animals such as ourselves, this basically means access to sufficient food, air, and nutrients; for plants, sufficient water, carbon dioxide, sunlight, and mineral nutrients; each other form of life has its own specific needs for survival. By looking more deeply into the metabolic pathways, we can draw further diagrams to show interrelationships, but firstly let's look at the two quite different purposes of metabolism. For convenience of study, metabolism may be divided into catabolism, which involves mainly oxidative, degradative pathways, and anabolism, which largely involves reductive, biosynthetic pathways (Figure 11.2).

Typically, catabolic pathways are involved in generating energy—they are exothermic—for use by the cell by transforming the chemical energy in foodstuffs and are usually oxidative in their chemistry. In contrast, anabolic pathways usually require the consumption of energy—they are endothermic—to build complex molecules required by the cell and are reductive in their chemistry. The synthesis of complex molecules is usually endothermic overall and is often achieved by the roundabout method of coupled reactions, as mentioned previously. Anabolic processes typically utilize NADPH as the coenzyme. Catabolic processes produce NADH, which is used as a source of chemical reducing power for the cell and much of this is ultimately transferred to the ETC, where it is used via oxidative phosphorylation to form ATP. The enzymes that utilize NADH vs NADPH can discriminate between the two coenzymes on the basis of the extra phosphate group in the latter. Catabolic processes typically produce ATP as their end product and, as we have seen, ATP is used to drive endothermic reactions.

The above descriptions deal with metabolism in very general terms. Starting with catabolism, let us be specific.

In the discussion that follows, the emphasis is on the overall significance rather than the individual steps to gain an appreciation of the pathways and cycles, and the ways in which they are interdependent. It is a problem that most of us have learned things, especially science, within a largely reductionist framework. When it is necessary to step outside the framework and take a more global view, we often have difficulty. Detail may obscure the broader patterns. To reduce this I shall dispense with chemical formulae as much as possible. Details are provided in appendices, as noted.

In a typical aerobic heterotrophic cell, most energy is derived from the catabolism of carbohydrates, proteins, and fats contained in food (Figure 11.3).

Stage 1 The metabolic pathways of these food materials begin separately, as they are broken down into their components, for example by the early stages of digestion.

Stage 2 In glycolysis, the components are further degraded, and converge to a common product, the two-carbon acetyl groups of acetyl-coenzyme A (acetyl-CoA).

Stage 3 Acetyl-CoA enters a metabolic pathway called the citric acid cycle (also known as the Krebs cycle or the tricarboxylic acid (TCA) cycle). The citric acid cycle plays a central and vital role in aerobic metabolism, feeding into the ETC, where oxidative phosphorylation generates large amounts of ATP. Glucose is the principal source of chemical energy for many life forms, from early anaerobes to today's mammals. For this reason I will concentrate on glucose pathways from now on and go into no further detail about the ways in which lipids and proteins (see Figure 11.3) are converted to pyruvate, to acetyl-CoA, or to compounds that can feed into the citric acid cycle.

11.2 Glycolysis

The metabolic pathway from glucose to pyruvate (pyruvic acid) is called glycolysis. Glycolysis itself is anaerobic, in keeping with its early evolutionary origins as a major pathway in anaerobic bacteria. It takes place in the cytoplasm

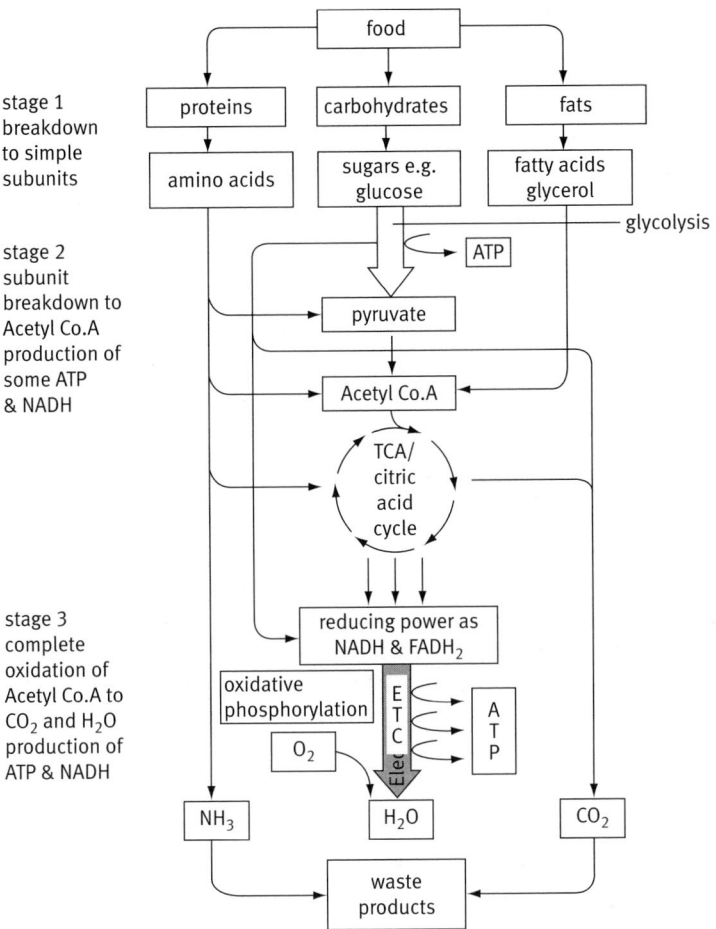

Figure 11.3. Catabolism. The three stages by which foods are broken down to generate ATP. ETP = electron transport chain/respiratory chain. (Illustration adapted from Pocock, G. and Richards, C.D. *Human Physiology*, p. 29. © 1999 Oxford University Press.)

of both prokaryotes and eukaryotes. It is believed to be the most ancient metabolic pathway, occurring in essentially all cells, having evolved before oxygen appeared in significant amounts in the atmosphere (Garrett and Grisham 1999 p. 609; Schopf 1999). Glycolysis has a dual role overall: to generate ATP and to produce intermediates that are used in a number of biosynthetic pathways, for example acetyl-CoA is the precursor for fatty acid biosynthesis. In glycolysis, two ATP are consumed and four ATP are generated, a net yield of two ATP per glucose molecule. Two NADH molecules are also formed and under aerobic conditions these yield further energy, as ATP, via the citric acid cycle, ETC, and oxidative phosphorylation (Figure 11.4).

The first phase of glycolysis actually consumes two ATP. It can be regarded as a preparatory phase that sets up the chemistry for the second, energy-producing phase.

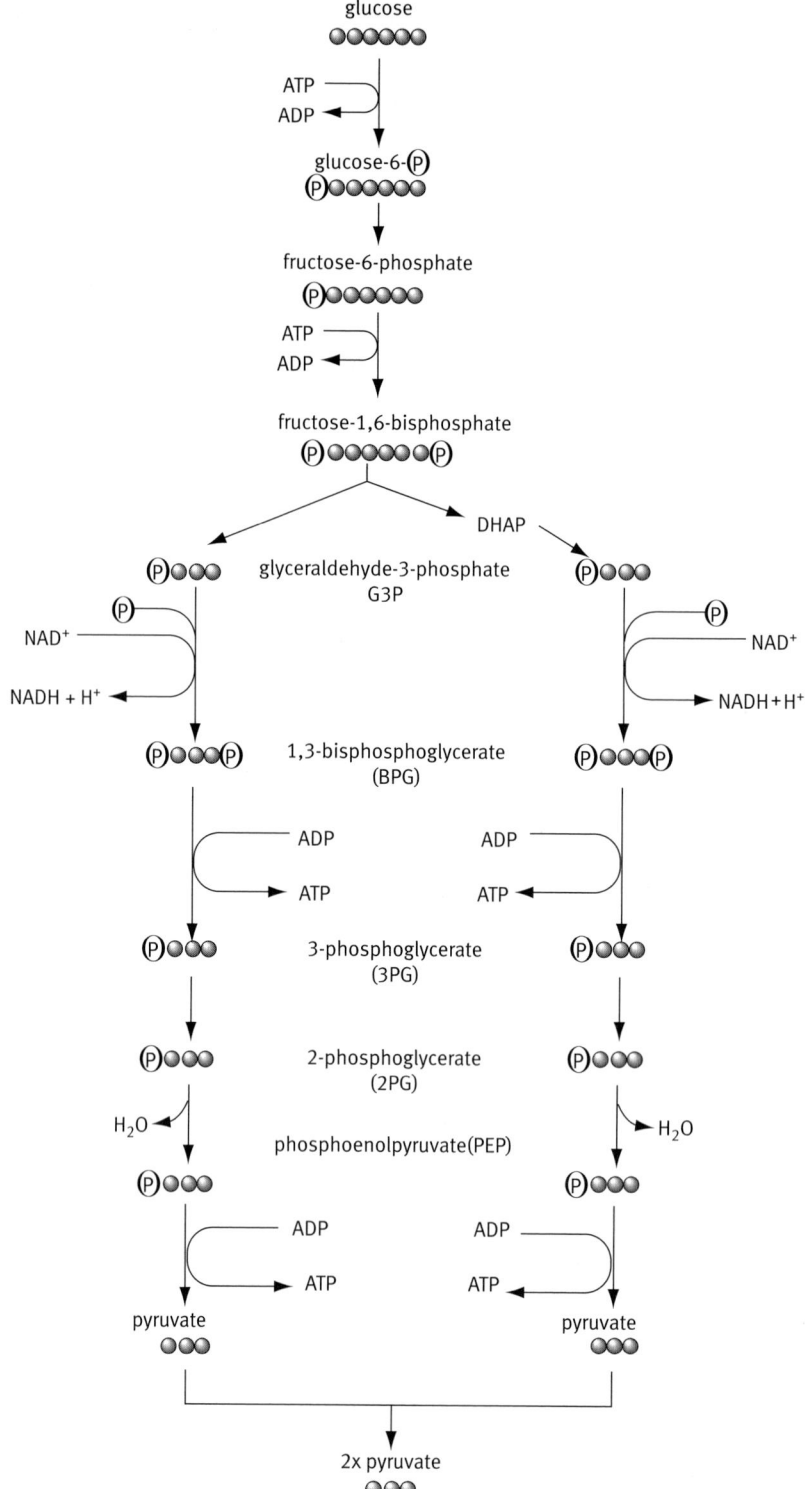

Figure 11.4. Glycolysis, showing the conversion of glucose to pyruvate. DHAP = dihydroxyacetone phosphate, which is immediately converted to G3P. ○ represents one carbon atom.

11.3 The reactions of glycolysis

1. Phosphorylation

$$\text{glucose} + \text{ATP} \xrightarrow{\text{hexokinase}} \text{glucose-6-}\textcircled{P} + \text{ADP}$$

($\Delta G^{0\prime} = -16.7$ kJ mol^{-1}; $\Delta G = -33.9$ kJ mol^{-1} under cellular conditions).

Phosphorylation of glucose keeps it in the cell as this ionized compound cannot diffuse through non-polar membranes. The large negative ΔG makes it a site for regulation of glycolysis.

2. Isomerization

$$\text{glucose-6-}\textcircled{P} \xrightarrow[\text{isomerase}]{\text{phosphogluco-}} \text{fructose-6-}\textcircled{P}$$

($\Delta G^{0\prime} = +16.7$ kJ mol^{-1}; $\Delta G = -2.9$ kJ mol^{-1} under cellular conditions).

The isomerization to fructose operates close to equilibrium in the cell (low ΔG). It produces the fructose molecule, which can undergo ready C_3–C_4 bond cleavage.

3. Phosphorylation

$$\text{fructose-6-}\textcircled{P} + \text{ATP} \xrightarrow[\text{kinase}]{\text{phosphofructo-}} \text{fructose-1,6-}\textcircled{P}_2 + \text{ADP}$$

($\Delta G^{0\prime} = -14$ kJ mol^{-1}; $\Delta G = -19$ kJ mol^{-1} under cellular conditions).

Fructose-1,6-bisphosphate is ideally set up to be cleaved into two three-carbon fragments that are already phosphorylated. Phosphofructokinase is inhibited allosterically by two compounds: ATP and even more strongly by its product, fructose-1,6-bisphosphate. The reaction is therefore an important regulation point in glycolysis.

4. Cleavage

$$\text{fructose-1,6-}\textcircled{P}_2 \xrightarrow{\text{aldolase}} \text{glyceraldehyde-3-}\textcircled{P} + \text{dihydroxyacetone-}\textcircled{P}$$

($\Delta G^{0\prime} = -24$ kJ mol^{-1}; $\Delta G = -0.23$ kJ mol^{-1} under cellular conditions).

This important reaction leads, after reaction 5, to two identical three-carbon, phosphorylated compounds, that is two glyceraldehyde-3-phosphate molecules. The favourable chemistry for C3–C4 bond cleavage is demonstrated by two reactions:

1) Free fructose is cleaved chemically in alkaline solution at room temperature to glyceraldehyde and dihydroxyacetone, just as in reaction 4.
2) Under similar alkaline conditions, a mixture of glyceraldehyde and dihydroxyacetone forms substantial amounts of fructose, plus some C_3 and C_4 epimers. This reversible reaction is well known to organic chemists as an example of aldol condensation, after which the enzyme aldolase is named. Note the small value of ΔG, showing that the reaction is close to equilibrium *in vivo*.

5. Isomerization

$$\text{dihydroxyacetone-}\textcircled{P} \xrightarrow{\text{triose phosphate isomerase}} \text{glyceraldehyde-3-}\textcircled{P}$$

($\Delta G^{0\prime} = +7.6$ kJ mol^{-1}; $\Delta G = -2.4$ kJ mol^{-1} under cellular conditions).

From this point on, we have two molecules of glyceraldehyde-3-phosphate, etc. from each molecule of glucose, as illustrated in Figure 11.4. I emphasize this by adding (×2) after each heading. The chemistry has converged to one type of compound, simplifying the rest of glycolysis.

6. Oxidation (×2)

$$\text{glyceraldehyde-3-}\textcircled{P} + \text{NAD}^+ + \text{P}_i \xrightarrow{\text{glyceraldehyde-3-P dehydrogenase}} \text{1,3-bisphosphoglycerate}$$

$+\text{NADH}+\text{H}^+$

($\Delta G^{0\prime} = +6.3$ kJ mol^{-1}; $\Delta G = -1.3$ kJ mol^{-1} under cellular conditions).

This can be considered as a two-stage reaction, involving an electron transfer oxidation and a phosphorylation that does not involve ATP. (The reaction involves a thioester intermediate in the active site of glyceraldehyde-3-phosphate dehydrogenase. See Box 11.1 for more on the importance of thioesters in metabolism.)

Despite the slightly positive ΔG, this reaction proceeds because of the overall favourable ΔG for glycolysis, and there is an advantage in saving energy for the next step in the form of the phosphate group on position 1 of 1,3-bisphosphoglycerate (BPG). This is in the form of an acid anhydride (specifically, an acyl phosphate) between phosphoric acid and the carboxyl group.

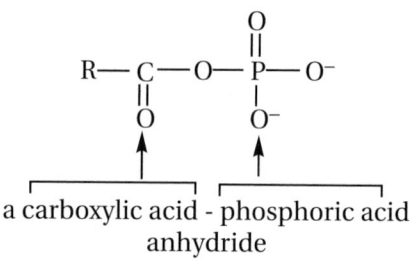

a carboxylic acid - phosphoric acid anhydride

As such, it has a high phosphate group-transfer potential, capable of synthesis of ATP in step 7.

(In comparison, the familiar terminal phosphate in ATP is a phosphoric acid anhydride, also having high phosphate group-transfer potential.)

7. Phosphate transfer (×2)

$$\text{1,3-bisphosphoglycerate} + \text{ADP} \xrightarrow{\text{phosphoglycerate kinase}} \text{3-phosphoglycerate} + \text{ATP}$$

($\Delta G^{0\prime} = -19$ kJ mol^{-1}; $\Delta G = +0.1$ kJ mol^{-1} under cellular conditions).

Box 11.1 Thioesters and coenzyme A

As we have seen in many examples so far, phosphate in the form of phosphorylated compounds is ubiquitous in metabolism, especially in energy metabolism. Phosphate was scarce in the prebiotic world and it has been proposed that other 'high-energy' compounds must have been present to 'drive' ancient metabolic pathways before phosphorylated compounds became established. There is considerable evidence for this in the literature on the origins of life. One strongly supported candidate for such a role is the thioester. Chemically, thioesters are formed by the condensation of a carboxylic acid and a thiol:

$$\text{R-COOH} + \text{HSR}' \longrightarrow \underset{\text{a thioester}}{\text{RCOSR}'} + H_2O \qquad (11.B1)$$

Thioesters have a $\Delta G^{0\prime}$ of hydrolysis much higher than ordinary esters, RCOOR.

In modern metabolic pathways, thioesters play a central role in the biological oxidation of carbonyl groups ($-C=O$) to carboxyl groups ($-COOH$). Biochemically, these reactions are of two types:

$$\underset{\text{aldehyde}}{\text{RCHO}} + \text{HSR}' \longrightarrow \text{RCOSR}' + 2e^- + 2H^+ \qquad (11.B2)$$

$$\underset{\substack{\|\ \| \\ O\ O \\ \alpha\text{-keto acid}}}{\text{RC-COH}} + \text{HSR}' \longrightarrow \text{RCOSR}' + 2e^- + 2H^+ + CO_2 \qquad (11.B3)$$

These reactions bring about the direct formation of a high-energy thioester bond. They are 'driven' by the energy released by the electron transfers shown in eqns (11.B2) and (11.B3). (In vivo, the electron acceptors for reactions (11.B2) and (11.B3) are NAD^+ or $NADP^+$.)

Thioesters RCOSR' have a high group-transfer potential for acyl groups, RC=O. Thus

$$\text{RCOSR}' + P_i \longrightarrow \text{RCO-P} + R'\text{SH} \qquad (11.B4)$$

$$\text{RCO-P} + \text{ADP} \longrightarrow \text{RCOOH} + \text{ATP} \qquad (11.B5)$$

$$\overline{\text{RCOSR}' + P_i + \text{ADP} \longrightarrow \text{RCOOH} + \text{RSH} + \text{ATP}} \qquad (11.B6)$$

Overall reaction (11.B6) is the sum of (11.B4) and (11.B5). It results in the formation of a carboxylic acid, and via a coupling mechanism mediated by an appropriate enzyme, in the formation of ATP. This is an example of a substrate level phosphorylation, mentioned previously. As such it could have been a source of ATP in ancient organisms, before the advent of oxidative phosphorylation. This type of ATP formation occurs without the complex electron carriers involved

Box 11.1 (continued)

in the ETC, and without the need for a proton gradient and the protonmotive force. Most importantly, thioesters are central to energy metabolism in modern organisms.

1. Thioesters in glycolysis: an example of the reaction type

$$\text{RCHO} + \text{HSR}' \longrightarrow \text{RCOSR}' + 2e^- + 2H^+ \quad (11.B2)$$

Reaction 6 of glycolysis (see above) is the oxidation of an aldehyde:

$$\text{glyceraldehyde-3-P} + \text{NAD}^+ + P_i \xrightarrow{\text{glyceraldehyde-3-P dehydrogenase}}$$

$$\text{1,3-bisphosphoglycerate} + \text{NADH} + H^+ \quad (11.B7)$$

This reaction, the exothermic oxidation of an aldehyde, is used to drive the synthesis of a high-energy compound, the acyl phosphate 1,3-BPG. The last step in the mechanism shows that an acyl thioester intermediate forms, which is attacked by the phosphate group to produce 1,3-BPG (Figure 11.B1) (Voet *et al.* 2002 p. 395).

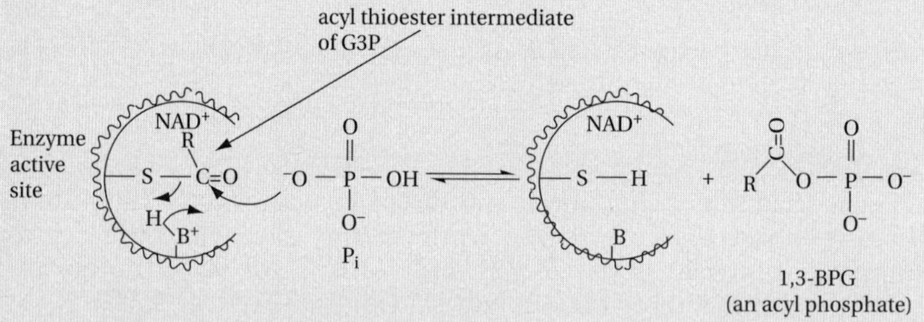

Figure 11.B1. The final step in the proposed mechanism for the oxidation of G3P to 1,3-bisphosphoglycerate (1,3-BPG), catalysed by glyceraldehyde-3-phosphate dehydrogenase (GAPDH). R = [(P)OCH$_2$CHOH].

Acyl phosphates have a large negative free energy for acyl group transfer. The next step in glycolysis (step 7 above) uses this to generate ATP from ADP:

$$\text{1,3-BPG} + \text{ADP} \xrightleftharpoons[\text{kinase}]{\text{phosphoglycerate}} \text{3-phosphoglycerate (3PG)} + \text{ATP}$$

The energy of aldehyde oxidation (of G3P) is conserved in two ways by steps 6 and 7 of glycolysis: the production of reducing power in the form of NADH, plus the formation of 1 mole of ATP. This is very efficient use of energy, which in glycolysis up to this point had actually consumed ATP. From step 7 onwards, glycolysis goes into its net energy-producing stage, as mentioned above in the discussion on glycolysis. The net formation of two ATP by glycolysis comes about by substrate level phosphorylation, which does not involve oxygen.

Box 11.1 (continued)

Glycolysis, as we have seen, can be used by anaeroboic organisms as their sole energy source. This is in keeping with glycolysis being an ancient metabolic pathway and the fact that a thioester is involved lends credence to the argument that thioesters may have been important in the emergence of early life.

2. Coenzyme-A: examples of the reaction type

$$\text{RC-COH} + \text{HSR} \longrightarrow \text{RCOSR'} + 2e^- + 2H^+ + CO_2 \qquad (11.B3)$$
$$\overset{\|\ \|}{\underset{O\ O}{}}$$

CoA is involved in two key steps involving the TCA cycle. In its uncombined form the abbreviation is CoA-SH because of the terminal thiol (SH) group. It has a complex molecular structure:

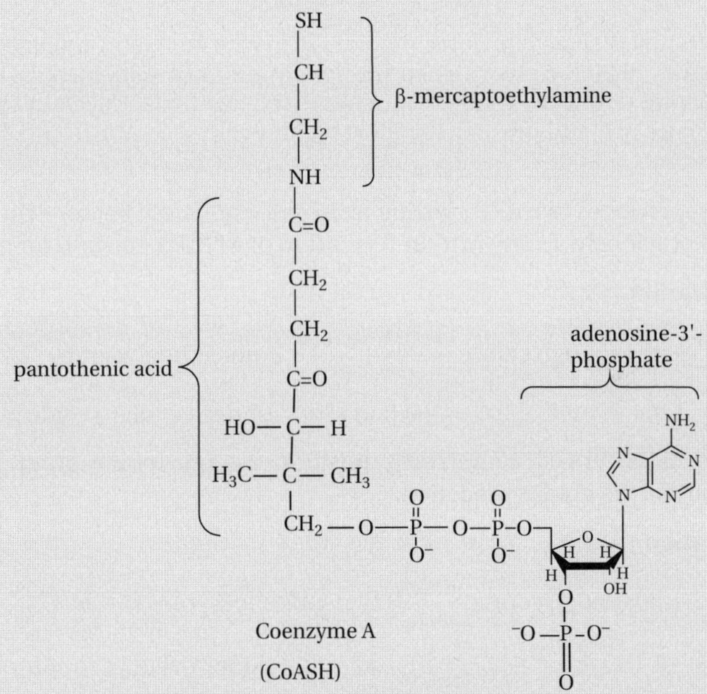

CoA-SH acts as a carrier of acetyl and other acyl groups. Acetyl CoA, or CoA-S-COCH$_3$, is a thioester of acetic acid having a $\Delta G^{0\prime}$ of hydrolysis of -31.5 kJ mol^{-1}, almost the same as that of ATP (-30.5 kJ mol^{-1}). Thus, formation of a thioester conserves energy that can be used in other reactions to drive endothermic processes. The linking of glycolysis to the TCA cycle allows access of pyruvate to oxidative phosphorylation. Instead of the formation of a mere two ATP per glucose molecule by glycolysis alone, a total of 32–38 ATP may be achieved. Significantly, the access of pyruvate to the TCA cycle involves not only thioesters. The conversion of pyruvate to acetyl CoA is carried out by a

> **Box 11.1 (continued)**
>
> multienzyme complex, pyruvate dehydrogenase. The complex contains three enzymes and carries out five sequential reactions. Overall:
>
> $$\text{pyruvate} + \text{CoA-SH} + \text{NAD}^+ \longrightarrow \text{acetyl-CoA} + \text{CO}_2 + \text{NADH} + \text{H}^+$$
>
> (11.7)
>
> Also involved are the coenzymes thiamine pyrophosphate (see Chapter 8), lipoamide, and FAD, making five coenzymes in all. This is an impressive array, reflective of the importance of this vital linkage between the two ancient metabolic pathways that played such a role in the development of modern metabolism.
>
> The third and final example is the involvement of thioester chemistry in the TCA cycle, which takes place at step 6:
>
> $$\text{succinyl-CoA} + \text{GDP} + \text{P}_i \longrightarrow \text{succinate} + \text{GTP} + \text{CoASH} \quad (11.12)$$
>
> $$\Delta G^{0\prime} = +4.4 \text{ kJ mol}^{-1}$$
>
> As mentioned above, the formation of GTP is the energetic equivalent of formation of ATP. Thus, we see the involvement of thioesters at three crucial points, one in glycolysis and two involving the TCA cycle, all at the very centre of metabolism in the most highly evolved organisms on Earth.

This step produces two ATP starting from glucose. These balance the two ATP consumed previously, so any further formation of ATP is a net gain for glycolysis.

8. Isomerization (×2)

$$\text{3-phosphoglycerate} \xrightarrow{\text{phosphoglycerate mutase}} \text{2-phosphoglycerate}$$

($\Delta G^{0\prime} = +4.4$ kJ mol^{-1}; $\Delta G = +0.8$ kJ mol^{-1} under cellular conditions).

This step is necessary to correctly position the phosphate group on C_2 in preparation for the enolase reaction.

9. Dehydration (×2)

$$\text{2-phosphoglycerate} \xrightarrow{\text{enolase}} \text{Phosphoenolpyruvate} + \text{H}_2\text{O}$$

($\Delta G^{0\prime} = +2.1$ kJ mol^{-1}; $\Delta G = +1.1$ kJ mol^{-1} under cellular conditions).

By rearrangement of the bonding in the molecule, this reaction produces a phosphate with high group-transfer potential, another so-called 'high-energy' phosphate, to be used in step 10 to produce two further molecules of ATP.

10. Phosphate transfer (×2)

$$\text{Phosphoenolpyruvate} + \text{ADP} \xrightarrow{\text{pyruvate kinase}} \text{pyruvate} + \text{ATP}$$

($\Delta G^{0\prime} = -31.7$ kJ mol^{-1}; $\Delta G = -23$ kJ mol^{-1} under cellular conditions).

Pyruvate kinase is an allosteric enzyme made up of four subunits. It is inhibited by ATP, to slow down production of the latter when it is in high concentration in the cell.

(The above *in vivo* thermodynamic data are from Minakami and Yoshikawa (1965). They refer to glycolysis in red blood cells. Data referring to heart muscle (Newsholme and Start 1973, reported in Voet *et al.* 2002 p. 407) give somewhat different values, but qualitatively show the same trends.)

The net reaction of glycolysis from glucose to pyruvate is:

$$\text{glucose} + 2\text{ADP} + 2\text{P}_i + 2\text{NAD}^+ \longrightarrow 2\,\text{pyruvate} + 2\text{ATP} + 2\text{NADH} + 2\text{H}^+ \quad (11.1)$$

The ATP formed during glycolysis is another example of the substrate-level phosphorylation that was mentioned in Chapter 10. The Gibbs energy to drive substrate-level phosphorylation comes from the group-transfer potential of certain phosphate groups. In glycolysis, these powerful phosphate donors to ADP are 1,3-BPG and phosphoenolpyruvate (PEP).

Note again the formation of two molecules of the three-carbon molecule pyruvate from each six-carbon molecule of glucose. Pyruvate may have one of three main fates (Figure 11.5).

In aerobic metabolism, it loses carbon dioxide and the remaining two-carbon fragment becomes linked to CoA to form acetyl-CoA. This enters the citric acid cycle. Two anaerobic fates are possible for pyruvate. It may be reduced to lactate (lactic acid). Some bacterial species such as *Lactobacillus*, responsible for the taste of sour milk, produce lactate and survive by anaerobic glycolysis. Animal examples include mammalian red blood cells, which rely on anaerobic glycolysis for energy, and muscle tissue during vigorous exercise, when oxygen cannot be supplied rapidly enough to the muscles. There is a limit to this; muscle pain occurs as lactate builds up, and one is forced to stop. Deep breathing eventually

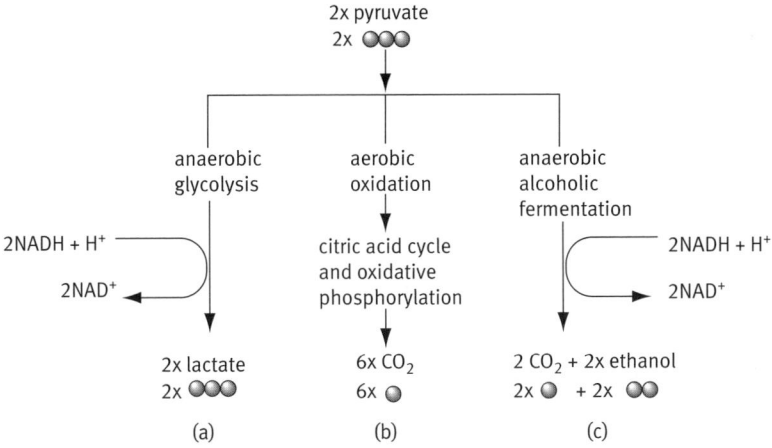

Figure 11.5. A summary of three fates of pyruvate derived from glucose. (a) Anaerobic formation of lactate, e.g. in exercising muscle tissue. (b) Formation of acetyl Co.A, entry to the citric acid cycle, oxidative phosphorylation to ATP, CO_2, and water. (c) Fermentation to form ethanol and CO_2. ○ represents one carbon atom.

allows aerobic metabolism to take over, the lactate is oxidized to pyruvate then converted to acetyl-CoA.

The other anaerobic fate of pyruvate is its conversion to acetaldehyde then ethanol in those organisms capable of alcoholic fermentation, such as yeasts. This is utilized by the brewing industry to produce a variety of alcoholic beverages, notably beer and wine. Of overall biological importance is the ability of anaerobic glycolysis to produce enough ATP to sustain a variety of living organisms. It is possible to make an estimate of the efficiency of anaerobic glycolysis in the utilization of glucose. For fermentation to lactate:

$$\text{glucose} \longrightarrow 2\text{lactate} + 2\text{H}^+ \quad \Delta G^{0\prime} = -196 \text{ kJ mol}^{-1} \quad (11.2)$$

For fermentation to alcohol:

$$\text{glucose} \longrightarrow 2\text{CO}_2 + 2\text{CH}_3\text{CH}_2\text{OH} \quad \Delta G^{0\prime} = -235 \text{ kJ mol}^{-1} \quad (11.3)$$

Each of these reactions is coupled to the formation of 2ATP, which requires $2 \times \Delta G^{0\prime}$ for the synthesis of ATP, or $2 \times 30.5 = 61$ kJ mol^{-1}. The 61 kJ mol^{-1} is the energy return to the cell for the release of 196 or 235 kJ mol^{-1}. The rest is dissipated as heat. Thus, the efficiencies of lactate and alcoholic fermentation are $(61/196) = 31\%$ and $(61/235) = 26\%$, respectively, under biochemical standard state conditions. Under cellular conditions, the efficiency has been calculated to be greater than 50% (data from Voet et al. 2002 p. 406). A comparison with oxidative phosphorylation, which yields up to 38 ATP per glucose (see below) reveals that anaerobic glycolysis is a much less efficient user of the free energy available from a mole of glucose. When oxygen became available to life courtesy of photosynthesis, it was soon 'chosen' by natural selection as the ultimate electron acceptor by a vast number of species.

On the other hand, it is fortunate that we mammals still have access to anaerobic glycolysis, as it can produce ATP up to 100 times faster than oxidative phosphorylation, a useful property in hard-working muscle tissue. The lactate is aerobically processed as soon as we are forced to slow down.

Close regulation of metabolic pathways is necessary to allow them to adapt to changing cellular conditions. Such control is often exercised near the start and end of a pathway, and sometimes at key intermediate steps. Regulation of glycolysis occurs at three points.

1) Conversion of glucose to glucose-6-phosphate (Figure 11.4) is catalyzed by hexokinase ($\Delta G^{0\prime} = -16.7$ kJ mol^{-1}, $\Delta G = -33.9$ kJ mol^{-1}). This reaction is inhibited by its product, glucose-6-phosphate. Kinases are enzymes that transfer the terminal phosphate group of ATP to nucleophilic acceptors. Hexokinase can use several hexoses such as glucose, fructose, and mannose as substrates. There are specific glucokinases in some tissues, for example mammalian liver and pancreas.
2) Conversion of fructose-6-phosphate to fructose-1,6-bisphosphate is catalyzed by phosphofructokinase ($\Delta G^{0\prime} = -14$ kJ mol^{-1}, $\Delta G = -19$ kJ mol^{-1}).

This enzyme is inhibited by ATP.

3) The last step in glycolysis, formation of pyruvate and ATP from PEP, is catalyzed by pyruvate kinase ($\Delta G^{0\prime} = -31.7$ kJ mol^{-1}, $\Delta G = -23$ kJ mol^{-1}). This reaction is also inhibited by ATP.

All three reactions have large negative $\Delta G^{0\prime}$ values, making them potentially useful as control points. Somewhat different values have been determined (Voet et al. 2002 p. 407) but the overall features are the same. Regulation at these points makes perfect logic, as high levels of all three inhibitors indicate that plenty of ATP is available; slowing down of its synthesis makes cellular sense.

11.4 The pentose phosphate pathway

Although ATP is the major energy source at any given instant, cells must have another essential 'currency' available. This is reducing power. As well as ATP, cells require a constant supply of NADPH for anabolic endothermic reactions such as the reductive biosynthesis of fatty acids and cholesterol. Metabolically, NADH and NADPH are not interchangeable. NADH uses the free energy of metabolite oxidation to synthesize ATP via oxidative phosphorylation, while NADPH uses the same energy source for reductive biosynthetic purposes. Both NADH and NADPH act as coenzymes for a variety of dehydrogenase enzymes. Each type of dehydrogenase is specific for either NADH or NADPH. They can discriminate between the two structures. Cells maintain their NAD$^+$/NADH ratio at about 1000/1. This favours oxidative, catabolic processes, as NAD$^+ \rightarrow$ NADH + H$^+$ is oxidative. The NADP$^+$/NADPH ratio is maintained at about 0.01/1 and this favours the reductive processes typical of biosynthesis.

NADPH is formed from glucose-6-phosphate by the pentose phosphate pathway, an alternative to glycolysis. Tissues such as liver, adipose tissue, and the adrenal cortex are involved in lipid biosynthesis and are rich in pentose phosphate enzymes. They are located in the cytosol, the site of fatty acid biosynthesis. The pathway is versatile, producing three-, four-, five-, six-, and seven-carbon sugars. In muscle tissue, which is mainly involved in energy production, the pentose phosphate enzymes are essentially absent.

The overall reaction for the pathway is:

$$3\text{Glc6P} + 6\text{NADP}^+ + 3\text{H}_2\text{O} \rightleftharpoons 6\text{NADPH} + 6\text{H}^+ + 2\text{F6P} + 3\text{CO}_2 + \text{G3P} \quad (11.4)$$

glucose-6-phosphate   fructose-6-P   glyceraldehyde-3-P

Although, as indicated, a number of sugars are formed during the cycle, I won't go into further detail. The important point is that reducing power in the form of 6NADPH is generated. One important sugar formed is ribose-5-phosphate (R5P), which can be directed to the synthesis of nucleotides and subsequently of DNA/RNA.

What regulates the level of activity of the pentose phosphate pathway? It is controlled by the activity of the first enzyme in the pathway, glucose-6-phosphate dehydrogenase. The activity of glucose-6-phoshate dehydrogenase is in turn regulated by the level of NADP$^+$:

$$\text{Glc6P} \xrightarrow{\text{Glc6P dehydrogenase}} \text{6-phosphogluconolactone} \quad (11.5)$$
$$\text{NADP}^+ \quad \text{NADPH} + \text{H}^+$$

$$\Delta G_{\text{reaction}} = -17.6 \text{ kJ mol}^{-1}$$

This reaction is well set up for a regulatory role. It has a large negative ΔG, so is essentially irreversible. As the cell consumes NADPH, the level of NADP$^+$ will increase, activate the enzyme, and stimulate the pentose phosphate pathway.

11.5 The citric acid or tricarboxylic acid cycle

The TCA cycle is much more than just an addition to the end of glycolysis. It is a central metabolic pathway for the conversion of several types of foodstuff to metabolic energy (see Figure 11.3). It also supplies the reactants for a number of biosynthetic pathways. The starting point for the study of the cycle is the formation of acetyl-CoA from pyruvate or from β-oxidation of fatty acids. The citric acid cycle, via a series of eight reactions, oxidizes acetyl-CoA to produce energy as guanosine triphosphate (GTP), the reducing power of NADH and FADH$_2$, plus CO$_2$. GTP is a purine nucleoside triphosphate, similar to ATP in structure and properties such as large negative ΔG of hydrolysis. Like ATP, it can be used to drive endothermic reactions via energy coupling, so in energy terms GTP = ATP. Oxidative phosphorylation and electron transport are tightly linked to the cycle (Chapter 10). The NAD/NADH redox system acts as a shuttle that carries electrons released from foods to the ETC, ultimately transferring them to oxygen and forming water. ATP is formed by oxidative phosphorylation, as described in Chapter 10.

Expressing this generally:

$$\text{AH}_2 + \text{NAD}^+ \longrightarrow \text{A} + \text{NADH} + \text{H}^+ \quad (11.6)$$

where AH$_2$ represents a typical substrate, for example $-$CH$_2-$ in a foodstuff, which is oxidized to A, and NAD$^+$ is reduced to NADH+H$^+$. The NADH+H$^+$ is reoxidized to NAD$^+$ when it transfers its reducing power to the ETC in mitochondria. This step is the reverse of eqn (11.2), and the electron acceptor A is oxygen. It recycles the precious NAD$^+$. The citric acid cycle components are located in the mitochondrial matrix of eukaryotes, except for succinate dehydrogenase, which is embedded in the inner mitochondrial membrane (Figure 10.2). The TCA cycle is found in the cytosol of prokaryotes. In eukaryotes, all the substrates must be formed in the mitochondria or be capable of being transported there. The

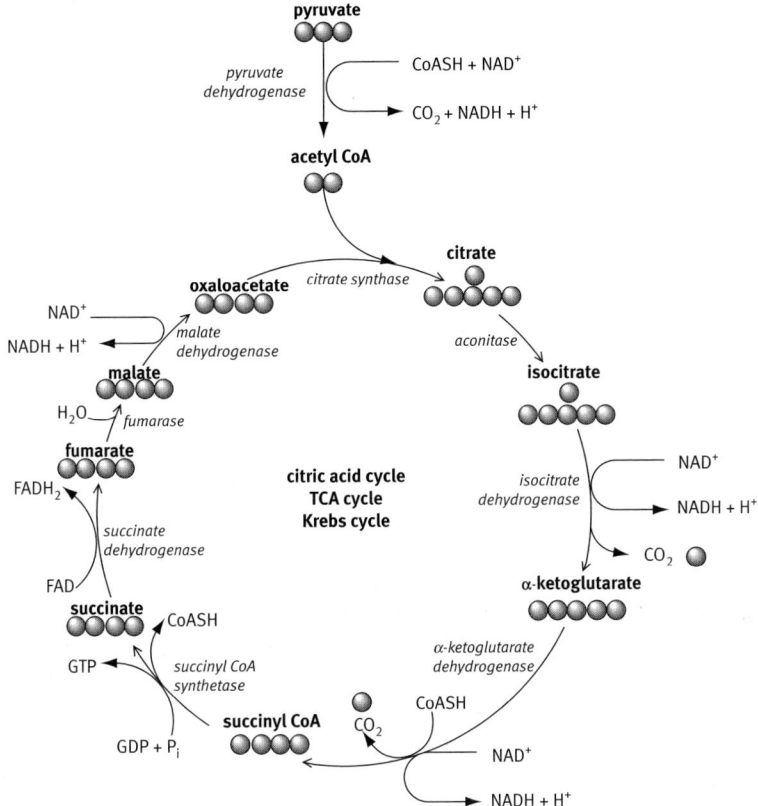

Figure 11.6. The citric acid or TCA cycle. One turn of the cycle oxidizes the equivalent of one acetyl group to two CO_2, while generating three ($NADH+H^+$), one $FADH_2$, and one GTP, and regenerating one oxaloacetate to continue the cycle. Two turns of the cycle are required to oxidize 1 mole of glucose, via input of two pyruvate from glycolysis. ● represents one carbon atom.

products are either consumed there or are capable of transport out of the mitochondrion to the cytosol.

In the following section, the $\Delta G^{0\prime}$ values are from Garrett and Grisham (2010 p. 572). The ΔG values shown are those pertaining to cellular conditions (Newsholme and Leech 1983).

1) Pyruvate from glycolysis in the cytoplasm enters the mitochondrion via a specific transporter (Figure 11.6). It undergoes activation to form a thioester with CoA (CoASH):

$$\text{pyruvate} + \text{CoASH} + \text{NAD}^+ \longrightarrow \text{acetyl-CoA} + CO_2 + \text{NADH} + H^+ \quad (11.7)$$

$$\Delta G^{0\prime} = -33.4 \, \text{kJ mol}^{-1}$$

This reaction is catalyzed by the pyruvate synthase complex, which in mammals is made up of five enzymes (see Box 11.1).

2) Catalyzed by citrate synthase, acetyl-CoA (two carbons) is linked to oxaloacetate (four carbons) provided by the cycle to form citrate (six carbons).

$$\text{oxaloacetate} + \text{acetyl-CoA} + H_2O \longrightarrow \text{citrate} + \text{CoASH} \qquad (11.8)$$

$$\Delta G^{0\prime} = -31.5 \text{ kJ mol}^{-1}$$

$$\Delta G = -31.5 \text{ kJ mol}^{-1}$$

3) Catalyzed by aconitase, citrate is isomerized to isocitrate (still having six carbons), a reaction that involves moving an OH group from one carbon atom to a neighbouring one. Aconitase requires Fe^{2+} for activity.

$$\text{citrate} \longrightarrow \text{isocitrate} \quad \Delta G^{0\prime} = +6.7 \text{ kJ mol}^{-1} \qquad (11.9)$$

$$\Delta G = +0.8 \text{ kJ mol}^{-1}$$

4) Catalyzed by isocitrate dehydrogenase, isocitrate is converted to α-ketoglutarate. This is an oxidative decarboxylation.

$$\text{isocitrate} + NAD^+ \longrightarrow \alpha\text{-ketoglutarate} + NADH + H^+ + CO_2 \qquad (11.10)$$

$$\Delta G^{0\prime} = -8.4 \text{ kJ mol}^{-1}$$

$$\Delta G = -17.5 \text{ kJ mol}^{-1}$$

This step is notable as it is the first in the cycle to achieve an oxidation of one net carbon to CO_2 with concurrent reduction of NAD^+ to $NADH + H^+$. The whole objective of the large expenditure on metabolites and energy in the citric acid cycle is for it to produce much more energy than it consumes. The production of reducing power in the form of $NADH+H^+$ largely achieves this objective. As we have seen, the channelling of electrons from species such as $NADH/H^+$ and $FADH_2$ into the ETC generates ATP.

5) Catalyzed by α-ketoglutarate dehydrogenase complex, another oxidative decarboxylation step produces more $NADH/H^+$.

$$\alpha\text{-ketoglutarate} + NAD^+ + \text{CoASH} \rightleftharpoons \text{succinyl-CoA} + NADH + H^+ + CO_2$$
$$(11.11)$$

$$\Delta G^{0\prime} = -30 \text{ kJ mol}^{-1}$$

$$\Delta G = -44 \text{ kJ mol}^{-1}$$

In accord with our previous observations on metabolic control, this highly exothermic reaction is another point of regulation of the cycle. The reaction is complex, similar to that forming acetyl-CoA from pyruvate. It takes place in several steps and there is a requirement for TPP, FAD, lipoic acid, and Mg^{2+}. The CO_2 is removed, making the cycle irreversible in vivo. To this point, two molecules of CO_2 have been formed, leaving a four-carbon acid that needs to be converted back to oxaloacetate to continue the next cycle. The carbon atoms in the two CO_2 molecules released do not come from the acetyl of the acetyl-CoA added in the first step, as shown by specific carbon-labelling experiments.

6) Catalyzed by succinyl-CoA synthetase, succinate is formed and the large negative ΔG of hydrolysis of the succinyl-CoA is coupled to the synthesis of GTP, making more energy available through this 'high-energy' compound.

$$\text{succinyl-CoA} + \text{GDP} + \text{P}_i \longrightarrow \text{succinate} + \text{GTP} + \text{CoASH} \qquad (11.12)$$

$$\Delta G^{0\prime} = -3.3 \text{ kJ mol}^{-1}$$
$$\Delta G = \sim 0 \text{ kJ mol}^{-1}$$

7) Catalyzed by succinate dehydrogenase, succinate is oxidized to fumarate, with the generation of FADH$_2$, which as we have seen feeds into the mitochondrial ETC (Figure 10.2).

$$\text{succinate} + \text{FAD} \longrightarrow \text{fumarate} + \text{FADH}_2 \qquad (11.13)$$

$$\Delta G^{0\prime} = +0.4 \text{ kJ mol}^{-1}$$
$$\Delta G = \sim 0 \text{ kJ mol}^{-1}$$

8) Catalyzed by fumarase, water is added to fumarate to form malate.

$$\text{fumarate} + \text{H}_2\text{O} \longrightarrow \text{malate} \qquad (11.14)$$

$$\Delta G^{0\prime} = -3.8 \text{ kJ mol}^{-1}$$
$$\Delta G = \sim 0 \text{ kJ mol}^{-1}$$

9) Catalyzed by malate dehydrogenase, malate is oxidized to oxaloacetate, completing the cycle with the formation of further reducing power as NADH/H$^+$.

$$\text{malate} + \text{NAD}^+ \longrightarrow \text{oxaloacetate} + \text{NADH} + \text{H}^+ \qquad (11.15)$$

$$\Delta G^{0\prime} = +29.7 \text{ kJ mol}^{-1}$$
$$\Delta G = \sim 0 \text{ kJ mol}^{-1}$$

This final step is endothermic, but overall the citric acid cycle from pyruvate is exothermic:

$$\text{pyruvate} + 4\text{NAD}^+ + \text{FAD} + \text{GDP} + \text{P}_i + 2\text{H}_2\text{O} \longrightarrow 2\text{CO}_2 + 4\text{NADH}$$
$$+ 4\text{H}^+ + \text{FADH}_2 + \text{GTP} \qquad (11.16)$$

For eqn (11.16) $\Delta G^{0\prime} = -77.7 \text{ kJ mol}^{-1}$

Starting from acetyl-CoA:

$$3\text{NAD}^+ + \text{FAD} + \text{GDP} + \text{P}_i + \text{acetyl-CoA} + 2\text{H}_2\text{O} \longrightarrow 2\text{CO}_2 + 3\text{NADH}$$
$$+ \text{H}^+ + \text{FADH}_2 + \text{GTP} + \text{CoASH}$$

$$\Delta G^{0\prime} = -40 \text{ kJ mol}^{-1}$$

From either starting point, the cycle will run *in vivo*.

Why go through such a complex process to oxidize acetyl groups $\mathrm{CH_3\text{-}\underset{O}{\overset{\|}{C}}\text{--}}$ to $2CO_2$, which is what the citric acid cycle really does? Direct oxidation of pyruvate would require cleavage of C−C bonds. In biochemical systems, the chemical mechanisms used cleaves C−C bonds (i) between carbon atoms α- and β- to a carbonyl group or (ii) α- to an α-hydroxyketone.

biological cleavage

$$R\text{—}C_\beta\text{—}C_\alpha\text{—}\underset{O}{\overset{\|}{C}}\text{—}R_1 \quad \text{or} \quad \text{—}\underset{HO}{C_\alpha}\text{—}\underset{O}{\overset{\|}{C}}\text{—}R_1$$

(i)   (ii)

Acetate has no β-carbon and chemically hydroxylation of acetate is unfavourable.

The end product of the evolutionary 'search' for a means to oxidize acetate is hardly a compromise, considering the versatility of the resulting citric acid cycle.

The overall yield of ATP from the oxidation of glucose via glycolysis, the citric acid cycle, and oxidative phosphorylation is as follows:

glycolysis: glucose to pyruvate	2 net ATP
citric acid cycle	2 net ATP (as GTP)
oxidative phosphorylation	28–34 net ATP
total per glucose molecule	32–38 net ATP

The above numbers include some consensus values used in the calculation of ATP yield in oxidative phosphorylation (Campbell and Farrell 2006 p. 563; Garrett and Grisham 2010 p. 623). Phosphorylation and the redox reactions are not directly coupled, so the NADH/ATP and $FADH_2$/ATP ratios are not whole numbers. The modern consensus values are 2.5 and 1.5 ATP produced from one NADH and one $FADH_2$, respectively. Varying metabolic conditions can alter the estimate of ATP. Other estimates propose a yield of 36–38 net ATP (Campbell et al. 2008 p. 176; Voet et al. 2002 p. 484).

The citric acid cycle also has a role in producing a range of precursors for various biosynthetic (anabolic) pathways, and is integrated with these. Examples include:

- from acetyl-CoA: steroids and fatty acids
- from α-ketoglutarate: some amino acids, purines
- from oxaloacetate and fumarate: some amino acids, pyrimidines (purines and pyrimidines are components of DNA and RNA)
- from succinyl-CoA: porphyrins (found in the structures of haemoglobin and chlorophylls).

Thus, the citric acid cycle is involved in both catabolic and anabolic metabolism, and is said to be amphibolic for this reason. It is important to remember that in living cells many of the metabolic pathways are often operating simultaneously. Every stage is under control to maintain optimal concentrations of substrates for the numerous reactions. It is just for the convenience of understanding that we discuss them separately.

11.6 Regulation of the citric acid cycle

The TCA cycle is situated between glycolysis and the ETC where oxidative phosphorylation occurs, that is, it is located at the heart of aerobic respiration in the cell and must be regulated carefully. If the cycle were to run uncontrolled, there would be an oversupply of ATP and NADH. If it were to run too slowly, there would soon be an undersupply of these components, and the energetic requirements of the cell could not be met. Control occurs at three points in the cycle, plus at the pyruvate to acetyl-CoA step. Why is control exerted at these points in particular?

Some of the reactions in the cycle operate close to equilibrium, that is $\Delta G \sim 0$. Thus, small changes in concentration of reactants or products could push these reactions either forwards or backwards. On the other hand, reactions that have a large negative ΔG under cellular conditions, if inhibited, can come to a complete stop but won't go into reverse. It is usually at these highly exothermic steps that metabolic control is manifested.

1) Pyruvate to acetyl-CoA: the pyruvate decarboxylase complex is inhibited by ATP and NADH.
2) Oxaloacetate to citrate: citrate synthase is inhibited by ATP, NADH, succinyl-CoA, and citrate. It is an allosteric enzyme. As the first reaction of the cycle it is also a logical candidate for regulation. This is commonly the case in metabolic pathways.
3) Isocitrate to α-ketoglutarate: isocitrate dehydrogenase is inhibited by ATP and NADH, and stimulated by ADP and NAD^+. ADP is an allosteric activator. There are many instances in which ATP and NADH inhibit enzymes of a pathway, while ADP and NAD^+ activate the same enzymes.
4) α-Ketoglutarate to succinyl-CoA: the 2-ketoglutarate dehydrogenase complex is inhibited by ATP, NADH, and succinyl-CoA. ADP and NAD^+ are activators.

Use of the inhibitors and activators above make perfect sense. When the ratios of ATP/ADP and NADH/NAD^+ are high, the cycle needs to be slowed, and vice versa. Some of the reactions are inhibited by their own products, which also makes perfect regulatory sense.

I should mention that in plants, plus some bacteria and algae, an extra pathway related to the citric acid cycle exists: the glyoxylate pathway. This pathway allows plants to use acetyl-CoA for the synthesis of carbohydrates, indeed for all the carbon-based compounds needed. This does not occur in animals. Animals can convert carbohydrates to fats, but not fats to carbohydrates. Although animals can produce acetyl-CoA from the catabolism of fatty acids, they do not possess the enzymes of the glyoxylate pathway to convert it to carbohydrates. While the citric acid cycle can provide some intermediates for biosynthesis, it cannot do so for carbohydrates. The citric acid cycle does not lead to a net synthesis of intermediates, as two CO_2 molecules are given off for each two-carbon acetyl-CoA fragment that enters it. Thus, it would be impossible to build up large amounts of storage molecules such as carbohydrates. The availability of the extra enzymes of the glyoxylate pathway allows plants, etc. to overcome this problem by utilizing acetyl-CoA directly.

The macromolecular carbohydrates are called polysaccharides and, like all polymers, consist of monomers, in this case simple sugars—the monosaccharides—joined by covalent linkages. Examples are starch in plants, and glycogen in the muscle and liver of animals. Both these polysaccharides have a branched structure and consist of glucose units, but they are joined in slightly different ways. Enzymes exist in cells and are made available as appropriate to break down the efficient energy storage molecules starch and glycogen to glucose. The paths of synthesis and breakdown of polysaccharides are regulated by feedback controls similar in principle to those described for glycolysis and the citric acid cycle. Glucose can also be obtained directly in the diet or from other simple sugars by enzyme-catalyzed reactions. Whatever its source, the glucose may enter the glycolytic pathway and undergo metabolism as described above.

It is by such means that living cells self-regulate at key control points in their metabolic pathways and maintain the flow of energy that is essential for their survival.

Finally, here is a reminder of the different locations in the cell of the components of glycolysis, the TCA cycle, the ETC, and the ATP synthase (Figure 11.7).

Glucose comes from outside the cell (from the blood in mammals) by active transport via a specific protein channel. Oxygen diffuses into the cell from the blood capillaries and CO_2 diffuses out to the blood. The enzymes of glycolysis are in the cytosol, but pyruvate from glycolysis (and other pathways) is fed into the mitochondria, where the citric acid cycle enzymes and the ETC are located.

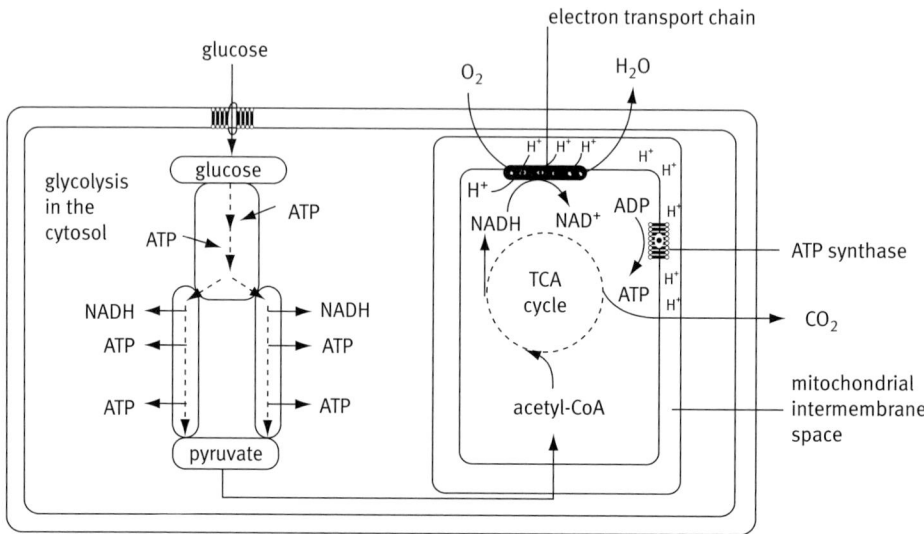

Figure 11.7. Location of the glycolytic and citric acid cycles, electron transport chain, and ATP synthase components within an aerobic, eukaryotic cell.

GENERAL REFERENCES

De Duve, C. (2005) *Singularities: Landmarks on the pathways of Life*. Cambridge University Press, Cambridge.

REFERENCES

Campbell, M.K. and Farrell, S.O. (2006) *Biochemistry*, 5th edn. Thomson Brooks/Cole, Belmont, CA, p.563.
Campbell, N.A., Reece, J.B., Urry, L.A., Cain, M.L., Wasserman, S.A., Minorsky, P.V., *et al.* (2008) *Biology*, 8th edn. Pearson Benjamin Cummings, San Francisco.
Garrett, R.H. and Grisham, C.M. (1999) *Biochemistry*, 2nd edn. Saunders College Publishing, Fort Worth, pp. 570, 609.
Garrett, R.H. and Grisham, C.M. (2010) *Biochemistry*, 4th edn. Brooks/Cole, Cengage Learning, Boston.
Minakami, S. and Yoshikawa, H. (1965) Thermodynamic considerations on erythrocyte glycolysis. *Biochemical and Biophysical Research Communications* **18**, 345–9.
Newsholme, E.A. and Start, C. (1973) *Regulation in Metabolism*. Wiley & Sons Inc, New York, p. 97.
Newsholme, E.A. and Leech, A.R. (1983) *Biochemistry for the Medical Sciences*. Wiley, New York.
Pocock, G. and Richards, C.D. (1999) *Physiology*. Oxford University Press, Oxford, p. 29.
Schopf, J.W. (1999) *Cradle of Life*. Princeton University Press, Princeton, NJ, p. 150.
Voet, D., Voet, J.G., and Pratt, C.W. (2002) *Biochemistry*, upgrade edn. Wiley & Sons Inc, New York, pp. 395, 404.

12

From Prokaryotes to Eukaryotes: Getting Ready for Multicellular Life

The earliest forms of life on Earth were unicellular. As a prelude to looking at the development of multicellular life, we now need to extend the discussion on cells to fill in some of the information gaps not covered previously. The first of these concerns the development of the nucleated type of cell, the eukaryotic cell, which led to the emergence of multicellular life forms. The broad evolutionary pathway of living cells now appears to have been agreed on. This differs somewhat from earlier ideas, which supposed that eukaryotes evolved in a 'linear' progressive fashion from the simpler prokaryotes, which were the earliest form of life. During the late 1970s and the 1980s research using molecular genetics techniques made it clear that the earlier classification was inadequate, revealing that there are two quite different kinds of prokaryotes. Carl Woese announced the results of work done on the comparative sequencing of ribosomal RNA, the so-called 16S rRNA (Woese and Fox 1977; Woese 1987). His work demonstrated that all bacteria were not members of the same family, as previously thought. Rather, they fell into two groups, which must have diverged very early in the development of life, now called the Bacteria (for a while, the term eubacteria, from the Greek *eu*, good or true, was proposed) and the Archaea (from the Greek *arkhaios*, ancient; formerly called the archaebacteria). Woese also showed that the eukaryotes, until then thought to have originated about 1 billion years ago, were actually nearly 3 billion years older, making them about as old as the Archaea and the Bacteria.

This caused confusion in evolutionary circles, as the timing and order of events was unclear. At present, opinion favours Woese's proposal, based on sequencing evidence, not only of RNA subunits but also on gene sequences and protein homology, that the universal ancestor of life is a prokaryote related to the hyperthermophilic (living at a very high temperature) Archaea. Not all Archaea are hyperthermophiles and the universal ancestor cannot have been a particular organism. In Woese's words:

'The universal ancestor is not an entity, not a thing. It is a process characteristic of a particular evolutionary stage.' (Woese 1998)

On the basis of Woese's proposal in 1990, today it is thought that organisms are better grouped into three lineages or domains (Woese 2000). These are the two prokaryote groups Archaea and Bacteria, plus all the rest, the Eucarya (Figure 12.1).

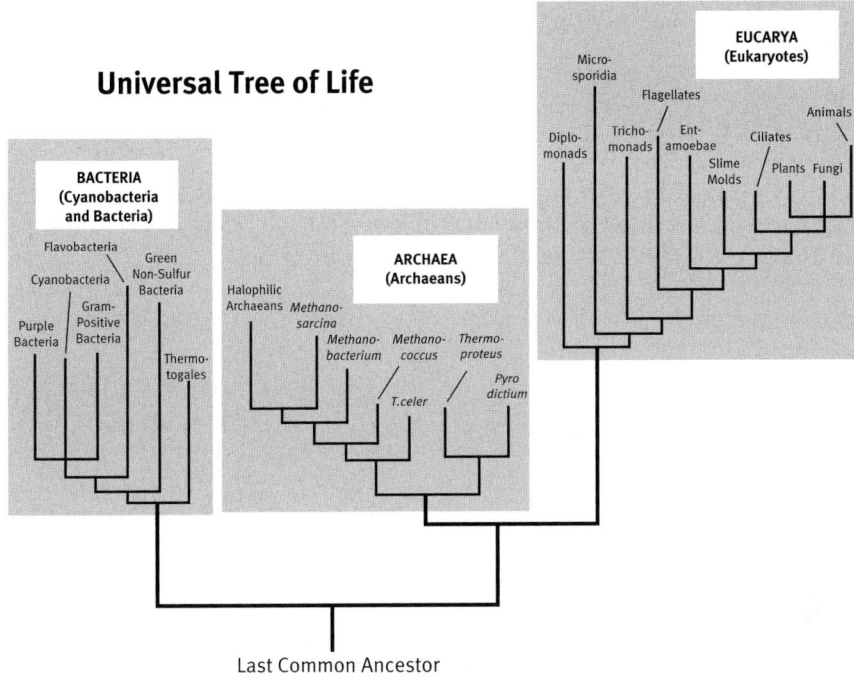

Figure 12.1. The tRNA Universal Tree of Life. (Illustration from Schopf, J. W. *Cradle of Life*. © 1999 Princeton University Press. Reprinted by permission of Princeton University Press.)

The Eukarya (commonly referred to as the eukaryotes) are usually divided into the microbial, usually single-celled, Protista (protists), plus the multicellular Plantae (plants), Animalia (animals) and Fungi (mushrooms, yeasts, and moulds). The formal names are those starting with capitals, and the more commonly used names are those without capitals.

Not all biologists accept the three-domain classification described above, with some notable researchers such as Ernst Mayr, Lynn Margulis, and Tom Cavalier-Smith dissenting on various grounds (Margulis 1996, 1997).

As new experimental evidence pours in, the details become less clear. This is nearly always the case in areas of science that are still under active investigation, as molecular evolution certainly is. There is a huge and increasing amount of data in the scientific literature on which to draw and what follows is a reasonable outline, speaking broadly, of what biologists believe might have occurred.

According to molecular DNA analysis, the first eukaryotes evolved from the Archaea, possibly from one such as *Thermoplasma*, a bacterium that lives in hot acidic springs. *Thermoplasma*, unlike most bacteria, has its DNA coated with proteins that resemble the histones supporting the chromosomes of eukaryotic cells. The early eukaryotes lacked mitochondria, but later apparently formed a symbiotic relationship with a group of non-photosynthetic bacteria, probably purple non-sulphur bacteria. It appears that initially the larger eukaryotic cell engulfed the purple non-sulphur bacteria as a source of food, or perhaps the latter actively invaded the larger cells. Somehow, some of the bacteria survived the ordeal and became resistant to digestion. Eventually, the engulfed bacteria became the mitochondria of early non-photosynthetic eukaryotic cells. This seems at first astonishing, and indeed it has taken more than a century for the so-called endosymbiosis theory, proposed by the Russian Konstantin Merezhkovsky (1855–1921), to become accepted. In the early 1970s, the endosymbotic theory was powerfully re-espoused in the USA by Lynn Margulis and several of her proposals have since become widely accepted by biologists (Margulis and Sagan 1997; Margulis 1999). Evidence for endosymbiosis includes the fact that mitochondria possess their own DNA (i.e. genes) independent from the DNA in the nucleus of the host cell. Margulis also proposed that somewhat later, in a similar symbiotic fashion, one type of the photosynthetic bacteria, probably an ancestral form of the cyanobacteria, became the chloroplasts of a group of early protists, from which modern plants eventually evolved. Again, in support of the endosymbiotic theory, chloroplasts have been shown to possess their own DNA. Both mitochondria and chloroplasts contain their own ribosomes, and their chromosomes resemble those of bacteria. Apparently, both mitochondria and chloroplasts over time lost some of their genes, which became unnecessary after they became organelles inside a host cell, rather than free organisms (Figure 12.2). Most of the lost genes now appear in the nuclei of the host cell.

The timing of the above events in the history of life is no coincidence. Before about 2 billion years ago there was essentially no free oxygen in the atmosphere. Oxygen was being produced by photosynthetic bacteria before this time, as long as 3.5 billion years ago, but it was removed by reaction with Fe^{2+} in the oceans to form mainly haematite, Fe_2O_3. Vast amounts of insoluble haematite were deposited in large basins, hundreds of kilometres across, to produce banded iron formations (BIFs) that today provide the bulk of iron ore for industry. Only after the removal of available Fe^{2+} and the further development of photosynthetic bacteria did the concentration of free oxygen began to increase (Schopf 1999). At this time, all life consisted of microorganisms, but most were anaerobic and for them oxygen was a poison. A strong argument has been made that the increase in 'poisonous' oxygen actually stimulated the development of respiration. Cells that had engulfed aerobic bacteria thereby developed a way to remove oxygen which happened to find its way inside and thus protected, they survived. The invaders became permanent residents in the form of mitochondria and the parent cells had a new source of energy. From such humble beginnings sprang the superathletes of today.

Similarly, the later addition of chloroplasts to the eukaryotic energy armoury led to the development of green unicellular or colonial protists, known to us as the various types of simple algae, the main components of the phytoplankton, microscopic organisms that are important photosynthesizers in the upper layers of the oceans. More complex algal forms, the brown and red algae, exemplified by the seaweeds and kelps, possess simple tissues and are not classified as protists. Subsequently, true multicellular plants evolved, which set the scene for the serious 'invasion' of the land.

A third endosymbiotic event may have provided the protists with their efficient means of propulsion, cilia, and flagella. Margulis proposes in the overall

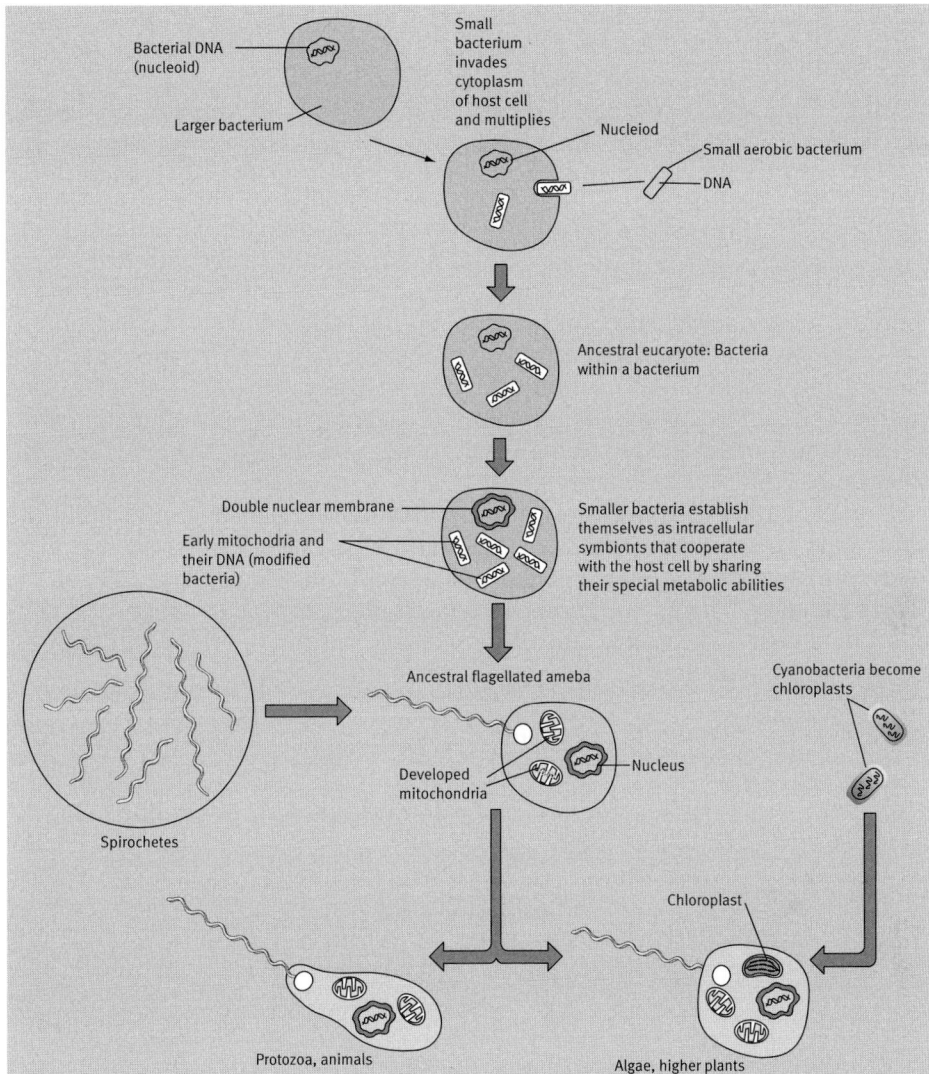

Figure 12.2. Possible symbiotic origins of eukaryotic cells. (Reproduced with permission from Talaro, K., Talaro, A. *Microbiology*, 2nd edn, p. 126. Copyright © 1996 Times Mirror Higher Education Group Inc.)

serial endosymbiosis theory (SET) that eukaryotic flagella and cilia are the consequence of endosymbosis between spirochaetes and early eukaryotic cells. Spirochaetes are helical-shaped bacteria that typically move by flexing and rotation. One genus contains the spirochaete *Treponema pallidum*, which causes syphilis. The proposals by Margulis went through stages of resistance, then reasonably wide acceptance, especially her ideas concerning mitochondria and chloroplasts, but the proposal of endosymbyotic involvement of spirochatetes to form flagella and cilia has few supporters today. Other details involving the mechanism of uptake of the prokaryote that became mitochondria are also disputed (de Duve 2005). Other theories continue to add intriguing possibilities to the debate, which seems to be far from settled in detail (Martin and Muller 1998; Dyall *et al* 2004).

These monumentally successful symbiotic events, preserved for the last 2 billion years, involve the capture of pre-existing, efficient energy-transducing organisms. The process has been termed 'prepackaged evolution' (Schopf 1999). It is highly significant, and again underlines the central role of controlled energy flow in the living world. The evolutionary advantages conferred by the mitochondria, chloroplasts, and other devices on the recipient cells become apparent when we note that modern eukaryotic cells are the only types that have allowed the development of multicellular organisms. This did not take place rapidly. The earlier eukaryotes remained unicellular, and only after the emergence of the modern eukaryotic cell with its internal structures, organelles, and other bodies did multicellularity flourish.

Notwithstanding the appearance of multicellular organisms, unicellular life forms have been enormously successful, and it has been estimated that even today prokaryotes and single-celled eukaryotes constitute about half of the Earth's biomass. This is staggering to contemplate, especially when one considers that most of the visible animals and plants living in the vast tracts of forest and grassland, plus those in the oceans, are multicellular.

Why then did multicellular life evolve? Perhaps, once life based on DNA had established itself on Earth, the eventual emergence of complex organisms was inevitable. We shall probably never know for certain, but we can be quietly confident that anything capable of evolving by natural selection, given favourable conditions and time (deep time), will evolve. Implicit in the phrase 'capable of evolving by natural selection' are the constraints imposed by heredity, genetic mutation, and genetic exchange, plus the vagaries of a changing and often hostile environment. In one sense, from unicellular to multicellular is the only available way for an organism to go. The first attempts at this probably consisted of a very few different types of cells, as in the protist *volvox*, a colonial green alga with an outer layer of ciliated cells and an inner one of reproductive cells. In one form of *volvox*, hundreds of single cells are united by strands of cytoplasm into a hollow sphere. This is not a multicellular organism in the accepted sense, rather an intermediate form called a colony in which the individual cells retain a high degree of independence.

Multicellular organisms typically contain specialized or differentiated cells, of various types. Differentiated cells are usually arranged in groups to form the various tissues and organs of a higher animal or plant. Each tissue and organ has its own function in assisting the survival of the whole organism. Specialization of cells allows a multicellular organism to protect itself, move about, regulate its internal systems, hunt prey, seek out mates, and engage in complex behaviour that is impossible for its unicellular cousins. The potential for differentiation allowed the eventual development of skeletal systems, blood vessels, brains, and immune systems. It is not surprising, considering the many advantages, that multicellular organisms have evolved independently many times. Evolution did not seek multicellularity, but when the opportunities arose, natural selection ensured that the advantages were seized.

In considering the development of multicellular organisms, two major questions arise:

1) What overall changes did prokaryotic cells undergo which led to the first eukaryotes?
2) How did unicellular eukaryotes develop into multicellular ones?

The crises the early prokaryotes faced, such as increasing oxygen levels, were overcome by attributes that the later eukaryotes lacked. Some bacteria, notably Gram-negative types, are able to transfer some of their genes readily and directly to other individual bacteria by a process called conjugation (from the Latin *conjugatus*, linked together). Eukaryotes cannot do this. Many texts have photographs of the conjugation process in operation (Talaro and Talaro 1996; Madigan and Martinko 2006). The donating bacterium develops a protein tube called a sex pilus (Latin for *hair*), which attaches itself to a recognition site on the receiving bacterium.

Intermolecular recognition and energy are involved here. Some of the donor's DNA, in the form of a plasmid, is replicated, and the replicated DNA is passed through or along the pilus to the recipient. The bacterium on the receiving end can use the acquired DNA to perform functions that its own original genes cannot manage. Antibiotic resistance genes may be transferred in such a manner. In principle, this rapid acquisition of genes by bacteria allows each of them access to much of the bacterial gene pool on Earth. Such capabilities allowed the early prokaryotic bacteria to produce the diversities of form and function that would lead to the first eukaryotic cells. They also complicate studies of molecular evolution, as in the appearance of 'unexpected' genes in certain organisms where they did not 'fit' into to the current evolutionary ideas (Woese 2002).

Apart from the nucleus, a major structural difference between prokaryotic and eukaryotic cells is the possession of a rigid cell wall by the former. The prokaryotic cell wall surrounds the lipid bilayer of the cell membrane, maintains the cell shape, and protects the cell, but is restrictive to the absorption of solid particles for use as food. To feed on solids, the prokaryote needs to secrete enzymes into the surrounding medium and absorb the soluble breakdown products. This is slow and wasteful, as the enzymes are presumably lost, costing the bacterium considerable expenditure of energy. Many eukaryotic cells, on the other hand, lack a cell wall entirely and are able to change their shape. They are capable of ingesting solid particles by a process called phagocytosis. The lipid bilayer of their membrane can distort around the particle and finally 'pinch off', forming a small membrane-bounded vesicle called a food vacuole inside the cell membrane. The food vacuole can then fuse with other lipid-based vesicles inside the cell, such as lysosomes, that contain digestive enzymes. In this way, no enzymes or nutrients are wasted. This is what seems to have occurred in primitive eukaryotes.

What might have caused the loss of the cell wall in the ancestral prokaryotes involved is still a matter for speculation, but once it occurred, the advantages were retained. Problems also arose, as without a protective cell wall the cell was much more fragile. These problems were solved on the one hand by developing a much more rigid cell membrane structure using different membrane molecules, leading to the Archaea. The other solution, which led to the eukaryotes and eventually to ourselves, involved the formation of an internal skeleton, the cytoskeleton (skeleton in the cytoplasm, or interior medium of the cell). The cytoskeleton consists of two major protein molecular classes, actin and microtubules. Actin filaments resist pulling or stretching forces, while microtubules are resistant to shearing and compressing forces. The macromolecular properties of these two components allow the cell to maintain its shape in the absence of an external cell wall. They also allow the cell to change its shape and to move things around inside it. This latter property is important to the function of modern eukaryotic cells.

Microtubules are used to guide the movement of particles and vesicles, to pull chromosomes apart during cell division, and are involved in movement devices such as the cilia and flagella, which by their rapid, whip-like movements drive cells through the surrounding liquid medium. Eukaryotic flagella are much more complex (they contain microtubules, for example) and larger than their prokaryotic counterparts. The acquisition was another important addition to the energy-transducing capacity of the host, this time providing an efficient form of locomotion. Locomotion is important in responding to stimuli and is a most useful attribute. Stimuli could include response to food, light, temperature, pH gradients, and chemical concentration gradients (including avoidance of harmful levels of these). Mammalian sperm are an example of this long-past encounter. Indeed, sperm are essentially a lump of motorized DNA equipped with a receptor, which recognizes the ovum. Having no future after their frantic race with millions of competitors ends in fertilization, sperm have very few mitochondria. This means that everyone inherits the vast majority of his or her mitochondria from the mother only, leading to the intriguing concept of the 'mitochondrial Eve' from which all human mitochondria descended.

Actin filaments are involved in phagocytosis and cell division. These activities involve movement, which requires the controlled expenditure of energy, the major source of which is ATP. This means that the proteins making up the actin filaments and the microtubules must be able to change their shape, to convert chemical energy into mechanical energy. The movements of the cytoskeleton are a long way from the twitching of your toe, but the connection exists nevertheless. Modern muscles are specialized organs consisting of specialized cells, adapted to maximize the conversion of chemical energy to mechanical energy. Considered in this way, energy is the parent of motion and motion is the quintessential theme of life, from unicellular to multicellular, from fish to frogs and fowl. The graceful movements of the ballet dancer, the explosive heave of the discus thrower, the jaw-breaking power of an uppercut all have their origins in deep time, when the building blocks of life as we now see it were being moulded.

How could molecules such as actin, tubulin, myosin, and the other parts of the cytoskeleton have evolved? They interact so precisely, often spontaneously, yet presumably evolved sequentially. Part of the answer, according to Nobelist Christian de Duve, lies in the idea of molecular complementarity of the proteins involved (de Duve 2002). Presumably after many unsuccessful attempts, mutations produced proteins which 'fitted together', that is could be held that way by secondary bonding to form bundles and helices, for example those that formed the early parts of the cytoskeleton, and which propped up the wall-less cell, allowing it to grow larger without collapsing. We have seen the power of molecular complementarity in the base pairing of DNA. A pairs with T, G pairs with C, by hydrogen bonding. Another related term is molecular recognition, as applied to cell-surface receptors that respond to messenger molecules such as hormones, to antibody–antigen interactions, to enzyme–substrate binding, and to allosteric effects. There is always energy exchange involved in such interactions, which may be provided by one or more of the usual non-covalent secondary forces—hydrogen bonding, ionic interactions, dipole interactions, van der Waals forces, hydrophobic effects, or desolvation effects. The complementarity of structure, where a 'key' on one molecule fits into a 'lock' on the other, is important. Only at very close contact distances do the above short-range non-covalent forces come into full effect. Let's pause once more to acknowledge the vital roles of protein macromolecules, with their richness of structural diversity that, under the influence of natural selection and with the assistance of deep time, have allowed the development of such unique and far-reaching functions.

Other differences between prokaryotic and eukaryotic cells have developed. Eukaryotic cells are up to 10,000 times the volume of their earlier prokaryotic counterparts, and in addition to the devices and organelles mentioned possess a system of internal membranes called the endoplasmic reticulum (ER). The ER in modern eukaryotes is a system of double membranes within the cell to which are attached the protein-synthesizing particles, the ribosomes. Various vesicles, such as the ribosomes and those which can fuse with the cell membrane to assist in growth, are also attached to the ER. These latter vesicles, with their enclosed enzymes, are 'budded-off' periodically and proceed (with the assistance of the microtubules) to where they are needed in the cell. The ER is continuous with the outer of the two membranes surrounding the nucleus. The presence of pores in the nuclear membrane facilitates the movement of gene products, such as messenger RNA, out of the nucleus into the cytoplasm, and of enzymes needed for replication into the nucleus. Other components of eukaryotic cells include the peroxisomes and the Golgi apparatus or Golgi bodies. Peroxisomes contain enzymes involved in the metabolism of hydrogen peroxide, a compound toxic to cells and one which is kept isolated if possible. The Golgi apparatus consists of membranous sacs found close to a part of the ER. It is involved in the secretion of proteins (and polysaccharides in plants) from the cell and, when required, in attaching sugars covalently to proteins.

This then is the basic eukaryotic cell. The details of how these cells evolved are still being finalized, using ever more sophisticated methods to analyse more species (Zimmer 2009b). Whatever the series of events was, the result triggered a revolution. Large, energy-efficient, sexually reproducing cells, underpinned by the potential diversity conferred by a genome based on DNA, were ready for the next great step in the journey of life on Earth.

Many other mutations no doubt occurred that were unsuccessful, but these disappeared from the gene pool. Subsequently, a long succession of mutations must have taken place, each with its evolutionary advantage surviving, slowly honing the eukaryotic cell to its present state of sophistication. New proteins had to be evolved and when they were, natural selection sorted them out from the motley ensemble generated by random mutation, but once again this took a long time, presumably with many 'failures' along the way. The number of mutations necessary is unknown, but we can gain some idea of the immensity of the task by looking at the time it took modern eukaryotes to evolve by natural selection. All the major functionalities for eukaryotic life had been evolved quite early by the prokaryotes, and the early eukaryotes emerged fairly soon after.

The early eukaryotes were unicellular and lived as independent organisms, the early protists. Over time some forms began to aggregate and form colonies, which still consisted of 'independent' cells. Eventually groups of cells within colonies must have become specialized, to perform some function that was advantageous to the entire colony such as movement, feeding, or reproduction. This led to the development of tissues consisting of specialized cells that had lost the ability to live independently. The colony had developed into a multicellular organism, which could only survive as a single unit. Existing eukaryotic organisms are not all multicellular, however. All modern protists, plus numerous algae and fungi, exist in unicellular or colonial form. The true multicellular organisms are found only in animals, plants, and some of the fungi and algae. Multicellular life is believed to have developed about 600–700 million years ago. There is no trace of it before this time.

The above description of the evolution of eukaryotes is just one of a number of possible pathways based on reasonably accepted evidence. It is far from complete in detail, as much of the detail has yet to be discovered. As to *why* such a pathway, or any particular pathway, should have been taken, an explanation is even further off. We can, however, make some educated guesses as to why eukaryotic cells, once they had developed, might have led to multicellular life.

Before proceeding with the development of multicellularity, I must deal with another development of the eukaryotes I have not mentioned previously; the newer way of cell division, or looking at the process more fundamentally, the newer way of replicating the genome. Prokaryotes, exemplified by the typical bacterium, undergo cell division by a process known as cell fission. Omitting the detail, this involves replication of the DNA in their single, circular chromosome by a process requiring its attachment to the cell wall in two places. Eukaryotes, lacking the strong cell wall, had to devise an alternative means of replication, which turned out to be the process now known as mitosis. The existence of a nucleus in eukaryotes also created special problems for cell division. In prokaryotes, after chromosome duplication, the cell initiates division by forming a simple furrow across its middle. This is followed by a 'nipping off' across the cell wall and membrane such that each of the daughter cells inherits one of the duplicated chromosomes. In eukaryote cells, the final division of the cytoplasm is by a similar mechanism, but before that happens the vital contents of the nucleus must undergo duplication. The chromosomes are firstly duplicated in the nucleus and the resulting chromosome pairs are compacted into filaments which, just before cell division, become visible under the microscope as the thread-like structures that gave this division process its name, mitosis (in Greek, *mitos* means 'thread'). The nuclear membrane and associated proteins (collectively called the nuclear envelope) are then dismantled and replaced by a complex of microtubules called the spindle, which pulls the pairs of duplicated chromosomes apart, with one member of each pair moving to opposite poles of the spindle. Thus, a full set of identical chromosomes is present at each pole of the spindle. New nuclear membranes then form around each set of chromosomes, the spindle is disassembled, and the cell divides into two daughter cells, each with its full complement of chromosomes.

The formation and dismantling of the mitotic spindle and the complex nuclear envelope must occur at each mitotic division of a cell and represent remarkable examples of molecular complementarity leading to the formation of very complex structures. Not surprisingly, the expenditure of energy in the form of ATP is involved in both sets of processes.

Note again the pattern: large molecules with specific structures, plus a source of energy. Without both of these, acting in cooperation, no life would ever have evolved, let alone the complex life forms characteristic of modern eukaryotes. The time, after time, after time reproducibility of cellular behaviour has been programmed, by the Ockham's razor of natural selection and the majestic sweep of deep time, into our deepest molecular structures.

Although the above description is brief, it is obvious that even the early eukaryotic cell, such as that typified by protists, is extremely complex, considerably more so than its prokaryote predecessors. Assuming the above development scenario to be broadly correct, what were the overall evolutionary advantages, conferred on the modern eukaryotic cell by natural selection, which ultimately allowed the development of multicellular organisms? One is the development of sexual reproduction in eukaryotes, with its potential for rapidly increasing biological diversity.

What is the essence of sexual reproduction in eukaryotes? To quote evolutionist John Maynard Smith, sex 'requires that DNA from different ancestors be brought together in a single descendent' (Maynard Smith and Szathmary 1999). How is this achieved in practice?

Cells having single copies of each type of chromosome are said to be haploid, for example the early unicellular protists. Now suppose a haploid organism were to reproduce sexually, according to Maynard Smith's requirements. This would lead to a doubling of the normal number of chromosomes, one set from each parent, which would be abnormal (for the organism involved) and would constitute a complex problem to be overcome. Some types did overcome the problems of extra chromosomes and sets of chromosomes, but this was not the best answer. We have seen that for the development of mitosis in eukaryotic cells certain changes in the manner of replication were necessary, such as the early development of microtubules (tubulin), which allowed the formation of the mitotic spindle. Sexual reproduction required further developments, such as the process of halving the normal number of eukaryotic chromosomes. This process is known as meiosis. Organisms with two copies of each chromosome type are called diploid, for example the cells of mammals. Meiosis turns diploid

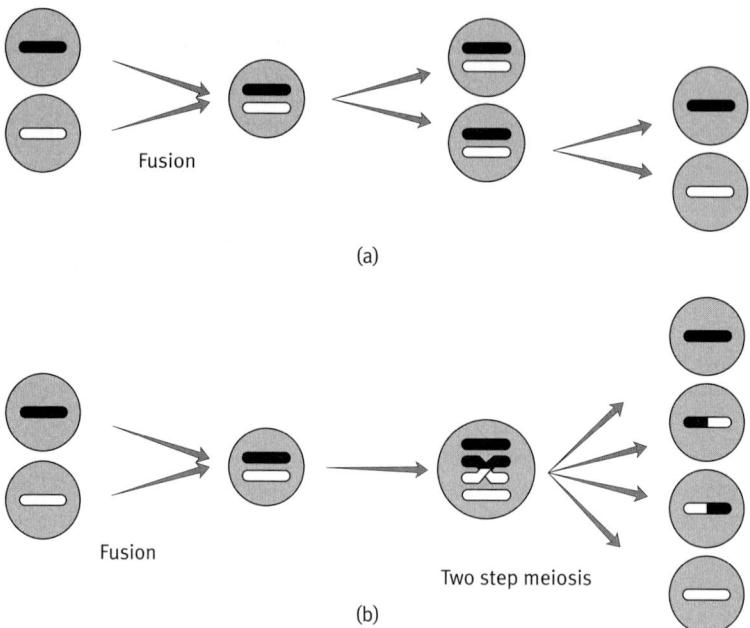

Figure 12.3. Possible stages in the evolution of meiosis and gamete formation. (a) Haploid fusion of gametes with one-step meiosis, without crossover. (b). The modern sexual life cycle. Gamete fusion involving two-step meiosis with crossover and recombination creates greater genetic diversity than (a). (Illustration adapted from Maynard Smith, J. and Szathmary, E. *The Origins of Life.* p. 88. © 1999 Oxford University Press.)

cells into haploids, although rather special ones. The advent of meiosis, perhaps 2 billion years ago (Zimmer 2009a) made possible the formation of special 'sex' cells, called gametes, each of which has half the number of chromosomes of a normal body (somatic) cell.

The two similar chromosome copies in diploid cells are called homologous pairs. Some proposed advantages of diploidy, that is of having homologous pairs of chromosomes are (i) damaged DNA can only be repaired if there is an undamaged strand to copy and (ii) as DNA is damaged more frequently in the presence of oxygen, diploidy may be a useful adaptation to an aerobic existence.

In sexually reproducing species, both plant and animal, the sperm is the male gamete and is haploid, while the ovum, the female gamete, is also haploid. When the two fuse, fertilization occurs, the normal diploid number of chromosomes for the species is restored, and the fertilized ovum, the zygote (from the Greek *zygotos*, 'joined together') proceeds to develop by standard mitosis into a multicellular embryo. In humans, a diploid species, there are 23 homologous pairs of chromosomes in each somatic cell and 23 single chromosomes in each sperm cell and each ovum, formed by meiosis during the reproductive lifetime of each male and female. A possible route to the evolution of meiosis is shown in Figure 12.3.

However meiosis developed, the result in terms of greater genetic diversity can readily be seen in the illustrations of the four simple gamete cells in Figure 12.3(b). One gamete has the same genetic makeup as the male parent, one has the same makeup as the female parent, but two have unique gene combinations not seen in either parent. Thus, one generation cycle can yield two unique gene combinations. This is very rapid generation of diversity, compared with asexual cell fission in bacteria, for example, where the two 'daughter' cells have an essentially identical genetic makeup. Approach (a), although showing true sexual reproduction, would yield gametes identical to those of the parents. Genetic diversity in bacteria arises from errors of copying, by mutations in the DNA caused by chemical or radiation damage, or by direct assimilation of genomic DNA from other bacteria by conjugation (referred to as horizontal gene transfer, rather than vertical transfer from parent to offspring).

In the diploid organism represented in Figure 12.3, suppose one parent has an unspecified genetic 'defect'. Stage (a) in Figure 12.3 would perpetuate this defect. In stage (b), however, two of the four gametes formed have a chance, by virtue of the shuffling of genes during crossover, of forming without the defect at all. When such chances occur, as they frequently do in reality, they help ensure that the 'better' gamete produces defect-free progeny, with an increased chance of survival. Such is the case for the simple example above. Humans possess about 20,000–25,000 genes (Human Genome Project Information). In humans, combine this with the fact that unique genetic combinations occur, as each sperm and each ovum is formed by multiple crossing over during meiosis (although there are certain limitations to the extent of crossing over) and you will see that the possible genetic diversity is enormous, even from the mating of one human couple. The process of crossing over during meiosis is a fascinating one. For a single crossing over and recombination event, corresponding regions of DNA in the homologous chromosomes must line up, each DNA thread is broken by an enzyme, and the two 'excised' pieces are spliced back into the gap left in the other thread, by yet other enzymes. The fidelity of such a process depends on multiple molecular recognition events, all exquisitely controlled. How could such a complex system evolve? Evolution is conservative and parsimonious. In this case the components evolved earlier, for a different purpose. The enzymes involved in recombination are the same type that repair damaged DNA, and as such possess the same precise DNA cutting and splicing properties.

In the development of cellular types, we have reached the stage of the complex, sexually reproducing eukaryotes, such as the unicellular protists. At the cellular level, essentially all the problems to confront multicellular life have been solved; what remains is their association into organisms capable of taking advantage of specialization and the division of labour. At the stage of life on Earth we have reached, all life still lived in water, from deep, hot vents, cold currents, and all the other ocean habitats to shallow lakes and streams. It is still a long way from multicellularity, let alone from 'conquering' the land and the air.

GENERAL REFERENCES

de Duve, C. (2002) *Life Evolving*. Oxford University Press, New York.

REFERENCES

de Duve, C. (2005) *Singularities: Landmarks on the pathways of life*. Cambridge University Press, New York, Ch 17, pp. 203–213.
Dyall, S.D., Brown, M.T., and Johnson, P.J. (2004) Ancient invasions: From endosymbiosis to organelles. *Science* **304**, 2553–7.
Human genome Project Information http://www.ornl.gov/sci/techresources/Human_Genome/project/info.shtml.
Madigan, M.T. and Martinko, J.M. (2006) *Brock Biology of Microorganisms*, 11th edn. Pearson Education Inc., Upper Saddle River, NJ, p. 278.
Margulis, L. (1996) Archael-eubacterial mergers in the origin of Eukarya: Phylogenetic classification of life. *Proceedings of the National Academy of Sciences USA* **93**, 1071–6.
Margulis, L. (1999) *The Symbiotic Planet*. Phoenix Books, London, p. 38.
Margulis, L. and Sagan, D. (1997) *Slanted truths*. Springer-Verlag, New York.
Martin, W. and Muller, M. (1998) The hydrogen hypothesis for the first eukaryote. *Nature* **392**, 37–41.
Maynard Smith, J. and Szathmary, E. (1999) *The Origins of Life*. Oxford University Press, Oxford, pp. 88–90.
Schopf, J.W. (1999) *Cradle of Life*. Princeton University Press, Princeton, NJ, pp. 148, 171–3.
Talaro, K. and Talaro, A. (1996) *Foundations in Microbiology*, 2nd edn. Times Mirror Higher Education group, Inc., Dubuque, IA, p. 95.
Woese, C.R. (1987) Bacterial evolution. *Microbiological Reviews* **51**, 221–71.
Woese, C.R. (1998) The universal ancestor. *Proceedings of the National Academy of Sciences USA* **95**, 6854–9.
Woese, C.R. (2000) Interpreting the universal phylogenetic tree. *Proceedings of the National Academy of Sciences USA* **97**, 8392–6.
Woese, C.R. (2002) On the evolution of cells. *Proceedings of the National Academy of Sciences USA* **99**, 8742–7.
Woese, C.R. and Fox, G.E. (1977) Phylogenetic structure of the prokaryotic domain: The primary kingdom. *Proceedings of the National Academy of Sciences USA* **74**, 5088–90.
Zimmer, C. (2009a) On the origin of sexual reproduction. *Science* **324**(5392), 1254–6.
Zimmer, C. (2009b) On the origin of eukaryotes. *Science* **325**(5941), 666–8.

13
Multicellular Life: The Last Hurdle?

13.1 Multicellularity arrives

Most of us have read various accounts of the ways in which life, after some 2.5 billion years essentially restricted to a water environment, emerged on to land and eventually flourished there. Emotive terms such as 'conquest of the land' and 'invasion of the Earth' were suitable for the Empire-building nineteenth century when they were coined. I prefer descriptions such as 'greening of the Earth' or even 'colonization of the land'. The former is not merely descriptive but also reflects the order in which life, in a substantial way, first emerged on the land. Of the complex forms of life, plants came first. In reality things were difficult and success was by no means assured.

Thus far we have dealt with individual cells and some life forms are, as we have seen, single celled to this day. They gain nutrients from and excrete waste products directly to their environment through their cell membrane (and cell wall, if any). They move by means of tail-like flagella or the concerted beating of many cilia. We probably are more familiar, however, with multicellular organisms that live on land. Single-celled life forms such as bacteria, cyanobacteria, and protists have been highly successful for billions of years, so how did multicellular organisms evolve? What extra problems did they need to overcome to survive, not only in water where they evolved, but also on land?

Imagine the land slightly less than 1 billion years ago. The continents, vastly different in shape and position from what we see today, were apparently barren, rocky, scoured by intermittent streams but unable for lack of plants and organic soil to maintain much moisture, baking under a scorching sun, but often freezing at night. A number of microorganisms undoubtedly existed in these environments, including the atmosphere, with sufficient nutrients to support them, but overall 'land' was most likely almost bereft of life. Oxygen levels were building and an ozone layer was developing that was to protect any emerging organisms from the worst of ultraviolet radiation.

What changed all this? The oceans were populated by prokaryotic bacteria and by unicellular eukaryotic organisms, the protists that had already begun to diversify. These eukaryotes comprised both phototrophic and heterotrophic types, some of which had adopted a sexual mode of reproduction. Of the many tentative associations formed by the protists, undoubtedly most did not survive, but those which did laid the biological foundations for the truly multicellular organisms that were to come. What made eukaryote cells first associate is unknown. Some of the associations, fortuitously, must have possessed some survival advantages. These reproduced for long enough to undergo natural selection, which resulted in further survival, further improvements, further survival, and so on.

The close association of many cells is the most obvious characteristic of multicellularity. Association is considered as the first of several principles to have been involved in the evolution of multicellular organisms. Presumably, simple association led to improved reproductive success and thus it persisted. Merely by staying alive, by persisting, multicellular organisms, like all others, were inevitably subjected to natural selection pressures. More than simple association was needed. Some form of cooperation was required, which meant intercellular communication and transmission of information. However, the cell membrane presents a significant barrier to water and other polar molecules. Fortunately, cell membranes are more than just a lipid bilayer. They can accommodate proteins, for example. Some 50–70% by weight in modern animal cell membranes is protein. Non-polar sections of membrane proteins span the membrane, leaving polar heads and tails in the cytoplasm and extracellular fluid. Such transmembrane proteins can move laterally in the membrane. They are also capable of allosteric interactions with smaller molecules to alter their conformation and hence their uptake properties. Such regulatable property variations lend themselves to the formation of specific channels and receptors that allow cross-membrane and thus intercellular information exchange by molecular messengers.

Some experimental (in the evolutionary sense) multicellular types would have disappeared fairly quickly, while others must have begun to develop specialized tissues, that is they took advantage of simple association to develop cell differentiation, the second principle of multicellularity. Differentiation enhanced survival and so off the multicellular organisms went, ultimately to become flowers and fungi, dinosaurs and mushrooms, frogs and humans. From the early simple colonies emerged the means for cells to remain associated. The so-called slime-moulds are an instructive example. Their unicellular forms developed surface molecules that allowed them to adhere to others by a complementary, lock-and-key type of interaction. They probably developed the early equivalent of cell-surface receptors, which held them to the viscous slime they secreted. These association principles remain in force in modern organisms, for example in the cell-adhesion molecules and substrate-adhesion molecules of mammals. Unicellular organisms that were to become plants and fungi most likely developed their cellular association within a structure that became the cell wall of modern plants and fungi. Plant cell walls are complex structures and structurally very strong. The size of some modern trees attests to this. Fungal walls are somewhat less complex, but are well suited to their scavenging lifestyle. Cell walls are absent in animals, which for land-dwellers places a limit on the size they can achieve and even affects their body shape. These limitations are compensated for by greater mobility—animals can run, fly, swim, hunt, and hide to avoid predators.

Differentiation can only occur and tissues be maintained if the cells possess a well-developed capability for gene regulation. Superb control mechanisms are required. Genes need to be turned on and off, and their regulation by controlling the transcription of DNA into messenger RNA (and hence into protein) is important in several ways. Each somatic cell of an organism contains all the genes (the genome) of that organism, but only a small proportion of these genes is ever used in a given cell type. All cells have what are called 'housekeeping' genes, which code for structural proteins, enzymes for intermediary metabolism, etc, that is all the general functions needed to run the cell from moment to moment. These genes must be capable of being active when needed, and are under the control of a group of regulatory proteins. Special proteins called transcription factors bind to specific sites on DNA and either activate or deactivate their respective genes. Not surprisingly, the genes that make the transcription factors are called regulatory genes. A transcription factor that inhibits DNA transcription is called a repressor, and one that stimulates transcription is called an inducer. Depending on the activity levels of these essential transcription factors, the genes are either transcribed or not transcribed into messenger RNA, and thence to protein. A model for control of gene expression in *Escherichia coli* was first proposed by the French researchers Jacob and Monod (Maynard Smith and Szathmary 1999).

In the cells comprising any given tissue or organ, apart from the housekeeping genes, the genes that can be activated are those which define the type of tissue. For example, in muscle cells the genes coding for the special functional proteins of muscle are often activated, but the genes coding for the specialized functions of, say, the pancreas or liver, although present in the muscle cells, are deactivated early in development, when the embryonic cells undergo differentiation. Such deactivation of certain genes after cell differentiation is not always absolutely irreversible. An example well known to gardeners is that certain plants may be grown from a cutting or shoot. The end of the cutting in the ground will often grow roots and eventually an entirely new, complete plant will develop. This could occur only if there is a complete set of genes in the cells of the part of the stem placed in the ground and if these genes are capable of being reactivated to produce root cells.

Regulation of genes by transcriptional control is especially important during the development and maturation of an organism. There are many examples, but let's just consider briefly the onset of puberty in humans. Specific genes in the cells of organs concerned with puberty are activated, such as those that induce facial hair in boys and the development of breasts in girls. Steroid hormones that trigger a response in the appropriate cells are released into the bloodstream and on reaching the appropriate cells interact with specific cell-surface receptors. This triggers the activation of regulatory proteins, which bind to the complementary segments of DNA in the genes to be activated. Transcription can now take place in these genes and the processes leading to the development of hair or breasts commence. At the gene level much the same kind of control is involved in other stages of development.

The regulation of gene activity is actually much more complex than the above outline suggests, but I hope you have grasped the principles involved.

Thus differentiation leads to cell specialization, and to division of labour between the tissues and organs of a multicellular organism. We will deal with some more of the advantages and problems associated with cell specialization soon, but before leaving gene regulation an important question must be answered: What is the role of energy in gene regulation?

I said above that regulatory proteins bind to specific portions of the genomic DNA. The binding occurs because it is energetically favourable under the prevailing conditions. Complementarity of shape between the regulatory proteins and the DNA regions to which they bind is involved and this allows the short-range secondary bonding forces to act cooperatively, holding the two portions together. This is fine for binding, and 'unwanted' genes may thus stay inactivated quite long term. What about the reverse, the removal of, say, a repressor protein to allow the gene to become activated? Removal of a bound regulatory protein may be achieved if a third molecule, such as a hormone or a metabolite, binds to the regulatory protein and alters its shape. This is another example of the allosteric effect. If the altered shape does not have the same complementarity with DNA as before, secondary bonds cannot be maintained and the regulatory protein/DNA complex dissociates. There are some subtle variations on this theme, brought about by the four different types of regulatory protein so far known. I won't discuss these here. During the entire, incredibly complex, regulatory process, the laws of thermodynamics are satisfied—there's thermodynamics in everything!

To outline the story of the evolution from protists to humans could take anything from a large book to a small library. My approach has been to look at the major 'problems' that organisms faced along the way, particularly those involving energy. Put another way, what had to be 'invented' to 'take advantage' of the 'potential opportunities' that were 'offered' by multicellularity? As with all evolutionary developments, the existing life forms didn't actively seek to solve such problems or to make such inventions—hence my use of so many quotation marks.

As mutations occurred, the variety of resulting new life forms sampled the available niches. We don't know how many attempts were made. Depending on how a particular life form managed to cope with all the conditions confronted, it either survived or disappeared. Ultimately, the three main multicellular kingdoms emerged independently and survived: the plants, the fungi, and the animals.

13.2 The evolution of plants

The early history of plants is reflected in some of their modern counterparts, the algae and the seaweeds. Seaweeds and indeed all algae do not develop from any kind of embryo and are thus not classified as plants. Three main evolutionary lines of algae emerged and in each case endosymbiosis appears to have been involved. Depending on the type of cyanobacterium involved in the symbiotic relationship, the red, brown, and green algae emerged. All these algae are phototrophic, all contain green chlorophylls, and all produce molecular oxygen as a result of their photosynthetic activity. The different colours are caused by the presence of different pigments. The pigments typically have at least two functions: they act as secondary absorbers of light and/or as antioxidants to protect the algal tissues from the high reactivity of molecular oxygen and its products. Pigments look coloured in sunlight precisely because they absorb some wavelengths of white light (sunlight) preferentially. The colour we see is what is 'left over' from white light. The absorbed light energy, as we saw in Chapter 10, drives the endergonic photosynthetic processes via an electron transport chain. Within the three groups of algae, there is considerable diversity of size, shape, metabolism, and reproductive styles. Common to all algae is the cell wall, which contains the glucose polymer cellulose as its main component, plus a variety of other polysaccharides such as alginates or carrageenans, depending on the species. Structurally, algae are quite simple as they contain only a few specialized cell types such as those which anchor them to the bottom, floats filled with air, and simple sex cells. Their gradual adaptation to a life on land occurred some 450 million years ago.

As always, natural selection was at work. One advantage was the acquisition of a waxy cuticle, which prevented plant desiccation in their newly adopted environment. Because the cuticle prevented the free flow of gases and nutrients, another acquisition must have been the development of special surface cells containing pores to facilitate the uptake of carbon dioxide from the atmosphere, rather than from water, and to release oxygen to it. These pores eventually evolved into the stomata of present-day plants. Stomata are variable apertures that allow control of the exchange of carbon dioxide, water vapour, and oxygen between the plant and its environment.

For some unknown reason, green aquatic algae were the only protists to evolve into plants which became fully successful on dry land, so use of the term 'the greening of the Earth' is appropriate. The successful ones eventually evolved into three phyla, the mosses, the liverworts, and the hornworts, whose members are known collectively as the bryophytes. The need for water at the critical fertilization stage largely restricted the bryophytes to moist regions in temperate or tropical habitats. Their growth was also restricted, thus bryophytes are typically ground-hugging and small. This success story of the transition from an aquatic to a terrestrial way of life is hardly one of invasion or conquest in the accepted sense. Rather it is an example of the immense power of environmental factors in exerting a genetic load and influencing the direction of evolution. Humans have often deliberately increased the genetic load on organisms by selecting the characteristics they want and 'forcing' the chosen plants or animals to adapt. Examples include the selective breeding of dogs from their wolf ancestors to an extent that almost defies belief, the development of enormous sunflowers in Russia, and increased yields of cereal grains (without the application of genetic engineering).

The first terrestrial plants were restricted to coastal areas, where high humidity assisted the transition from water to land. The remaining vast terrestrial spaces were still barren of life, their surface layers largely dry and deserted. To spread further, the plants needed to develop root systems that could penetrate the ground to draw up precious moisture and nutrients. The aerial parts needed to grow larger and longer, to be more efficient in harnessing solar energy for conversion into larger and larger structures, as conditions permitted. A characteristic of land plants, the cuticle, would have prevented excessive evaporation from the early land plants and protected the tissues from damaging ultraviolet radiation.

The land plants were forced by their physical environment into developing two distinct zones. Above ground were the green, light-seeking leaves, which also functioned as gas exchangers with the atmosphere. Below ground were the colourless roots, lacking in chlorophyll. Joining these was a stem, which allowed the passage of water and nutrients from the roots upwards and of sugars formed by photosynthesis downwards. What was lacking in the bryophytes, but which was needed desperately for plants to have the best of both worlds, was an efficient transport system connecting all parts of the plants.

The vascular (vein) system, a fluid-conducting system that plants evolved, consists of the xylem tissues and phloem tissues. These were a vast improvement. The xylem is the tissue through which most of the water and minerals are circulated, whereas the phloem is the food-transporting system in vascular plants. The vascular systems of most land plants consist of these tissue types. Xylem tissues comprising the vascular system consist mainly of dead, heavily lignified cells and living phloem cells. The early vascular plants, such as the ferns and horsetails, lacked seeds and possessed spores only. The leafy ferns we see today are the dominant, adult, diploid sporophyte stage. Under the leaves, haploid spores are produced by meiosis, are discharged, and germinate into gametophytes. Omitting much of the detail, male and female gametophytes produce sperm and eggs, respectively, and when there is sufficient water present the sperm are released and swim into the female archegonium, where fertilization occurs, as in the bryophytes. The (now diploid) zygote then develops into the mature sporophyte stage.

Cell differentiation into xylem and phloem systems was one advance that allowed the vascular plants to solve the problem of efficient food and nutrient transport. Another, which helped start plants on their upward pathway towards the sun, was the ability to synthesize lignin. Lignin is a chemically complex, amorphous material that is deposited in the intracellular space as plant cells, particularly those of the stem, mature. It adds rigidity to the carbohydrate components of the cell wall, allowing some plants to grow to considerable heights. Land plants needed to cope with the Earth's gravitational field to support their stems and branches of increasing span. Large trees are heavily lignified, especially in the regions that support the large branches. On the other hand, the smaller, more flexible grasses contain less lignin. The presence of lignin makes the carbohydrate of plant cell walls inaccessible to the digestive enzymes of many grazing animals, which tend to prefer the grasses for a meal.

The ancient plants gradually evolved more highly differentiated, efficient systems suited to life on land.

About 400 million years ago, oxygen levels were about 10% of what they are today. Plants began to colonize the land on an increasing scale. Climatic changes assisted by making the atmosphere more humid, and once the plants took firm hold they also contributed to this by drawing up moisture from the soil and releasing it into the air through their stomata by the process of transpiration. Huge tropical swamps were formed, where large trees grew rapidly, and when these and other plants died, they decayed and were transformed to form vast deposits of carbonaceous material, which are mined today as coal. Some 50 million years later these swamps began to dry out as the continental masses then present drifted together to form a single massive land mass, Pangaea, a large portion of which was located over the South Pole, and as a result was covered in a thick layer of ice. Interior parts contained large deserts. The climate became colder, possibly because volcanic eruptions caused dust clouds, which blocked out much of the energy from the sun. More ice-sheets and glaciers formed. The large areas of snow and ice reflected more of the energy from the Sun back into space; more ice formed, the sea-level dropped, and the Earth entered its most severe Ice Age. These rapid changes in the energy balance of the Earth had a profound effect, especially on land species. As a result, the great Permian Extinction, which wiped out vast numbers of both land and marine species, took place.

We might be tempted to look on these events aghast as yet another example of disasters which seem periodically to affect life on our planet in a negative way. This would not be only wasted emotion, but in this case (and others) a decidedly misguided one. As with many events in the history of life on Earth, the Permian Extinction presented evolutionary opportunities for those organisms which were, fortuitously, at a stage to exploit it. One of the results was the emergence of true seeds and ultimately the flowering plants that give humans (again fortuitously) so much pleasure today. The flowering plants—Class Angiospermae—have steadily increased their influence on the living world since the early Cretaceous, some 130 million years ago. Since about 90 million years ago they have been dominant in the plant world. Today, some 90% of plant species are angiosperms. They range in size from the gigantic eucalypts of Australia to tiny duckweeds with 1 mm leaves. Angiosperm flowers range from pinhead size to those of the giant Indonesian *Rafflesia*, up to a metre across. The grains, fruits, and vegetables we eat are angiosperms, as are the plants that provide us with cotton and linen. Domestication of angiosperms (and animals) had enormous influence on the early history of humans, which was extended firstly by conventional plant breeding, through to today's genetically modified crops. Although angiosperms reproduce both sexually and asexually, it is aspects of the former method that have helped their evolution and fascinated humans. Flowers and fruits are two unique characteristics of angiosperms. With the aid of animals, in particular insects, flowers have helped plants to overcome one of the problems they encounter because of their immobility—their inability to reproduce efficiently with distant members of their own species. Rather than allowing the wind to carry pollen randomly, many angiosperms have developed complex relationships with insects. These range from the mutualistic interaction between bees and flowering plants, where both participants benefit, to the outright bizarre mating behaviour of male wasps of the species *Campsoscolia ciliata*, which attempt to copulate with the flowers of the Mediterranean orchid. The male wasp is apparently fooled into this act by a physical resemblance of the flower to the female wasp and also, more remarkably, by chemical compounds with a scent similar to that of a receptive female wasp. The wasp carries pollen to other orchid flowers and the species is efficiently fertilized. There is no apparent benefit to the wasp, in contrast to the efforts of bees, which gather nectar (and other materials to produce their propolis) as well as spreading pollen. Some plants and their attendant insects can truly be said to have coevolved, so much have their ways of life become enmeshed. Insects view the world at wavelengths different from ours. Studies have shown that bees perceive ultraviolet as a distinct colour, but don't perceive red. Bee-pollinated flowers are usually blue or yellow, with distinct markings that can be seen by ourselves if a photograph is taken by a camera sensitive to ultraviolet light. The most important plant pigments, which give flowers their colour, are flavonoids, complex organic compounds that block far-ultraviolet light and thus protect the plants, while selectively admitting blue-green and red light for photosynthetic purposes. An important class of flavonoids, the anthocyanins, are major determinants of flower colour. Their colour can range from red to blue depending on the pH. Some red, orange, and yellow flowers contain carotenoids, which also are found in leaves and stems. Various combinations of pigments give rise to the variety of colours we see in plants and their flowers. Reduced to the prosaic level of chemistry, all colours we perceive arise from the interaction between electromagnetic energy and the electrons in organic molecules. A rose by any other name.

13.3 The evolution of fungi

Difficult though it may be to believe, fungi are more closely related to animals than to plants, according to recent molecular evidence. Animals and fungi appear to have diverged from a common ancestor, possibly from a protist resembling a member of the colony-forming choanoflagellate family. Choanoflagellates are heterotrophic, aerobic protists with a single flagellum. The earliest fungi-like fossils are filaments from about 540 million years ago (Lower Cambrian).

Fungi are to be found almost everywhere on Earth. Their spores float in the air and settle on our faces, our food, and our fences. Well in excess of 70,000 fungal species have been named and it has been estimated that the total number of species exceeds 1.5 million, second only to the insects. Although they are principally land dwellers, about 500 known species are marine. I mention these numbers to emphasize the huge potential for fungi to influence their environment. Fungi obtain their food by secreting enzymes directly into their immediate environment, thereby digesting it externally rather than ingesting it as animals do. This way of feeding makes the fungi so distinctive that they have been assigned their own kingdom. They are characteristically multicellular, but a few, including yeasts, are unicellular. Although once believed to be primitive plants which lacked chlorophyll, the fungi in fact share little with the plant kingdom, except in general appearance and their lack of mobility. Fungi and plants both possess cell walls, but these are chemically different. Fungal cell walls consist mainly of the polysaccharide chitin, the same material found in the exoskeletons of insects, spiders, crabs, and shrimps. Chitin is more resistant to degradation by microbes than cellulose, the major component of plant cell walls.

Fungi are formed of slender filaments called hyphae. Typically, hyphae are divided into cells by septa, but these do not usually form a complete barrier between cells. Rather, the septa have large pores that let the cytoplasm stream through quite freely, allowing proteins synthesized anywhere in the hyphae to pass to the tip, which is the point of growth. As a result of this freedom of transport for essential molecules, the growth of hyphae may be very rapid during times when food and water are plentiful. It has been estimated that some individual (presumably large!) fungi can produce up to a kilometre of new hyphae within 24 hours. This astonishing growth rate reflects an equally impressive rate of metabolic energy flow and is only possible because of the structure and arrangement of the hyphae. Individual hyphae, which are barely visible to the naked eye, are arranged in a mass called a mycelium (from the Greek for fungus, *myketos*; the study of fungi is called mycology). The mycelium of a fungus may be many metres long. It penetrates into the surroundings, or substratum, and because of the filamentous form of the hyphae gives the fungus a relationship with its environment that is unique in muticellular organisms. The surface-area-to-volume ratio is very high, bringing fungi into intimate contact with their surroundings such that no somatic cells are more than a few microns from the environment, with only a thin cell wall and the plasma membrane separating them. The cell wall is nevertheless rigid, preventing microorganisms and particles from being engulfed directly. Instead, the fungi secrete a battery of enzymes into a substratum containing food and absorb the smaller molecules, which are released, mainly at or close to the growing tip of the hyphae. All fungi are heterotrophic, obtaining their food supply by growing on top of or inside their food source. Depending on the relationship with their food supply, fungi function as saphrophytes (which live on organic compounds from dead organisms), as parasites (living off other live organisms, usually harming them but giving nothing in return), or in mutual symbiotic association, where the association benefits both organisms. Ecologically, fungi are important as decomposers of the living and dead tissues of other organisms. In this role they release nutrients and other compounds to the environment, such as carbon dioxide to the atmosphere and nitrogen compounds to the soil. Fungi thus help to maintain the cycles that keep the limited amounts of these essential materials constantly in circulation in the biosphere. Keeping the balance of nature is a role not to be underestimated and the fungi are fully involved. They attack cellulose and lignin in wood and other plant material, but also impact negatively on many human activities. Fungi are great destroyers of human foods, as they grow on bread, fruit, meat, and vegetables. Amazingly, their digestive repertoire extends to cloth, paint, paper, rubber, leather, and even petroleum. Valuable crops are attacked by over 5000 species of fungi, causing billions of dollars in losses each year, while many trees and wild plants also succumb to pathogenic types. Over 150 species are involved in causing serious diseases in humans and domestic animals.

Other fungal types are commercially valuable. Yeasts, such as *Saccharomyces cerevisiae* (baker's yeast), produce alcohol and carbon dioxide, and are widely used in baking, brewing, and winemaking. A number of antibiotics, including penicillin, have been isolated from fungi, and in 1979 the 'wonder drug' cyclosporin became available. Continuing tissue rejection problems up to the late 1970s almost resulted in the cancellation of attempts develop organ transplants in humans. Cyclosporin was found to suppress immune responses causing rejection and allowed the programme to proceed. Truffles, morels, and some 20 types of mushrooms are enjoyed as food by people around the world. The ability of fungi to break down all kinds of substances has led to their use in toxic waste cleanup and also in non-toxic waste disposal.

What needed to be 'invented' by the ancestral protists that evolved into the fungi? Considering the relative simplicity of structure, and their way of obtaining food (and thereby energy), not a great deal. Development of a basic set of enzymes allowed fungi to take advantage of the many food sources that existed. The components of the basic set changed as new fungal species evolved along with new plant species and after animals became available as a food source on land. The development of a cell wall was essential for terrestrial organisms to maintain structures that were rigid enough to support them against gravity and to penetrate the substratum. The fungal cell wall does not need to be as structurally strong as in plants, as fungi don't need to grow large above-ground structures. Much of the fungus is in the form of mycelia, which are buried close to or well inside the food source. There is no need for a large light-collecting canopy, as in a tree or shrub, to maximize photosynthesis.

These then are the fungi, with a kingdom to themselves and deservedly so. As an integral part of the biosphere, their influence is often overlooked, but they are all around us, quietly helping to maintain the energy balance of the environment that we more spectacular organisms so readily take for granted.

13.4 The evolution of animals

The sources of cellular energy in multicellular organisms are essentially the same as those in unicellular life forms. The basic biochemistry of the cell is the same and ATP is still the major energy source. Mammals, as examples of complex multicellular organisms, have added layers of structural complexity compared with their unicellular cousins. Mammalian bodies are often described and studied in terms of the eight or so systems of which they are constructed. Their body complexity leads to extra requirements, such as the means to distribute and regulate energy supply in the mammal as a whole. To achieve this, the energy requirements of all cells, moment by moment, need to be monitored, collected, and collated, and the resulting information distributed appropriately. This distributed information is directed to various feedback mechanisms, which act to supply individual cells with their required amount of energy. The whole integrated ensemble of processes is working all the time to maintain the delicate balance of energy demand and supply essential for healthy cell life.

To reiterate: animals are heterotrophic; they depend for their energy on food derived from other living organisms. The search for food dominates the behaviour of animals in a much more visible way than it does that of green plants. The behaviour of animals is more obvious as they are relatively large, mobile, and readily catch our interest. Plants are autotrophic, and as such require water, nutrients, CO_2, and sunlight, all of which are readily accessible on much of the land and in the surface layers of oceans and other bodies of water. At about the same time as autotrophic protists began tentative steps along the multicellular pathway that was to lead to modern plants, early heterotrophic protists, the ancestors of the animals with which we are all familiar, also 'experimented' via mutation and natural selection, with the advantages and drawbacks of multicellularity. This exquisite sorting process, over time, generated an enormous variety of ways to improve the vital processes of feeding and reproduction. Success, at least that measured by the survival of enough organisms to provide today's astonishing diversity of animals, was achieved through the cooperative association of cells. As with plants, no doubt there were many trials and failures. First, cells formed into simple tissues; some individuals survived and succeeded. Ultimately, came the complex and specialized organs that we see today in the domesticated and wild animals around us. One of the differences between animal groups which strikes us immediately is that of body shape.

We could spend quite some time on the topic of animal body shapes, or, as evolutionary biologists would say, body plans. As an example, consider the dog. All dogs have the same body plan, but it is obvious that all dog breeds do not look (or behave) alike. All existing dogs (family Canidae) evolved from the wolf, and humans over thousands of years have had a major hand in the accelerated development, under domestication, of the major breeds. The story of the domestic dog is an example of evolution not by natural selection, but of selective pressure exerted by humans. Much the same can be said of the other animals domesticated by humans—pigs, cattle, sheep, goats, poultry, horses, pigeons, and all the rest, plus the domesticated plants. All have been subjected to selective breeding, sometimes with astonishing success.

Evolution has in general tended to favour conservation of animal body plans, rather than replacement of them. Many examples were noted by Darwin, and in the mammals these include the familiar comparison of the bones comprising the wing of a bat, the flipper of a whale, the forelimb of a mouse, and the human arm. These are examples of structural homology, or structures built from the same basic anatomical feature. The overall plan has been retained by the animals, despite considerable differences in their respective ways of life. Such similarities indicate a common ancestor and constitute one very powerful type of evidence for evolution, and for its ability to produce diversity within a basic plan—remember the principle of common descent.

The more closely two organisms are related, the more characteristics they share, both morphologically (i.e. in body shape and body plan) and genetically. The animal kingdom as a whole has a rather limited number of characteristic body plans, which developed broadly as follows. Early association of cells to form primitive animals and fungi could well have started with the aerobic protist family called the choanoflagellates. These organisms probably formed hollow, spherical structures initially, which, via mutations, gradually flattened into a pancake shape consisting of two layers of cells. The cells of the back or dorsal side were thinner than those of the bottom (ventral) layer. The ventral layer was thicker and used for movement and food gathering. Over time, the ventral layer developed a cavity, which evolved into a groove that served as a primitive digestive region. Eventually the groove sealed over, forming an early version of what is now, in animals ranging from worms to mammals, called the alimentary canal. These early animals had two distinct layers of cells: those on the 'outside' became the ectoderm, while those lining the alimentary canal became the endoderm. The process also formed a body cavity, or coelom, which had no contact with the outside world. In time the coelom became lined with a third layer of cells, the mesoderm. Thus arose the three-layered body plan of the triploblasts, which most of the animal world, including ourselves, subsequently adopted. The body became elongated in the direction of the alimentary canal and thus became bilaterally symmetrical, that is it possessed a right side and a left side, which were reflections of one another, as well as the dorsal and ventral aspects. The two ends of the alimentary canal eventually developed distinct functions and became the mouth and the anus, with food passing unidirectionally from the former to the latter, being digested and absorbed along the way. In order to control an increasingly complex body, the nervous system evolved in parallel, ultimately leading to a head and brain located in the anterior (front) region near the mouth, where they were best places to house other sensory systems such as those for sight, smell, hearing, taste, and touch. The sensory organs allowed the animal to keep in constant contact with many aspects of the surrounding environment, and allowed it to respond rapidly to stimuli received from that environment. Later still, a backbone evolved in some animals, along the path of the nervous system which ran the length of the body, and the vertebrates, ranging from fish to amphibians, reptiles, birds, and mammals, came into being. We have seen a number of times previously that 'successful' genetic mutation, as the main mechanism underlying evolution, usually results in relatively small changes in an organism at a time. An accumulation of small changes, over many successive generations, can result in large differences. Body plans that have proven successful need not change drastically to achieve quite different functions, as shown by our example of wings, arms, and flippers above, so the question arises: how are the basic animal body plans maintained, yet allowed to differ in detail, that is to generate the enormous number of animal shapes we see around us? For part of the answer, we must probe a little into the mysteries of animal development.

There is a period in the development process of an animal called a 'phylotypic stage'. Before the phylotypic stage the 'decision' as to which phylum the animal will belong to has not been made. Before the phylotypic stage, the threshold of commitment has not been crossed. Despite this lack of decision, there has been already programmed into the developing organism a number of blocks of undifferentiated cells, arranged relative to one another just as they will be arranged in the mature adult. The basic body plan is already in place, preserved because it has been advantageous, so in the evolutionary sense it does not 'need' to change much. As development proceeds, there comes a stage where these blocks of cells are allowed to differentiate more or less independently, leading eventually to an individual belonging the phylum to which its own parents belonged. The all-important decision at the phylotypic stage is, not surprisingly, determined by specific regulatory genes characteristic of the phylum.

For example, the animal Phylum Chordata includes the Subphylum Vertebrata, whose members possess a body plan characterized by having the familiar left/right (bilateral) symmetry, a head with a skull and brain, a tail, and a segmented vertebral column (the backbone). Depending on the class to which our example belongs, its phylotypic stage may become a jawless fish, a cartilaginous fish such as a shark, a bony fish such as a herring, a frog, a reptile, a bird, or a mammal; the basic body plan is the same. This plan and its inherent flexibility allow the development of fins, legs, wings or arms, to name a few easily observed variations on the theme. Although we may view elephants and toads as vastly different, their body plans are similar; they belong to the same phylum (Chordata) and subphylum (Vertebrata), and begin to differ in classification only at the class level (*Mammalia* and *Amphibia*, respectively).

Built into the genome of every multicellular species are the instructions to produce the body plan characteristic of that species. Body plans are the result of a combination of up to several hundred cell types, all of which are formed by division and differentiation from a single cell, a fertilized egg in the case of animals (or, in some plants, a spore). With great certainty, we can assert that pea plants will produce pea plants and ducks will produce ducks. This certainty reflects our experience, and underpinning it all is the regulation of gene expression, from the fertilized egg to the mature adult containing cells in their billions. If something should upset the regulatory mechanism, disaster for the individual is usually the result. For humans, malnutrition, disease (rubella), some drugs (smoking, heroin), or other chemicals can upset the exquisitely balanced sequence of events that culminate in birth. Those humans who are born 'normal' are fortunate, considering the plethora of hazards they face leading up to that event. Other species face similar problems.

New discoveries in developmental genetics that depend on the marvellous techniques of molecular biology are rapidly advancing our knowledge, taking us further and further into the evolutionary past to reveal some most surprising relationships. In the fruit fly *Drosophila*, the classic choice of classical geneticists, there was discovered a series of genes, called Hox genes, each of which is active in a different section of the embryo, from front to back. Each gene appears to act as a master switch that activates a series, or cascade, of other genes that are needed for the development of the respective regions of the embryo. The Hox genes have been sequenced comparatively recently. The surprise was the finding of a similar series of genes in the mouse and subsequently in other groups of animals. In summary, it is now believed that the common ancestor of all bilaterally symmetrical animals, such as flies, segmented worms, fish, and mammals, possessed a series of Hox genes which acted on different regions of the body and controlled the development of structures in those regions. Whereas flies have one set of Hox genes, the more complex vertebrates have four sets, each slightly different from the others. These genes have been conserved for about 500 million years and it has been suggested that their possession should be the defining characteristic of all animals. Natural selection seems to work on the principle 'if the system works, why change it, especially if there are hazards in doing so?' Perhaps some early 'attempts' by animals did sample other systems, but these didn't survive. The widespread occurrence of the Hox system was so unexpected because there is nothing in common between the animal structures it controls in different groups of animals. There is nothing in cats or in pigeons which corresponds to the six legs and two pairs of wings of the insect thorax. In retrospect, it was realized that the control mechanism, or signalling system, was being conserved. The Hox genes signalled 'build something here', while at the next level of control the structural genes responded, producing a fin, a wing, or a leg. Thus, the actual genes being controlled were the variables which, over time, evolved to produce the different detailed structures we see in animals today—fin, wing, or leg. The Hox signalling system itself was what evolution conserved at a fundamental level.

13.5 The organization of the mammalian body

Unless it is necessary, in the following discussion I won't distinguish between humans and other mammals, as many aspects of their physiology are common. The basic building blocks of the mammalian body are the familiar cells. In humans there are some 210 different cell types. Cells of the same type are grouped into tissues, the main tissues being: epithelial, connective, lymphoid, nervous, and muscular. Aggregations of various tissues form the familiar organs of the body. In turn, the organs and tissues are traditionally classified into a number of functional systems, of which there are eight.

1) The *musculo-skeletal system* comprises the bones of the skeleton, skeletal muscles, and a number of associated tissues, whose main functions are to provide a means of movement and support for the internal organs. The system is especially important in land animals, which also need to support themselves against gravity.
2) The *cardiovascular system* is necessary to transport oxygen and nutrients to all the cells in the body. This is usually done by blood, pumped by some kind of heart. Simpler, single-celled animals, which are in direct contact with their environment, do not need such a system.
3) The *respiratory system* (lungs and associated organs) extracts oxygen from the air and delivers it to the bloodstream, and removes carbon dioxide from the deoxygenated blood for expiration to the atmosphere. As such it is of primary importance in energy processing in mammals, all the cells of which are dependent on oxygen, but only a few of which are in contact with it. Once again, unicellular organisms don't require such a system.
4) The *digestive system (gastrointestinal tract, GIT)* is involved in extracting nutrients from the diet, that is it processes ingested food. The GIT is involved in the degradation and absorption of foodstuffs, and the removal of the depleted food residues. It is another specialized system not required by unicellular organisms.
5) The *urinary tract and the kidneys* are involved in the removal of the soluble, non-volatile waste products of metabolism. This system is also involved in the control of the composition and volume of the fluid surrounding the cells (extracellular fluid).
6) The *endocrine and nervous systems* act to coordinate and regulate the various bodily systems. Without this regulation, the body as a whole would not be able to function. The nerves use electrical signals for information transmission, whereas the endocrine system largely uses chemical messengers, such as hormones, which travel from their source via the bloodstream to the cells where their activity is needed. Each type of hormone usually has its own

receptor on the target cell surface, with which it interacts to elicit a response inside the cell.
7) The *reproductive system* produces the specialized sex cells, or gametes. In mammals these are the sperm (male) and the ova (female), which each contain half (the haploid number) the number of chromosomes normally found in the organism (the diploid number).
8) The *immune system* is the body's defence system against infection. It not only kills invading organisms and eliminates some foreign chemicals, but also destroys damaged or diseased cells that may be hazardous to healthy cells in the body.

All the organ systems listed above are unique to multicellular animals. As multicellularity developed from the simple association of similar cells, natural selection, as ever, conserved the 'good ideas' from the many that were undoubtedly generated, gradually honing each system as it arose to be increasingly efficient. Natural selection can only work on what has come before. There are therefore limits, both qualitative and quantitative, on what it can do in the short term.

Before mammals could evolve, other 'simpler' animals needed to emerge as intermediates on the journey to the complexity that was ultimately achieved by mammals. The interrelatedness and complexity of mammalian bodily systems require that they be suitably regulated and controlled. These are some of the defining characteristics of complex organisms, and of life itself. The two major regulatory systems in mammals are the endocrine system and the nervous system. One aspect of control in mammals that heavily involves the endocrine system is the maintenance of a stable environment in which each group of cells may work. The internal environment of a mammal consists of groups of cells in tissues and organs, surrounded by extracellular fluid. Both the cells and the surrounding fluid must be stabilized against rapid changes. The maintenance of a stable internal environment is called homeostasis. I don't mean to imply that the environment for all cells types is the same. Far from it. What is important is that the environment for any type of cell be maintained close to the optimum for that type of cell, under the conditions prevailing at the time. For example, for the heart muscle to continue beating strongly and regularly, the heart cells must continue to contract rhythmically. This depends on electrical signals, which depend on the relative concentration of sodium and potassium ions in the intracellular and extracellular fluids. An excess of potassium ions in the extracellular fluid may cause the cardiac muscle cells to contract irregularly instead of rhythmically, with potentially fatal results. Fine control of the extracellular potassium and sodium ion concentration is clearly essential.

Homeostasis is a property of unicellular organisms as well. Their internal environment must also be kept quite close to the optimal levels. The cytoskeleton, various organelles, and other components must have their essential gradients and concentrations carefully maintained for optimum activity. In general, monocellular organisms have little control over such variables as temperature or the chemical composition of their immediate environment. Multicellular organisms have the same problems, and the development of solutions to these problems has taken considerable evolutionary time.

One solution involved the adoption of an optimal extracellular environment in which the various cells could carry out their functions without having to continuously 'worry' about this aspect of their existence. Mammals also need to coordinate their various organs and tissues into an integrated whole. The eight mammalian organ systems listed above took some 500–600 million years to evolve. This seems a short time compared with the 3 billion years of life that preceded them and indicates that many of the necessary control mechanisms and principles had already been evolved by bacteria. Not a great deal needed to be 'invented' by natural selection, but what was invented made possible multicellular life as we see it today.

The evolutionary development of each organ system has been studied intensively across the whole range of animal groups. Although it is convenient for many purposes to recognize the individual functions of organs, the body functions as a whole. No organ or system is more important than any other in maintaining the overall health and wellbeing of an individual. If one part fails, it will affect the rest. For the brain, which must consume large amounts of energy continuously, glucose and oxygen depletion can be tolerated for only a short time before severe damage and death occur. This is one price we pay for our biological complexity, but at least we are capable of realizing, understanding, and coping with it. Medical experts often specialize in one of the systems, such as the nervous system or the GIT, or may concentrate on a very limited aspect, such as the heart or the kidneys. The complexity of the mammalian body precludes a single person being an expert in all its workings. There is still much to learn, even for the experts of today.

The study of biology is a never ending journey, whatever your view of life.

GENERAL REFERENCES

Campbell, N.A., Reece, J.B., Urry, L.A., *et al.* (2008) *Biology*, 8th edn. Pearson Benjamin Cummings, San Francisco.

Raven, P.H., Evert, R.F., and Eichhorn, S.E. (1999) *Biology of Plants*, 6th edn. W.H. Freeman &Company/Worth Publishers, New York.

Tudge, C. (2000) *The Variety of Life*. Oxford University Press, Oxford.

REFERENCES

Maynard Smith, J. and Szathmary, E. (1999) *The Origins of Life*. Oxford University Press, Oxford, p. 112.

Appendix A

Electromagnetic Radiation

Electromagnetic and gravitational forces are the ultimate sources of the energy that most influences life on Earth. We have discussed some of the effects of the gravitational force in Chapter 3. Here I deal with the aspects of electromagnetic radiation as a background for the discussion in Chapter 5 of chemical structure, chemical stability, chemical reactions, and the energy associated with them.

Scottish physicist James Clerk Maxwell (1831–1879) developed the electromagnetic theory of radiation between about 1855 and 1873 (Maxwell 1873; Laidler 1993). Maxwell's equations described the nature of light mathematically, rather than conceptually. These equations were based on the wave theory of light, which had recently become accepted by leading physicists. Maxwell proposed that changing electric currents or accelerated electric charges gives rise to magnetic and electric fields, jointly called electromagnetic radiation (sometimes termed electromagnetic waves). It was not until 1888 that Heinrich Hertz (1857–1894) demonstrated experimentally the existence of electromagnetic radiation in the form of waves.

Electromagnetic waves simultaneously posses an electric field component and a magnetic field component. The electric field acts on charged particles, whether they are stationary or moving. The magnetic field acts only on moving charged particles. These two fields vibrate as waves in mutually perpendicular planes (Figure A.1).

The velocity (c) of any wave is the product of its wavelength (λ) and its frequency (ν):

$$c = \lambda \nu \quad \text{or} \quad \nu = c/\lambda$$

The SI unit for λ is the metre, but the nanometre (10^{-9} m) is commonly used. The SI unit for frequency ν is hertz (Hz), where 1 Hz = 1 s^{-1}. The frequency is the number of waves passing a given point per second, for example 10 Hz means 10 full waves passing per second.

Electromagnetic waves, like water and sound waves, are transporters of energy. In contrast to water and sound waves, electromagnetic waves require no medium for their transmission, that is they can travel through a vacuum. Sunlight and starlight reaching Earth effectively do so through the near-vacuum of space.

The velocity of visible light (c) and of all other wavelengths of electromagnetic radiation is 2.9979×10^8 m s^{-1} in vacuum. In air it is very nearly the same, but in water, which is much more dense, c is less, at about 0.66 times that in air, or about 2×10^8 m s^{-1}. The velocity of light in a vacuum is one of the so-called universal constants, and the realization of this by Einstein was essential to his special theory of relativity.

In 1900, the German physicist Max Planck (1858–1947) published results of his work on the emission of blackbody electromagnetic radiation. A blackbody is defined as a perfect absorber and perfect emitter of radiation. Planck was trying to explain some anomalies in the emission of blackbody radiation that had defied solution by classical physics.

As part of his approach, Planck proposed that blackbody solids behave as a series of oscillators, one for each frequency of emitted light, in which the emitted electromagnetic energy (E) was limited to discrete values, according to the expression:

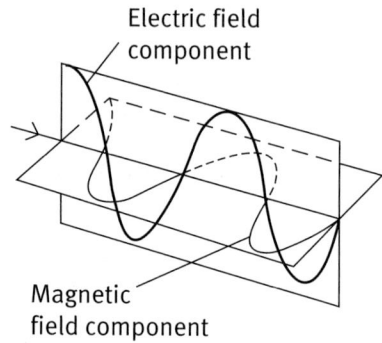

Figure A.1. Electromagnetic radiation comprises an electric field component and a magnetic filed component, as waves that oscillate in mutually perpendicular planes. The two components have the same wavelength and the same frequency, and thus the same velocity: the velocity of light. Each wavelength has a characteristic energy associated with it.

$$E = nh\nu$$

where ν is the fundamental frequency of the oscillator concerned, n is a whole number, and h is a constant (later called Planck's constant).

This implied that the permitted energies of the oscillator occurred as discrete packets (later to be called quanta) in simple multiples of $h\nu$. Thus, such an oscillator of frequency ν can possess only the energies 0, $h\nu$, $2h\nu$, $3h\nu$, etc. On the basis of his assumption Planck was able to derive a mathematical expression that precisely fitted the experimental curve for the distribution of energy density in blackbody radiation, thus neatly explaining the anomalies mentioned above.

Finally, if we take

$$E = h\nu \text{ and } \nu = c/\lambda \text{ from above, then}$$
$$E = h\nu = hc/\lambda$$

Thus, the energy of electromagnetic radiation becomes greater with increasing frequency ν and decreasing wavelength λ. Planck's constant has a value of 6.626×10^{-34} joule.second (Js), and c a value of 2.9979×10^8 m s^{-1} in a vacuum.

Why did Planck use the model of an oscillator in his work on electromagnetic radiation? A simple analogy can be used to explain his reasoning. One way to discuss the absorption and emission of light is to consider the electrons in an atom as electrons on helical springs, oscillating around their equilibrium positions. The mathematics are identical and a great deal of insight can come from the analogy.

Planck's studies lay the groundwork for quantum theory, which (unanticipated by Planck at the time) applies not only to oscillators, but also to electromagnetic radiation, to atomic and subatomic particles, and indeed to all types of energy.

At first, many physicists thought quantum theory applied only to material oscillators, such as solids, as initially proposed by Planck. In 1905, Albert Einstein (1879–1955) proposed that electromagnetic radiation itself was propagated as a series of discrete packets of energy he called quanta, rather than continuously, as had been previously believed.

Quanta (plural) and quantum (singular) derive from the Latin *quantum*, meaning 'how much'. The term was used by Einstein in one of his famous 1905 papers, where he noted that in some experiments electromagnetic radiation behaved as a beam of particles, which he called quanta. In 1926, G.N. Lewis introduced the term photon as the name for a 'particle' of electromagnetic radiation (Laidler 1993).

Until the proposals of quantum theory, physicists believed that energy was continuous, that is it could take any arbitrary value one cared to assign it. Quantum theory overturned all that, particularly at the atomic and subatomic scale. An important and revolutionary proposal of quantum theory is that energy is quantized, not continuous.

An analogy is a solid ball rolling down a set of steps, that is from higher to lower potential energy. The ball can lose potential energy only in a series of discrete steps. It cannot fall by a half step or a quarter step. Its energy behaviour is restricted to a series of discrete levels, each determined by the height of the step. This description is analogous to an atom emitting quanta of electromagnetic radiation as excited electrons relax from higher- to lower-energy states. I will return to this analogy when discussing the behaviour of electrons in atoms and molecules.

Quantum theory, developed into the highly mathematical quantum mechanics, is able to explain many aspects of atomic and molecular behaviour. It literally revolutionized physics and was complete in its essentials by about 1927.

Another of the strange proposals of quantum theory is that electromagnetic radiation has the properties of both waves and solid particles, and that moving solid particles have the properties of waves. The latter idea was proposed by Louis de Broglie in 1923.

Quantitatively, the wavelength, λ, associated with a moving particle is given by

$$\lambda = \tilde{h}mv$$

where h is Planck's constant, m is the particle mass, and v is its velocity.

The phenomenon (or should I say phenomena?) is/are known as wave-particle duality. It is one of the most puzzling aspects of quantum theory (certainly to non-experts). Suppose in an experiment we want to detect a beam of electrons. If we set up an experiment to detect a wave, we'll detect a wave, for example a diffraction pattern caused by the wavelike behaviour of the electrons (Figure A.2).

If we then set up an experiment to detect a particle, for example by using a scintillation counter, we'll detect a flash of light, caused by the electron behaving as a distinct particle. Paradoxically, both results are generated by a beam of electrons! Equally as strange, a beam consisting of other discrete particles such as protons, neutrons, and atoms has been shown experimentally to exhibit wavelike behaviour, such as diffraction and interference.

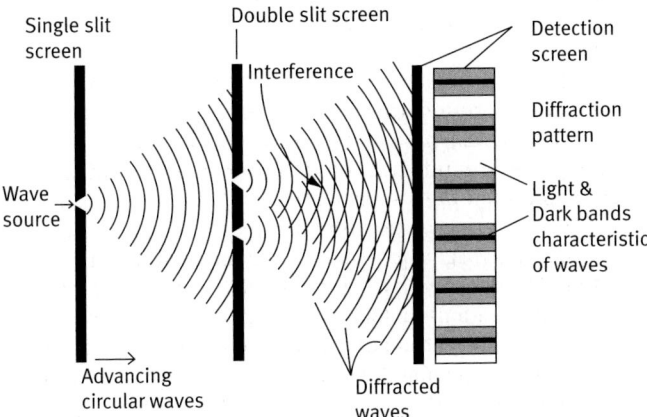

Figure A.2. Diffraction pattern formed by interference of waves passing through two small slits. The source could be electrons, visible light, X-rays, or neutrons.

Similarly, a ray of light that we normally consider as electromagnetic waves of frequency v can be shown to behave also as a stream of particles (photons), each having an energy hv. Many aspects of quantum mechanics are difficult to understand and counter-intuitive, but in the end they have to be accepted. Nature just happens to be that way. Several eminent physicists, from Niels Bohr (1885–1962) to Richard Feynman (1918–1988), have said essentially that. We cannot 'explain' quantum mechanics in everyday terms. To quote a recent physical chemistry text:

'As with any theory in science, quantum mechanics is accepted by scientists because it *works*. It is ... one of the most successfully tested theories devised by science.' (Ball 2003)

All forms of energy turn out to be quantized, but quantum effects only become significant at about the molecular level and below. As far as macroscopic, everyday objects such as footballs, people, or motor cars are concerned, their energy states are essentially continuous, in that the 'steps' between them are so small as to be undetectable. In practice, quantum effects are insignificant for such large masses.

By now, some readers might be thinking, why go into so much detail about the physics of electromagnetic radiation, plus the bit about quantum mechanics? Actually, the above description barely scratches the surface detail. I hope it gives sufficient background for what is to come.

Electromagnetic radiation is absorbed by atoms and molecules in very specific ways that are now well understood. Similarly, electromagnetic radiation is emitted by atoms and molecules in ways that reflect what is happening inside them. It is not exaggerating to say that almost everything we know about the structure and properties of matter has come from studies involving the interaction of electromagnetic radiation with matter.

Our understanding of atomic and molecular structures, their stability, and eventually the energy characteristics of chemical reactions, depends largely on experiments involving light and electrons.

Electromagnetic radiation is most familiar to us as visible light, but visible light is only the small part of the overall electromagnetic spectrum between about 400 and 770 nm (Table A1.1). The exact limits of the various regions of the spectrum are not universally agreed on, so there might be some differences between what I list below and values from other sources.

For visible light, the electromagnetic radiation is generated during changes in energy of the outer electrons in atoms or molecules. Infrared radiation (also known as thermal radiation) has wavelengths ranging from about 770 nm to 1 mm. It is emitted as a result of the rotational and vibrational motions of atoms and molecules, and some by outer electron transitions.

Microwaves (1–250 mm) and radio waves (>250 mm) complete the longer wavelength/lower energy end of the spectrum. Their sources are summarized in Table A1.1.

Going to the higher energy/shorter wavelength end, we have ultraviolet (UV; 4–400 nm), mainly involving outer electronic transitions in atoms and molecules. X-rays (about 1 nm–1 pm) are generated by bombarding metals with high-energy electrons. They involve transitions of inner electrons. Gamma rays (γ rays; $<10^{-12}$ m) are generated by transitions involving the nuclei of atoms. Gamma rays are the most energetic, and the most damaging to living organisms.

For atoms or molecules to emit electromagnetic radiation they must first be stimulated by an appropriate source of electromagnetic energy. For example, electrons in the atom will absorb energy and be raised from their ground-state energy levels to higher-energy levels. The emission of radiation we observe occurs (as photons) when the electrons in atoms or molecules fall back to their ground states (or to other permitted lower-energy states) by processes collectively called relaxation.

Table A1.1. The atomic and molecular origins of the electromagnetic spectrum.

Radiation type	Frequency range (Hz)	Wavelength range[1]	Atomic/molecular transition causing the radiation
Gamma rays	10^{20}–10^{24}	$<10^{-12}$ m (<1 pm)	Nuclear excitation
X-rays	10^{17}–10^{20}	1 nm–1 pm	Inner electron excitation
Ultraviolet	10^{15}–10^{17}	400 nm–4 nm	Outer electron excitation
Visible	4–7.5×10^{14}	770 nm–400 nm	Outer electron excitation
Near-infrared	1×10^{14}–4×10^{14}	2.5 µm–770 nm	Outer electrons/molecular vibrations
Infrared	10^{11}–10^{14}	1mm–2.5 µm	Molecular vibrations
Microwaves	3×10^{11}–10^{13}	250 mm–1 mm	Molecular rotations, electron spin*
Radio waves	$<3 \times 10^{8}$	>250 mm	Nuclear spin*

[1] m = metre; mm = millimetre; µm = 10^{-6} m; nm = 10^{-9} m; pm = 10^{-12} m.
*Electrons and atomic nuclei behave as though they are spinning on an axis. The spin generates a magnetic field, the strength of which can be detected and used to give information about molecular structure.

The ground state for a particular type of electron (call it E1) in an atom refers to the lowest energy level it can occupy. The ground state is usually the electron energy state that is most heavily represented in a collection of identical atoms at 'normal' temperatures. As the temperature is increased, the electron E1 will absorb energy and the higher-energy levels become more populated, at the expense of the respective ground states. All electrons are selective in the radiation they can absorb. Depending on their ground state energy and position in their atom, they will absorb energy in a series of particular wavelengths only. This will boost them into ever-higher energy states. Electron energy levels are quantized, hence the use of the step analogy. There is no stopping between steps.

If the energy supplied reaches a certain level, called the first ionization energy, for the type of atom involved, the electron will be removed from its atom. The energy levels of an electron become closer together at higher energy levels (Figure A.3). Figure A.3 can be used to rationalize the emission spectrum of atomic hydrogen.

Early in the history of spectroscopy, scientists studied the emission of electromagnetic radiation from heated gases, in particular elemental gases such as hydrogen, helium, and oxygen, inert gases such as neon, and some elements that were sufficiently volatile when heated, such as mercury or sodium. The emission spectrum of the simplest element, hydrogen, was important in the development of atomic theory. In contrast to heated metal rods and sunlight, hydrogen exhibits a line emission spectrum. Hydrogen and other elements will only produce a detailed line emission spectrum if heated to high temperatures, for example in a discharge tube, in which an electrical discharge is generated in a tube containing a small quantity of the element. At normal temperatures, most electrons in atoms are in the ground state and there will be virtually none in high-energy states available to emit photons as they relax.

The line emission spectrum of hydrogen was puzzling to early spectroscopists. Why was light emitted only at discrete wavelengths? We now know the explanation, thanks to quantum theory. Light is emitted from heated atoms as photons, when electrons

308 Introducing Biological Energetics

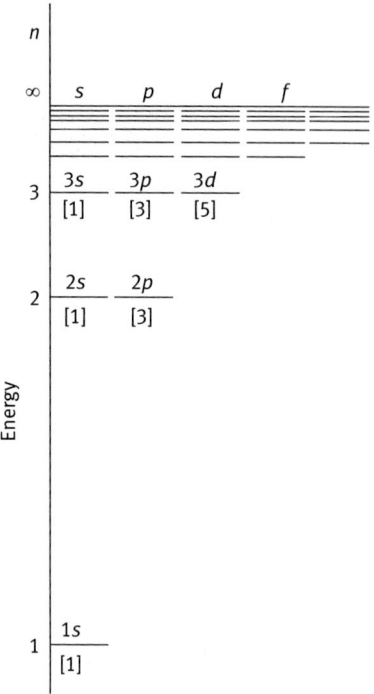

Figure A.3. Relative electron energy levels of hydrogen atomic orbitals. Orbitals with the same principal quantum number n have the same energy. For values of $n > 1$ there are subshells, designated s, p, d, and f. The numbers of orbitals in each subshell are shown in brackets. The energy levels become closer together at higher energies. The electron is removed when the first ionization energy level is reached. (Illustration adapted from Atkins, P. and de Paula, J. *Atkins' Physical Chemistry*, 7th edn, p. 373. © 2002 Oxford University Press.)

relax from higher-energy orbitals to lower ones. There is a strictly limited number of energy states that electrons are allowed to occupy. Each state has its own characteristic energy, and as the electrons jump between them a line emission spectrum is generated, one line for each permitted electron jump. The jump or step analogy is apt, as there are no intermediate states between each step. Each type of atom in the periodic table has its own unique number and arrangement of electrons, so each will emit its own characteristic line emission spectrum. This allows the identification of elements even in a mixture, using a technique called atomic emission spectroscopy. Hydrogen and helium have been identified in the Sun by detection of their emission spectra. Similarly, astronomers can detect elements in stars by spectroscopic means. Temperatures and other properties of stars can be determined by the appropriate type of spectroscopy.

Consider an iron rod heated to 'white heat' in a furnace. What we observe is the visible emission spectrum of the heated rod. Most of the wavelengths of visible light are emitted (plus infrared, which we feel as heat) as the electrons relax back to lower-energy levels. As the rod is removed from the furnace and begins to cool, it changes colour through red, eventually back to its initial room temperature colour. The change of colour reflects the loss of energy to the surroundings, as radiation, via electron relaxation processes. The rod

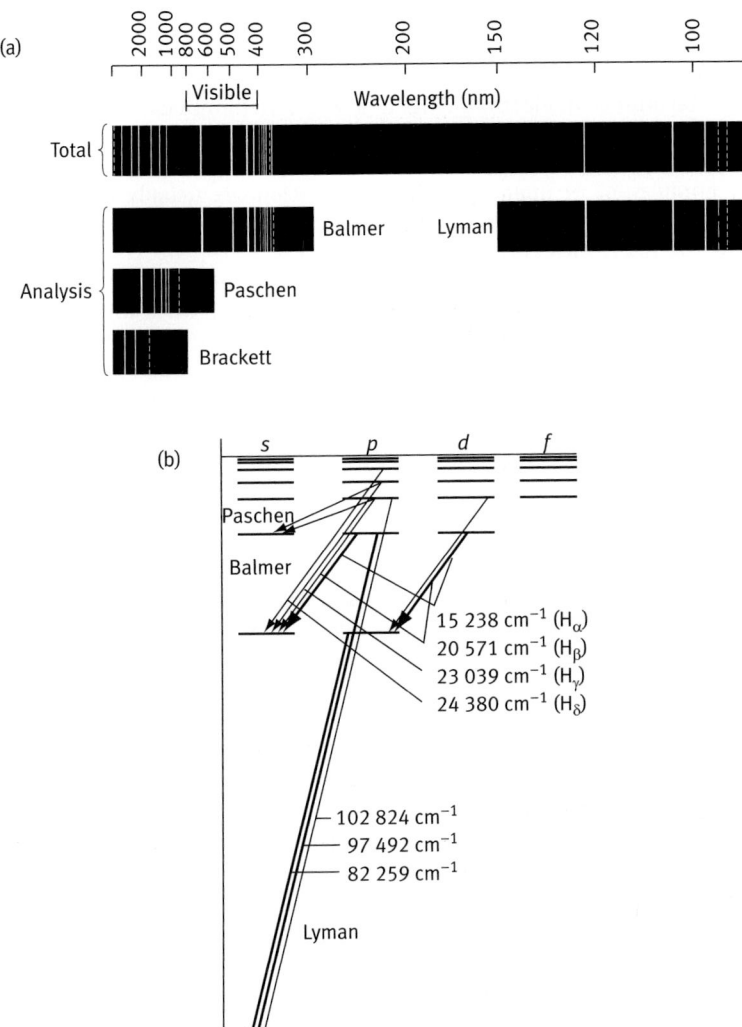

Figure A.4. (a) A line emission spectrum of atomic hydrogen. The total spectrum shows the combined series Balmer, Lyman, Paschen, and Brackett, named after the scientists who first observed them. The Balmer series of lines occurs in the visible region of the spectrum. (b) The electronic transitions involved in the formation of the hydrogen spectral lines in (a). An electron in a high-energy state can relax back to the ground state via two or more permitted intermediate energy states. Numbers refer to the wavenumber of each spectral line. Wavenumber = $1/\lambda$ = 1/wavelength. The thicker the line, the more intense the transition, and the brighter the observed spectral line. (Illustration adapted from Atkins, P. and de Paula, J. *Atkins' Physical Chemistry*, 7th edn, pp. 366, 383. © 2002 Oxford University Press.)

also cools, reflecting a reduction in the atomic vibrations in the metal, manifested as infrared radiation to the surroundings. Finally, the electrons in the iron rod, having lost the boost of energy from the furnace, will mainly occupy their ground states.

The visible spectrum is the region to which human visual pigments are sensitive. Other animals can detect other wavelengths: bees 'see' in the UV and some snakes can sense infrared (thermal) radiation (Cengel 1997).

Experimental studies in electromagnetic radiation depend heavily on various types of spectroscopy (literally, 'seeing the spectrum'). Although several types of spectroscopy were well developed before the arrival of quantum theory, spectroscopy is really an important application of quantum mechanics. Scientists usually classify the type of spectroscopy according to the region of the electromagnetic spectrum involved—UV, visible, infrared, microwave, radiowave.

Spectroscopy is widely used for analytical purposes. The absorption of UV and visible light by solutions of various compounds can be used both qualitatively to identify the compounds by their characteristic wavelength and quantitatively to measure their concentration. For elements, their emission spectra when heated can be used qualitatively and quantitatively. Several elements were discovered using emission spectroscopy, for example caesium and rubidium in 1861, by Kirschoff and Bunsen (Laidler 1993).

Nuclear magnetic resonance spectroscopy (NMR) and electron spin resonance (ESR) spectroscopy involve radiation in the microwave region. They depend on the magnetic properties of molecules. NMR is a powerful tool for the determination of chemical structure and ESR is useful in experiments involving free radicals.

Astronomy and cosmology derive their experimental information from the electromagnetic radiation received from space, using optical, infrared, UV, X-ray, radio- and microwave spectroscopy. Gamma radiation also provides information. Studies of the behaviour of light from stars have contributed to proof of some postulates of the special theory of relativity.

The light we observe from a white hot rod, or from the Sun, is called a continuous emission spectrum because essentially all visible wavelengths are emitted. To our eyes, the rod emits a single colour, but as we know, such white light can be dispersed by a prism into the familiar red, orange, yellow, green, blue, indigo, and violet. Sunlight can be dispersed similarly.

GENERAL REFERENCES

Maxwell, J.C. (1873) *A Treatise on Electricity and Magnetism*, W.D. Niven (ed.). Oxford University Press, Oxford.

REFERENCES

Ball, D.W. (2003) *Physical Chemistry.* Brooks/Cole-Thomson Learning, Pacific Grove, CA, pp. 273, 461.
Cengel, Y.A. (1997) *Introduction to Thermodynamics and Heat Transfer.* McGraw-Hill Inc., New York, pp. 632–3.
Laidler, K.J. (1993) *The World of Physical Chemistry.* Oxford University Press, New York, pp. 150–1, 314–22, 179.

Appendix B

Glycolysis: Chemical Structures of Intermediates

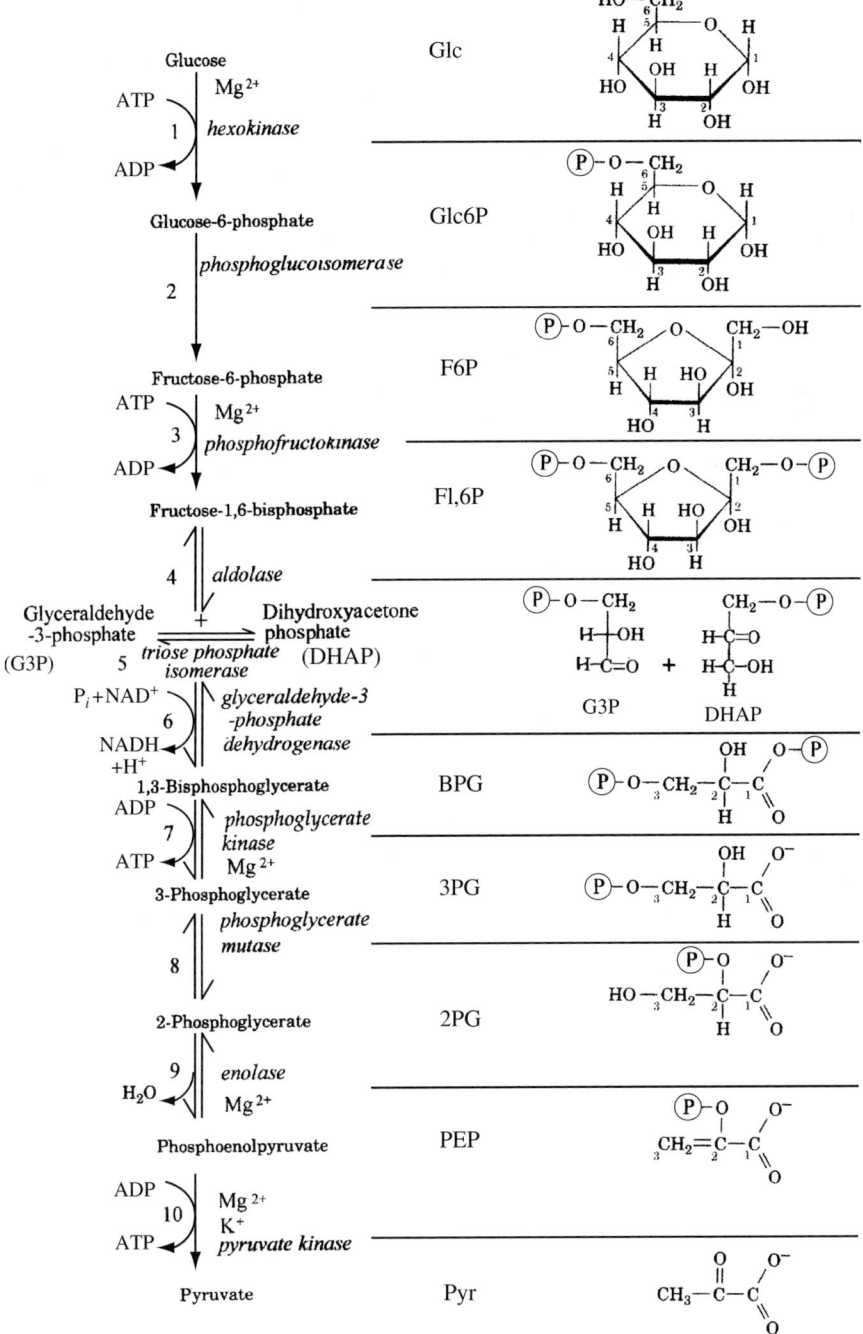

Appendix C

TCA Cycle: Chemical Structures of Intermediates

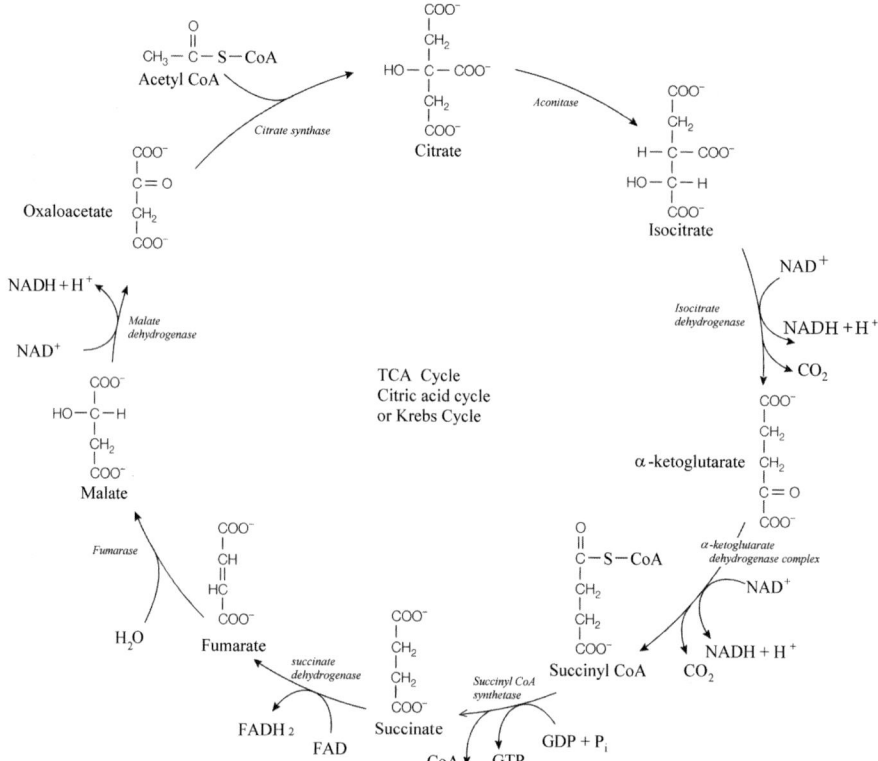

Appendix D

The Calvin Cycle in Photosynthesis, Showing Chemical Structures of the Intermediates, and the Enzymes Involved in each Step

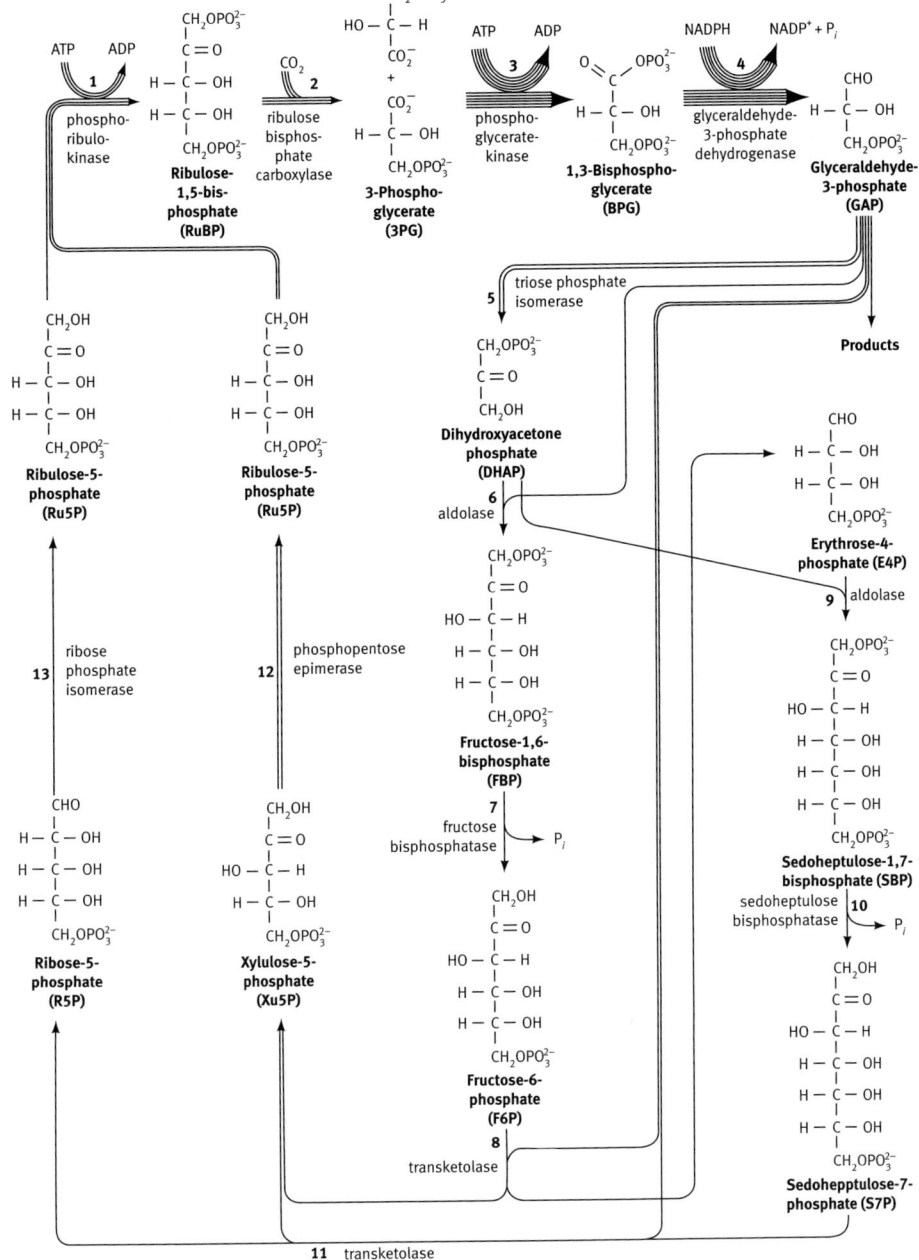

(Illustration from Voet, D., Voet, J.G., and Pratt, C.W. *Fundamentals of Biochemistry*, upgrade edn, p. 551. © 2002 John Wiley & Sons, Inc.)

Table D.1. Standard and physiological free energy changes for the reactions of the Calvin Cycle.

Step[a]	Enzyme	$\Delta G^{0\prime}$ (kJ · mol^{-1})	ΔG (kJ · mol^{-1})
1	Phosphoribulokinase	−21.8	−15.9
2	Ribulose bisphosphate carboxylase	−35.1	−41.0
3+4	Phosphoglycerate kinase + glyceraldehyde-3-phosphate dehydrogenase	+18.0	−6.7
5	Triose phosphate isomerase	−7.5	−0.8
6	Aldolase	−21.8	−1.7
7	Fructose bisphosphatase	−14.2	−27.2
8	Transketolase	+6.3	−3.8
9	Aldolase	−23.4	−0.8
10	Sedoheptulose bisphosphatase	−14.2	−29.7
11	Transketolase	+0.4	−5.9
12	Phosphopentose epimerase	+0.8	−0.4
13	Ribose phosphate isomerase	+2.1	−0.4

Source: Bassham, J.A. and Buchanan, B.B., (1982) in Govindjee (ed.), Photosynthesis, Vol. II, p. 155, Academic Press (1982).

Appendix E

Amino Acid Structures

Table E.1. Covalent structures, properties and abbreviations of the 'standard' amino acids of proteins.

Name, three-letter symbol, and one-letter symbol	Structural Formula[a]	Residue mass (D)[b]	Average occurrence in proteins (%)[c]	pK_1 α-COOH[a]	pK_2 α-NH_3^+[a]	pK_R Side-chain[a]
Amino acids with non-polar side chains						
Glycine Gly G	H—C—H with COO⁻ and NH_3^+	57.0	7.2	2.35	9.78	
Alanine Ala A	H—C—CH₃ with COO⁻ and NH_3^+	71.1	7.8	2.35	9.87	
Valine Val V	H—C—CH(CH₃)₂ with COO⁻ and NH_3^+	99.1	6.6	2.29	9.74	
Leucine Leu L	H—C—CH₂—CH(CH₃)₂ with COO⁻ and NH_3^+	113.2	9.1	2.33	9.74	
Isoleucine Ile I	H—C—C*(CH₃)(H)—CH₂—CH₃ with COO⁻ and NH_3^+	113.2	5.3	2.32	9.76	

(cont.)

Table E.1. (*Continued*)

Name, three-letter symbol, and one-letter symbol	Structural Formula[a]	Residue mass (D)[b]	Average occurrence in proteins (%)[c]	pK_1 α-COOH[d]	pK_2 α-NH$_3^+$[d]	pK_R Side-chain[d]
Methionine Met M	H—C—CH$_2$—CH$_2$—S—CH$_3$ (COO$^-$, NH$_3^+$)	131.2	2.2	2.13	9.28	
Proline Pro P	(cyclic structure with COO$^-$, H, C, N, CH$_2$ groups)	97.1	5.2	1.95	10.64	
Phenylalanine Phe P	H—C—CH$_2$—(phenyl) (COO$^-$, NH$_3^+$)	147.2	3.9	2.20	9.31	
Tryptophan Trp W	H—C—CH$_2$—(indole) (COO$^-$, NH$_3^+$)	186.2	1.4	2.46	9.41	

Amino acids with uncharged polar side chains

Serine Ser S	H—C—CH$_2$—OH (COO$^-$, NH$_3^+$)	87.1	6.8	2.19	9.21	

Amino Acid	Structure	MW	pI	pKa₁	pKa₂	pKa R
Thr T	H—C(NH₃⁺)(COO⁻)—C*H(OH)—CH₃					
Asparagine[e] Asn N	H—C(NH₃⁺)(COO⁻)—CH₂—C(=O)—NH₂	114.1	4.3	2.14	8.72	
Glutamine[e] Gln Q	H—C(NH₃⁺)(COO⁻)—CH₂—CH₂—C(=O)—NH₂	128.1	4.3	2.17	9.13	
Tyrosine Tyr Y	H—C(NH₃⁺)(COO⁻)—CH₂—C₆H₄—OH	163.2	3.2	2.20	9.21	10.46 (phenol)
Cysteine Cys C	H—C(NH₃⁺)(COO⁻)—CH₂—SH	103.1	1.9	1.92	10.70	8.37 (sulfhydryl)

Amino acids with charged polar side chains

Amino Acid	Structure	MW	pI	pKa₁	pKa₂	pKa R
Lysine Lys K	H—C(NH₃⁺)(COO⁻)—CH₂—CH₂—CH₂—CH₂—NH₃⁺	128.2	5.9	2.16	9.06	10.54 (ε-NH₃⁺)
Arginine Arg R	H—C(NH₃⁺)(COO⁻)—CH₂—CH₂—CH₂—NH—C(=NH₂⁺)—NH₂	156.2	5.1	1.82	8.99	12.48 (guanidino)

(cont.)

Table E.1. (Continued)

Name, three-letter symbol, and one-letter symbol	Structural Formula[a]	Residue mass (D)[b]	Average occurrence in proteins (%)[c]	pK_1 α-COOH[d]	pK_2 α-NH$_3^+$[d]	pK_R Side-chain[a]
Histidine[f] His H		137.1	2.3	1.80	9.33	6.04 (imidazole)
Aspartic acid[e] Asp D		115.1	5.3	1.99	9.90	3.90 (β-COOH)
Glutamic acid[e] Glu E		129.1	6.3	2.10	9.47	4.07 (γ-COOH)

[a] The ionic forms shown are those predominating at pH 7.0 (except for that of histidine[f]) although residue mass is given for the neutral compound. The C$_\alpha$ atoms, as well as those atoms marked with an asterisk, are chiral centres with configurations as indicated according to Fischer projection formulas. The standard organic numbering system is provided for heterocycles.
[b] The residue masses are given for the neutral residues. For the molecular masses of the parent amino acids, add 18.0 D, the molecular mass of H$_2$O, to the residue masses. For side-chain masses, subtract 56.0 D, the formula mass of a peptide group, from the residue masses.
[c] Calculated from a database of non-redundant proteins containing 300,688 residues as compiled by Doolittle, R.F. (1989) in Fasman, G.D. (ed.), *Predictions of Protein Structure and the Principles of Protein Conformation*, Plenum Press.
[d] Data from Dawson, R.M.C., Elliott, D.C., Elliott, W.H., and Jones, K.M. (1986) *Data for Biochemical Research*, 3rd edn, pp. 1–31, Oxford Science Publications (1986).
[e] The three- and one-letter symbols for asparagine *or* aspartic acid are Asx and B, whereas for glutamine *or* glutamic acid they are Glx and Z. The one-letter symbol for an undetermined or 'non-standard' amino acid is X.
[f] Both neutral and protonated forms of histidine are present at pH 7.0, since its pK_R is close to 7.0.

(Illustration from Voet, D., Voet, J.G., and Pratt, C.W. *Biochemistry*, upgrade edn, p. 80. © 2002 John Wiley & Sons Inc.)

Index

Note: page numbers in *italics* refer to Figures and Tables.

16S rRNA 218
24-hour clock analogy, geological timescales *27*

absolute zero (thermodynamic zero) 6
acceleration, definition 33
acetate ions, resonance forms 130
acetic acid
   dissociation constant and pK_a *148*
   dissociation in water 142
acetyl CoA
   chemical structure *312*
   conversion to citrate 268
acetyl phosphate, free energy of hydrolysis 129
acetylcholinesterase, effect of pesticides 197
acetylene, sp hybrid orbitals *92–3*
acidity constant (acid ionization constant) 142
acids 142
aconitase 268
actin 280
activation energy 110–12
   effect of catalysts 112–13
   effect of enzymes *180*
activation of enzymes 190
   chymotrypsin 186
activation steps 132
active site, enzymes 182, 186
active transport 159, 161
adenine 206–7
adipocytes 158
aerobic metabolism, fate of pyruvate 263
alanine, structure and properties *315*
alcohol dehydrogenase, classification 198
alcoholic fermentation *263–4*, 264
alcohols 91
   energy value *106*
aldehydes 91
aldolase 257
algae 290–1
alimentary canal, evolution 297, 298
alkalis 142
alkanes 91
alkenes 91
allopurinol 184
allosteric effects
   in gene regulation 290
   haemoglobin 175–7
allosteric enzymes *190–2*
α-amino groups 164

alpha helix 166, *167*
α-ketoglutarate
   chemical structure *312*
   conversion to succinyl-CoA 268
α-ketoglutarate dehydrogenase 268
   regulation 271
alternative energy sources 223
Alvarez, Luis 18
amines *91*
amino acids 164–5
   structures and properties *315–18*
   triplet code 213
aminoacylation reaction 214
ammonia, lone pairs 98
ammonium ion, dissociation constant and pK_a *148*
ammonium nitrate, reaction with water 59
AMP (adenosine monophosphate) *211*
amphiphilic (amphipatic) substances 156–8
   phospholipids 158
amylopectin, relative molecular mass 163
anabolism *253*, 254
anaerobic bacteria 22, 24
anaerobic metabolism, fate of pyruvate *263–4*
Anfinsen, Christian 177
angiosperms (Angiospermae/ flowering plants) 293–4
animals (Animalia) 275
   evolution 296–300
antenna complex, photosystems 241, *243*
anthocyanins 294
antibiotics 295
   resistance to 24
antibonding orbitals 86, *87*
anticodons 214
aquatic life 140–1
Archaebacteria (Archaea) 218, 274
arginine, structure and properties *317*
Arrhenius equation 188, *189*
asparagine, structure and properties *317*
aspartic acid, structure and properties *318*
astronomy, detection of electromagnetic radiation 310
atmosphere
   carbon dioxide levels 17–18
   early 14
   effect of plants 292
   origins of oxygen 24, 276

atmospheric energy balance 223
atomic mass 83
atomic number 75
atomic orbitals 75–6, 77–8
  energy levels 78–9
atomic structure 75–6, 77
atoms 74
  elementary particles 75
  hydrogen 75–6
  size 75
atorvastatin (Lipitor) 184
ATP (adenosine triphosphate)
  in aminoacylation 214
  coupling of reactions 123
    effect on position of equilibrium 124–6
  daily requirements 234
  in DNA replication 210
  in formation of biopolymers 162
  hydrolysis, Gibbs energy change 124–5, 129–30
  in protein folding 173
  regeneration 123–4, 127
  resonance stabilization 131
  structure and roles 128–32
ATP synthase 233
  in photosynthesis *248*
  rotary engine 234–5
  three-site alternating binding site mechanism *236*
ATP synthesis 228–9, 232, 233, 250
  in citric acid cycle 261
  creatine phosphate hydrolysis 237
  efficiency of fermentation 264
  efficiency of glucose oxidation 238
  during fermentation 264
  during glycolysis 255, *256*, 262, 263
  regulation 233–4
  substrate-level phosphorylation 259, 260
ATP yields of metabolic pathways 270
autotrophs (producers) 10, 251
azide poisoning 235

backbone, evolution 298
bacteria 218, 274
  adaptability 24
  anaerobic 22, 24
  conjugation 279
Bak, Per, *How Nature Works* 26
banded iron formations 276
barbiturates, action on electron transport chain 235
base pairs, DNA 207, *208*
bases 142
  DNA 206–7
  RNA 212
bees, pollination 294
beta decay 43
beta-strand (beta-pleated sheet), proteins 166, *167*, 168

bicarbonate ions
  dissociation constant and pK_a 148
  importance in living systems 95
bilayer formation, phospholipids *158*
biochemistry 74
biogeochemical cycles *252–3*
biological chemistry 74
biopolymers
  common features 162–3
  see also DNA (deoxyribonucleic acid); polysaccharides; proteins
biosphere 19–20
biotechnology 139–40
biotin 195
1,3-bisphosphoglycerate
  chemical structure *311*, *313*
  effect on haemoglobin 176
  free energy of hydrolysis 129
  phosphate transfer 258, 262
Black Holes 12
blackbody radiation 303–4
blood, buffers 148
body plans, animals 297–9
body size, relationship to gravity 37
body weight 47
Bohr, Niels 75–6
Bohr radius 75
boiling points 155
  hydrocarbons 155–6
Boltzmann, Ludwig 62–3
Boltzmann's constant 63
bond energies 86, 105
bond length 86
bonding orbitals 86, *87*
bonds
  carbonyl group 99
  covalent 85–8
    hybridization 88–90, 92–4
    pi (π) bonds 92, *93*
    sigma (σ) orbitals 86, *87*
  hydrogen bonding 96–8
  ionic 94–5
  polar 96
boundary processes 52
brain, development 204–5
brown adipose tissue 237
bryophytes 291, 292
BSE (bovine spongiform encephalopathy/mad cow disease) 173
buffers 144–6
  in biological systems 148
  pH ranges 146, *147*
butane 156

C_3 pathway *see* Calvin cycle
C_4 (Hatch–Slack) pathway 248–9
calcium chloride, reaction with water 59

calcium ions, importance in living systems 95
calorimeter experiments 56–7
Calvin cycle (C_3 pathway) 245, *246–7, 313*
   comparison with glycolysis 247
   regulation 247
   standard and physiological free energy changes *314*
Cambrian Explosion 25
Campsoscolia ciliata, pollination of orchids 293
cancers, mutations as cause 217
carbohydrates
   in cell membranes 159
   energy value *106*
carbon
   hybrid orbitals 88–90, 92–4
   isotope ratios, evidence of living organisms 21
   isotopes 83–4
      carbon-14 43–4
carbon cycle *252*
carbon dioxide
   bond dissociation energies 105
   as greenhouse gas 21
   sp hybrid orbitals 94
carbon dioxide fixation 238, 239, 245
   see also dark reactions, photosynthesis
carbon monoxide poisoning 235
carbonic acid, dissociation constant and pK_a 148
carbonyl groups 99
   oxidation 259
carboxylase reactions, role of biotin *195*
carboxylic acids *91*
cardiovascular system 300
Carnot, Sadi 59–60
Carnot cycle 51, 60–1
carotenoids 294
catabolism *253, 254, 255*
catalase 188
catalysts 112–13
cell division
   meiosis 283–5
   mitosis 282–3
   in prokaryotes 282
cell membranes 159–60
   transmembrane proteins 288
cell wall
   algae 291
   fungi 288, 294, 295, 296
   plants 288
      lignin 292
   prokaryotes 279
      evolutionary loss 279–80
cells, replacement of 133
cellulose 46
Celsius temperature scale 6
chair form, molecular structure 102, 103

chaperonins 173
chemical potential energy 7, 35
chemical reactions, work 50
chemistry 73–4
chemotrophs *251*
chitin 294
chlorofluorocarbons (CFCs), ozone hole 28
chloroform *150*
chloromethane, reaction with hydroxide ion 110, 114
chlorophylls
   chlorophyll P680 241–2
   chlorophyll P700 242–3
   light absorption 241
chloroplasts 239
   origins 276, *277*
   structure *240*
choanoflagellates 294, 297
chromosomes 200, 201–2
   homologous pairs 284
chymotrypsin
   mode of action 185–8
   structure 186
chymotrypsinogen 186
cilia of protists, endosymbiosis theory 276–7
cis configuration 92, *93*
citrate
   chemical structure *312*
   isomerization 268
citrate synthase 268
   regulation 271
citric acid cycle (Krebs cycle, tricarboxylic acid cycle) *107*, 254, 266–71, *267*
   ATP yield 270
   chemical structures of intermediates *312*
   location within cells *272*
   regulation 271
   as source of anabolic precursors 270
   thioesters 261–2
clathrate hydrates *154*
   methane 155–6
Clausius, Rudolf 61
climate, interaction with life 18
closed systems 51, *52*
CMP (cytosine monophosphate) *211*
coal deposition 292–3
cobalt-60 43
codons 213
coelom 298
coenzymes, 120–1, 178, 192
   coenzyme A 261–2
   coenzyme Q (ubiquinone) 229, 231
   FMN, FAD *193–4*
   group-transferring 194–5
   pyridine nucleotides 192–3
   supply to metabolic pathways 190
   vitamins as precursors *196*

coevolution, plants and insects 293–4
cofactors 178, 192
collagen, structure 168–9, *170*
colony formation, unicellular organisms 282
colour of flowers 294
combustion, heat of 56–7
comets 14
　orbits 39
common descent, principle of 29, 203
common ion effect 142–3
　and pH 143–4
competitive enzyme inhibitors 183–5, *184*
compounds
　difference from elements 83
　molecular 88
　naturally occurring 85
computer-designed drugs 103
concentration
　effect on position of equilibrium 119–21
　relationship to reaction rates 113
concentration gradients, Na^+ and K^+ ions 153
condensation reactions 162, 163
　amino acids 165
conjugate bases 142, 143
conjugation (bacterial) 279
conservation of energy, law of 5, 9, 34–5, 52–3, 55–6
consumers (heterotrophs) 10, 251
continental drift 14–*17*
contractile proteins 164
control mechanisms 8
cooperative binding, haemoglobin 175–7
copper sulphate, reaction with zinc 225–6
copying errors, DNA 216
coupling of reactions 67, 122, 250
　ATP regeneration 123–4
　glutamine metabolism 122–3
　involvement of enzymes 124
　redox reactions (redox systems) 225, 227–9
covalent bonds 86
　average bond enthalpies 87–*8*
　strength *173*
covalent compounds, crystal structures 95
creatine phosphate, hydrolysis reaction 237
Creutzfeldt–Jacob disease 173
Crick, Francis 201
critical micellar concentration (CMC) 157
crossing over, meiosis 285
crystal lattices 95
crystallization, Hbs 219
curly arrows, reaction mechanisms 100
cuticle, plants 291–2
cyanide poisoning 235
cyanobacteria 24
cyclic photophosphorylation 244
cyclic processes, state functions 59
cyclosporin 295–6

cysteine
　dissociation constant and pK_a 148
　structure and properties *317*
cytochrome oxidase 229, *230*, 231
cytochrome oxidase inhibitors 235
cytochromes 231
　cytochrome b_6f 242, 244, *248*
　cytochrome bc1 complex 229, *230*
　cytochrome c 229
　　in regulation of electron transport chain 234
　cytochrome P450 systems 178
cytosine 206–7
cytoskeleton, evolution 280–1

Dalton, John 74, 83
daltons (Da/ atomic mass units) 83
dark reactions, photosynthesis 240, 245–8
de Duve, Christian 280–1
death 132, 133, 134, 224
decomposers 10
　fungi 295
'deep hot biosphere' hypothesis 19
'Deep Time' 27
ΔG (change in Gibbs energy) 65–6, 65–7, 72
ΔS (change in entropy) 61
ΔU (change in internal energy) 54–5
denaturation of proteins 161, 171, 177–8
deoxyribonucleoside phosphates 210–*11*
Descartes, Rene 32
development 204–6
　regulation 299
dicoumarol, action on electron transport
　chain 235, 237
diesel engines, efficiency 60
differentiation 202, 278, 288–9
　in plants 292
digestive system 300
dihydroxyacetone phosphate
　chemical structure *311*
　isomerization 258
2,4-dinitrophenol, action on electron transport
　chain 235, 237
dinosaurs, mass extinction 18–19, 26
dipeptides 165
diploid organisms 283–4
directionality of polymers 163
　DNA and RNA 215
disulfide bonds, proteins *167*, 171–2
diversity
　effect of atmospheric oxygen 24, *25*
　number of living species 22
DNA (deoxyribonucleic acid) 200, 201–2
　determination of structure 138
　directionality 215
　hydrogen bonding 152
　non-coding regions (junk DNA) 218

plasmids 279
polymerase chain reaction (PCR) 220–2, *221*
repair 215–16, 217
replication 203, *209*–12
replication errors 216–17
structure 206–9, *208*
synthesis *210*
transcription 212–13
DNA gyrase 210
DNA polymerases 210
DNA profiling 218
Dobzhansky, Theodosius 5
dogs, selective breeding 297
domestication of animals 297
dopamine receptors, development in the brain 204–5
double bonds 87
  average bond enthalpies *88*
  sp^2 hybrid orbitals 90, *92–3*
double helix structure, DNA 207, *208*
Drosophila, Hox genes 299
drug design, computer modelling 103
dynamic energy balance (steady state) 7–9, 116, 127, 133

Earth
  continental drift 14–*17*
  effect of radioactive decay 44–5
  formation 12–14
  state of change 17–18
  structure 12, *13*
EC (Enzyme Catalogue) numbers 197–8
ectoderm 297
efficiency
  of fermentation 264
  of glucose oxidation 238
  of heat engines 60
Einstein, Albert 44, 304
elastic potential energy 35
electrical energy 42–3
electricity flow 43
electromagnetic force 35, 40–3
electromagnetic radiation 303, 306
  emission spectra 307–9
  energy of 304
  spectroscopy 310
  wave–particle duality 305–6
electromagnetic spectrum *40*, 240, 306, *307*
electron configurations 77
  elements 1 to 20 *79*
  in periodic table 82
electron density distribution, hydrogen atom 75–*6*, 77
electron shielding 97
electron spin resonance (ESR) spectroscopy 310
electron transfer channels 189

electron transfer reactions 224–5
  redox reactions 108
  standard reduction potentials 225–7
electron transport chain (ETC) 229–33, *230*
  action of drugs and poisons 235
  changes in redox potential 231–*2*
  location within cells *272*
  in photosynthesis 239, 241, 242–3
  regulation 233–4
  uncoupling 237
electronegativity 95–6
electrons 75
  wavelike properties 86
elements
  difference from compounds 83
  in human body 73, *74*
  origins 11, 76
  periodicity 79–80
  symbols for 74–5
emission spectra 307–9
endocrine system 300–1
endoderm 297–8
endoplasmic reticulum (ER) 281
endosymbiosis theory 275–8, *277*
endothermic processes 52, 55
  position of equilibrium, effect of temperature 119
energy 32
  chemical 7
  definition 3–4
  and entropy 64–5
  internal 53
  in living organisms 46
  release in nuclear reactions 44
energy changes 4–5, 35
energy conservation law 5, 9, 34–5, 52–3, 55–6
energy exchange, intensive and extensive factors 67–8, 69
energy flow 6, 70
  heat 7
  and origins of life 224
energy levels, atomic orbitals 78–9
energy storage, plants 239, 248
energy supply, mammals 296
energy transducers, living systems as 133–5
enthalpy (H) 53–4, 55–6
  standard enthalpy of formation 57
enthalpy change (ΔH), Hess's law 56–7
enthalpy of fusion 149
entropy (S) 53–4, 61
  nature of 63–5
  Schrodinger on 137
  second law of thermodynamics 135, 136
  and spontaneity of reactions 61–2, 65
environment, influence on development 204–6
enzyme kinetics 180–3

enzyme-catalyzed reactions
  reaction rate vs. concentration curves *191*
  selectivity 100
enzymes 113, 164, 178
  catalytic properties 179–80
  classification 197–8
  complexity 189
  in coupled processes 124
  deficiencies 196–7
  inhibitors 183
    competitive 183–5, *184*
    irreversible 185
    non-competitive 185
  as markers for disease 197
  membrane-bound 159
  mode of action 185–8
  pH-dependency 148
  reaction rate enhancement 188
  regulation of activity 189–92
  secondary bonds 185
  specificity 178
equilibrium 50–1, 114–15, 223
  dynamic 7–9, 116, 127, 133
  effect of ATP coupling 124–6
  and Gibbs energy 66
  influencing factors
    concentration 119–21
    temperature 118–19
  Le Chatelier's principle 118
  and living systems 132–3
  reversibility of reactions 115–16
  standard free energy change 117
equilibrium constant 117
  relationship to Gibbs energy change 118
erythrose-4-phosphate, chemical structure *311*
Escherichia coli 22, 24
esters *91*
  base-catalyzed hydrolysis 100–1
ethanol 151
ethylene, sp^2 hybrid orbitals 90, 92–*3*
Eubacteria (Bacteria) 218, 274
  see also bacteria
eukaryotes (Eukarya) 24, 200, 274–5
  cell size 281
  colony formation 282
  cytoskeleton 280–1
  evolution 279–82
  mitosis 282–3
  organelles 281
  origins of 275–8
  sexual reproduction 283–5
evolution 200
  of animals 296–300
  of cytoskeleton 280–1
  endosymbiosis theory 275–8
  of eukaryotes 275–6, 279–82

of fungi 294–6
loss of prokaryotic cell wall 279–80
of mammals 302
of meiosis *284*
of multicellular organisms 278–9, 282
    association of cells 287–8
    cell differentiation 288–9
  natural selection 8, 22, 137
  of plants 290–4
  role of mutations 217
  Carl Woese's work 274
evolutionary relationships, role of protein homology studies 198–9
exciton transfer (resonance energy transfer) 241, *244*
exothermic processes 52, 55
    position of equilibrium, effect of temperature 119
extensive factors in energy exchange 67–8, 70
extinctions *see* mass extinctions

$FADH_2/FAD$ redox system 232–3
fat-soluble vitamins 196
fats 158
  energy value *106*
fatty acids, amphiphatic nature 156–7
feedback inhibition 191, 204, 251, 296
  of electron transport chain 233–4
  of gene expression 203, 206
fermentation 264
ferns, spore formation 292
Feynman, Richard 64
fibroin, structure 166, 168, *169*
fission, nuclear 44–5
flagella
  eukaryotic 280
  protists, endosymbiosis theory 276–7
flavin coenzymes 193–*4*
flavonoids 294
flavoproteins 194
  role in electron transport chain 231
Flemming, Walther 200
flowering plants 293–4
flowers, colour 294
fluid mosaic model 159–*60*
FMN *see* flavin coenzymes
foetal development 204–6
foetal haemoglobin (HbF) 176
folding energetics, proteins 171
folding mechanisms, proteins *172*–3
foods
  calorific value 57
  metabolism 8
forces 32, *35*
  definition 33
  electromagnetic force 40–3
  gravitational force 36–40

Newton's laws 38–40
strong nuclear force 43
formaldehyde 99
  oxidation 106
formic acid, oxidation 106
frequency of a wave 303
fructose-1,6-bisphosphate
  chemical structure *311*, *313*
  cleavage 257
fructose-6-phosphate 264, 264–5
  chemical structure *311*, *313*
  free energy of hydrolysis 129
  phosphorylation 257
    regulation 264–5
fumarate
  chemical structure *312*
  conversion to malate 269
functional genomics 197
functional groups *91*, 98–9
functional systems, mammals 300–1
fungi 275
  evolution 294–6
fusion
  latent heat of (enthalpy of) 149
  nuclear 45–6
    in the Sun 76

Galileo Galilei 31, 32
gametes (sex cells) 200, 284–5
gamma rays *40*, 42, 240, 306, *307*
gene expression 202–3, 206
gene regulation 288–9
  during development 299–300
  energetics 290
gene therapy 202
genes 200, 201–2, 290
genetic code 213, *214*
genetic defects 216–17
genetic engineering 203
genetic load 291
genetic variation 217
genetics 200
  classical 201
  Mendel's experiments 201
genome 202, 205
  recipe analogy 205–6
geological timescales 20, 21, 22, 23
  24-hour clock analogy 27
  'Deep Time' 27
  questions arising 27
  year analogy 22, 24–6
Gibbs energy 53–4, 65
  changes during a reaction *115*–17
    in physiological conditions 118
  coupled reactions 123
  and spontaneity of reactions 66–7

Gibbs energy change
  in hydrolysis of ATP 124–5
  relationship to equilibrium constant 118
  role in protein structure 171
  and standard redox potentials 227
Gibbs energy change of hydrolysis, 'high energy'
    compounds 127, 129
global warming 21
globular proteins 161
D-glucopyranose, space-filling models *104*
glucose
  bond dissociation energies 105
  chemical structure *313*
  oxidation, heat of reaction 57–8
  phosphorylation 257
    regulation 264
  structure 102–3
glucose oxidation, efficiency 238
glucose-1-phosphate, free energy of
    hydrolysis 129
glucose-6-phosphate
  chemical structure *313*
  free energy of hydrolysis 129
  isomerisation 257
glucose-6-phosphate dehydrogenase,
    regulation 266
glutamic acid, structure and properties *318*
glutamine
  coupled reactions 122–3
  structure and properties *317*
glyceraldehyde-3-phosphate
  chemical structure *311*, *313*
  oxidation 258, 260
  production in Calvin cycle *246*–7
glycerol (glycerine) 151
glycerol-3-phosphate, free energy of
    hydrolysis 129
glycine 165, *315*
glycogen 272
glycolysis 248, 254–6
  ATP yield 270
  chemical structures of intermediates *311*
  comparison with Calvin cycle 247
  fates of pyruvate 263–4
  location within cells 272
  reactions
    cleavage of fructose-1,6-diphosphate 257
    dehydration of 2-phosphoglycerate 262
    isomerisation of glucose-6-phosphate 257
    isomerization of 3-phosphoglycerate 262
    isomerization of dihydroxyacetone
      phosphate 258
    oxidation of glyceraldehye-3-phosphate
      258, 260
    phosphate transfer from
      1,3-biphosphoglycerate 258, 262

glycolysis (*cont.*)
   reactions (*cont.*)
      phosphate transfer from
         phosphoenolpyruvate 262–3
      phosphorylation of fructose-6-phosphate 257
      phosphorylation of glucose 257
   regulation 264–5
   role of thioesters 260–1
glycomics 197
glycoprotein peptidase, inhibition by penicillins 185
glyoxylate pathway 271–2
GMP (guanosine monophosphate) *211*
God, and thermodynamics 135
Gold, Thomas 19
Golgi apparatus (Golgi bodies) 281
Gondwanaland 16, *17*
gradients 224
gravitational force *35*–8
   and Newton's laws of motion 38–40
gravitational potential energy 34, 36
gravity, acceleration due to 36, 37
greenhouse effect 21
'greening of the Earth' 287, 291
ground state 307
group transfer potential 132
groups, periodic table 80
group-transferring coenzymes 194–5
growth rate, fungal hyphae 295
GTP (guanosine triphosphate) 266
   production in citric acid cycle 269
guanine 206–7

haematite, formation 276
haemoglobin 161
   changes in shape 175–7
   mutations 219–20
   relative molecular mass 163
   in sickle cell anaemia (Hbs) 219
   structure *167*, 174, *175*
half-life, radioactive isotopes 43–4
haploid cells 283
heart attack, lactate dehydrogenase levels 197
heat
   energy flow 7
   from solar energy 3
   specific heat capacity 71–2
heat energy 6
heat engines 59–60
heat transfer 49, 52
   direction of energy flow 68, 70
   Hess's law 56–7
heats of reaction 56–7
   oxidation of glucose 57–8
helicases 210
helium 82–3
   formation in the Sun 76

Henderson–Hasselbalch equation 143, 145
Hess's Law of heat summation 56–7
heterotrophs (consumers) 10, 251
   animals 296
   classification *251*
hexane 156
hexokinase 264
'high-energy' compounds 126–7, 129
   doeoxyribo- and ribonucleic acids 212
   thioesters 259–61
Hiroshima, atomic bomb 44
histidine, structure and properties *318*
HIV (human immunodeficiency virus), rate of evolution 218
HMG-CoA reductase, inhibition by statins 184
Holmes, Arthur 15
homeostasis 301–2
homologous pairs of chromosomes 284
homology, structural 297
house-keeping genes 289
How Nature Works, Per Bak 26
Hox genes 299–300
Hsp70 (heat-shock protein 70) 173
Human Genome Project 203, 205
hurricanes 223
hybridization 90
   sp hybrid orbitals 92–4
   sp^2 hybrid orbitals 90, 92–*3*
   sp^3 hybrid orbitals 88–9
hydration shells 152–*3*
hydrocarbons
   boiling points 155–6
   oxidation 106–8
hydrochloric acid, dissociation in water 142
hydrogen 87
   electron energy levels of orbitals *308*
   emission spectrum 307–8, *309*
   H_2 molecule 85–6, 87
   nuclear fusion in the Sun 76
   reaction with oxygen 111–12
   redox potential 225
hydrogen atom 75–6, 77
hydrogen bonds 96–8
   in DNA 207
   hydroxyl groups 151
   between proteins 152
   in proteins *167*, 171, *174*
   strength *173*
   in water 149–*50*
hydrogen peroxide, decomposition, effect of catalysis 188
hydrolysis, phosphorylated compounds 127
hydrophilic compounds 151
hydrophobic interactions 155, 157
   in membranes 159

in proteins 161, *174*
strength *173*
hydrophobic pockets 139
hydrophobic substances 154
hydroxide ion, reaction with chloromethane 110, 114
hydroxyl group, hydrogen bond formation 151
hyperbolic curve, reaction rate vs. concentration curves *191*
hyperthermophiles, protein stability 171
hyphae 294–5

ice 149
    structure *150*
Ice Ages 293
imidazole, dissociation constant and pK_a *148*
immune system 301
    effect of ultraviolet radiation 42
inborn errors of metabolism 196–7
inducers 289
induction of enzymes 190
infrared radiation *40*, 41, 240, 306, *307*
    solar 3
inheritance, Mendel's experiments 201
inhibitors of enzymes
    competitive 183–5, *184*
    irreversible 185
    non-competitive 185
insects, relationships with flowering plants 293–4
insulin production, feedback mechanism 206
intensive factors in energy exchange 67–8, 69, 70
interconversion control of enzymes 190
intermediary metabolism 251
intermediate species 114, 115
internal energy (U) 53–4
    changes in (ΔU) 54–5
ionic bonds 94–5
    in proteins *174*
    strength *173*
ionic compounds, dissociation in water 95, 152–*3*
ionization energies 307
    relationship to pK_a 146–8
irreversible enzyme inhibition 185
irreversible processes 64
isocitrate
    chemical structure *312*
    conversion to α-ketoglutarate 268
isocitrate dehydrogenase 268
    regulation 271
isolated systems 51–*2*, 51–2
isoleucine
    biosynthesis, feedback inhibition 191
    structure and properties *315*
isotopes 83–4

joule (J) 33
Jupiter, acceleration due to gravity 37

Kelvin temperature scale 6
keratins, structure 168
ketones *91*
    reduction, reaction mechanism 99–100
kidneys 300
kinetic energy 5, 6, 33–4
    of molecules 110–12, *111*
kinetics 110
Krebs cycle *see* citric acid cycle
kuru 173

lactate
    oxidation to pyruvate 120–1
    production 263–4
lactate dehydrogenase 100, 120
    as marker for disease 197
lactic acid, dissociation constant and pK_a *148*
land, colonization by life 287, 291
latent heat of fusion 149
lattice enthalpies 95
Laurasia 16, *17*
Le Chatelier's principle 118
Leitch, Eric and Vasisht, Gautam 19
leucine, structure and properties *315*
Lewis, G.N. 304
life
    development
        24-hour clock analogy *27*
        year analogy 22, 24–6
    oldest remains 21–2
    origins 3, 10, 11, 22, 140–1, 224
ligands 190–1
light *40*, 41, 240, 306, *307*
    velocity of 303
light absorption, chlorophyll 239, 241, *243*
light reactions, photosynthesis 239–40, 241–4
lignin 292
lipases 101
lipids 158
Lipitor (atorvastatin) 184
lipoamide 262
liquefied petroleum gas (LPG) 156
liver, detoxification reactions 178
living systems, energy transduction 133–5
locomotion 280
lone pairs 98
lysine, structure and properties *317*

magnesium ions, importance in living systems 95
malaria resistance, sickle-cell anaemia 220
malate
    chemical structure *312*
    oxidation to oxaloacetate 269

mammals
  evolution 302
  functional systems 300–1
  homeostasis 301–2
Margulis, Lynn 276–7
mass 36–7
  relationship to energy 44–5
mass extinctions 3, 18–19, 25–6
  Permian 293
mass number 75
mathematical basis of science 31–2
mathematics 74
Maxwell, James Clerk 67, 303
Maynard Smith, John 283
McLintock, Barbara 201
McPhee, John, 'Deep Time' 27
mechanical energy 35
mechanical equilibrium 50–1
mechanical work 4, 36
mechanics 32
meiosis 283–4
  crossing over 285
Mendel, Gregor 201
mesoderm 298
messenger proteins 163
messenger RNA (mRNA) 212
  triplet code 213
metabolic classification 251
metabolic pathways 7, 178, 250–1
  citric acid cycle 266–71
  glycolysis 254–6
    reactions 257–65
  glyoxylate pathway 271–2
  pentose phosphate pathway 265–6
  regulation 190–2
metabolism 7, 250
  anabolism and catabolism 253–4
  of food 8
metabolomics 197
metal ions, cofactors 178
meteorite strikes 12, 18–19, 26
methane 150
  as greenhouse gas 21
  molecular shape 89–90
  oxidation 106
  sp^3 hybrid orbitals 88–9
methane clathrates 155–6
methanol, oxidation 106
methionine, structure and properties 316
micelles 157
Michaelis–Menton model, enzyme kinetics 180–2
microstates 63
microtubules 280
microwaves 40, 306, 307
mitochondria
  electron transport chain (ETC) 229–33

location of citric acid cycle components 266
origins of 275–6, 277
structure 230
mitochondrial DNA 213, 276
  inheritance 280
mitosis 282–3
mixed acid anhydrides 92
molecular biology 138
molecular compounds 88
molecular diseases 207
molecular evolution 203, 217
molecular formulae 83
molecular graphics 103
molecular kinetic energy 110–12, 111
molecular medicine 219
molecular orbital (MO) theory 85–6
molecular recognition 281
molecular turnover 133–4, 135–6
molecules 83
  thermal motion 49
  three-dimensional structure 102–3
monolayer formation, fatty acids 157
Montreal Protocol (1997) 28
moon
  acceleration due to gravity 37
  formation 12
  tidal effects 12–13
Morgan, Thomas Hunt 201
motion, Newton's laws 32, 38–40
mountain ranges, formation 16
Mullis, Kary B. 220, 222
multicellular organisms, evolution 278–9, 282
  association of cells 287–8
  cell differentiation 288–9
musculo-skeletal system 300
mutations 216–17, 218
  Hbs 219
mycelia 295, 296
myoglobin, structure 166, 168

$NaBH_4$ reactions 99–100
$NAD^+/NADH + H^+$, redox reaction 228
  in oxidation of lactate 120–1
$NAD^+/NADP^+$ 192–3
NADH 100
  electron donation to electron transport chain 231
  production during glycolysis 255, 256
NADH dehydrogenase 229, 230
$NADP^+$, reduction in photosynthesis 239–40, 243–4
NADPH, pentose phosphate pathway 265–6
natural selection 8, 22, 137
near-activity conformations 179–80, 182
neighbouring group effect 99
nervous system 300

neuraminidase, inhibition by Tamiflu 185
neutralization, heat of 56
neutrons 75, 82
Newton, Isaac, laws of motion 32
    first law 38–9
    second law 39
    third law 40
nitrogen, N_2 molecule 87
nitrogen cycle 252
noble gases 82–4
non-competitive enzyme inhibition 185
non-cyclic photophosphorylation 242
nuclear forces 35
    strong nuclear force 43
    weak nuclear force 43
nuclear magnetic resonance (NMR) spectroscopy 310
nuclear power stations 45
nuclear reactions 44
    fission 44–5
    fusion 45–6
    in the Sun 76
nucleic acids 162
    see also DNA (deoxyribonucleic acid)
nucleophilic reagents 99
nutrient cycles 252–3
    role of fungi 295
Nuttall, George 217

oceans, formation of 14
octet rule 84, 84–5
oil, mixture with water 154–5
'oil-drop effect' 154, *155*
open systems 9, 51, *52*
order
    in human achievement 136–7
    maintenance in living organisms 133, 134, 135
organic chemistry 88
overweight 9
ovum 284
oxaloacetate, chemical structure *312*
oxidation of hydrocarbons 106–8
oxidation process 108
oxidative phosphorylation 127, 229, 250
    ATP yield 270
    comparison with photosynthesis 244
    see also electron transport chain (ETC)
oxidoreductases (redox enzymes) 192
oxygen
    atmospheric
        effect on diversity of life 25
        origins 24, 276
    O_2 molecule 87
    production in photosynthesis 239, 242
    reaction with hydrogen 111–12

oxygen cycle *252*
ozone hole 28, 42

p-type orbitals 77–8
Pangaea 16, *17*, 293
parasites, fungi 295
passive transport 159
path-independence, state functions 54
Pauling, Linus 96
penicillin antibiotics, mechanism of action 185
pentane 156
pentose phosphate pathway 265
    regulation 266
PEP *see* phosphoenolpyruvate
peptide linkage 165, 166
    hydrolysis, chymotrypsin mechanism *187–8*
periodic table of the elements 80–4, *81*
periodicity of elements 79–80
Permian extinction 293
peroxisomes 281
pH 141–2
    buffers 144–6
    and common ion effect 143–4
    effect on haemoglobin 176
    and enzyme activity 148
phagocytosis 279
phenotype 205
phenylalanine, structure and properties *316*
phenylketonuria (PKU) 196–7
phloem 292
phosphate esters *91*
phosphate ion, resonance forms 130
phosphoanhydride bonds, ATP 129
phosphoanhydrides 92
phosphocreatine, free energy of hydrolysis 129
phosphoenolpyruvate (PEP) 124, 237, 249
    chemical structure *313*
    free energy of hydrolysis 129
    phosphate transfer 262–3
    regulation 265
phosphofructokinase 257, 264–5
phosphoglucoisomerase 257
2-phosphoglycerate (2-PG)
    chemical structure *313*
    dehydration 262
3-phosphoglycerate (3-PG)
    chemical structure *311, 313*
    isomerization 262
    production in photosynthesis 245
phospholipids, bilayer formation *158*
phosphoric acid, dissociation constants and pK_a *148*
phosphorus cycle 252
phosphoryl group transfer potential 128
phosphorylation
    of fructose 257
    of glucose 257

photoautotrophs, classification 251
photography 41
photoheterotrophs 251
photons 304
photorespiration 248–9
photosynthesis 10, 24, 28, 103, 238–9
   C_3 pathway 248–9
   comparison with oxidative phosphorylation 244
   dark reactions 240, 245–8
   energy storage 248
   light reactions 239–40, 241–4
   regulation 247
   spatial relationship between components 248
photosystems 241
   energetics 242
phylogenetic studies 198–9
phylogenetic tree 217–18
phylotypic stage 298
physiological conditions 121, 139
phytol group, chlorophylls 241
pi (π) bonds 92, 93
pigments
   in algae 290–1
   in flowers 294
pK_a 142
   and buffers 146
   relationship to standard free energy of ionization 146–8
Planck, Max 303–4
Planck's constant 304
planets
   formation 12
   orbits 39
plants (Plantae) 275
   evolution 290–4
plasmids 279
plastocyanine 242
plastoquinone 242
platinum, use as a catalyst 188
Plimer, I. 19
polar bonds 96
polar compounds, water-solubility 151
polar ice-caps 140
polarity, relationship to boiling point 155
pollination 293–4
polymerase chain reaction (PCR) 220–2, 221
polymers 162
polypeptides 165
polysaccharides 162, 272
porphyrin groups, chlorophylls 241
potassium ions
   concentration gradients 153
   importance in living systems 95
potassium-40 44

potential energy 34, 35
power 33, 38
prepackaged evolution 278
pressure, as a state function 53–4
primary bonds 95
primary structure, proteins 166, *167*, 170
primers, PCR 221
prions 173
processes, definition 4
producers (autotrophs) 10, 251
prokaryotes 22, 200, 218
   cell fission 282
   cell wall 279
proline, structure and properties *316*
propane 156
prosthetic groups 178, 192
protective proteins 163–4
protein homology studies 198–9
protein kinases, interconversion control 190
protein sequencing 218–19
protein synthesis 212–15, *216*
proteins 162, 163
   amino acids 164–5
   as buffers 148
   in cell membranes 159, *160*, 288
   changes in shape 174
      haemoglobin 175–7
   denaturation 177–8
   determinants of shape 170
      disulphide bonds 171–2
      folding mechanisms *172*–3
   energy value *106*
   folding energetics 171
   hydrogen bonds 152
   maintenance of structure 173–4
   role in living organisms 163–4
   structure 161, 165–6, *167*
      collagen 168–9, *170*
      fibroin 166, 168, *169*
      keratins 168
      myoglobin 166, *168*
      quaternary 169
   see also enzymes
proteomics 197
Protista 275
   origins of cilia and flagella 276–7
protons 75
puberty, gene expression 289
pyridine nucleotides 192–3
pyridoxal phosphate 194–5
pyruvate
   chemical structure *312*, *313*
   conversion to acetyl CoA 261–2
   formation from lactate 120–1
   metabolism *263*–4
   reduction to lactate 100

pyruvate dehydrogenase 267
  regulation 271
pyruvate kinase 263

quanta 304
quantification 31
quantum theory 304–6
quaternary structure, proteins *167*, 169

radio waves *40*
radioactive decay 43, 82
reaction mechanisms
  hydrolysis of an ester 100–1
  reduction of ketones to form alcohols 99–100
reaction profiles *110*
  enzyme-catalysed reactions *181*
reaction rate vs. concentration curves,
    enzyme-catalysed reactions *191*
reaction rates
  activation energy 110–15
  catalysts 112–13
  effect of enzymes 188
  relationship to reactant concentrations 113
  reversible reactions 116
reactions, Newton's third law 40
recipe analogy, genome 205–6
red giant, Sun as 11
redox enzymes (oxidoreductases) 192
redox potentials 225–7
  changes along electron transport chain *231–2*
redox reactions (redox systems) 108, 224–5
  ATP synthesis 228–9
  coupled systems 227–8
  $NAD^+/NADH + H^+$ 228
reducing power, pentose phosphate
    pathway 265
regulatory enzymes 190
regulatory genes 289
regulatory proteins 163
Relenza (Tamiflu) 103, 185
replication of DNA 203, *209–12*
repression of enzymes 190
repressors 289
reproductive system 301
resonance forms 130
resonance stabilization, ATP 131
respiration, enthalpy of reaction 103–4
respiratory chain *see* electron transport chain (ETC)
respiratory system 300
reversibility of reactions 115–16
reversible processes 51, 65
  in living systems 127
  reaction rates 116
ribonuclease, renaturation 177
ribonucleoside triphosphates *211*
ribonucleosides 212

ribose-5-phosphate 265
  chemical structure *311*
ribosomes 212, 281
  protein synthesis 213–15, *216*
ribulose-1, 5-bisphosphate, chemical structure *311*
ribulose-5-phosphate, chemical structure *311*
Rift Valley 15
RNA (ribonucleic acid)
  16S rRNA 218
  directionality 215
  messenger RNA 212
  structure 212
  transfer RNA (tRNA) 213–14
rotary engine, ATP synthase *234–5*
rotation around bonds 92
rubella, effect on foetal development 204
rubisco (ribulose biphosphate carboxylase/
    oxygenase) 245
  regulation 247

s-type orbitals *77–8*
Sagan, Carl, on quantification 31
salts 143
  of fatty acids 157
Sanger, Frederick 218
saponification 101
saprophytes, fungi 295
satellites, orbits 39
schizophrenia, possible developmental
    abnormalities 204–5
Schrodinger, Erwin, on entropy 137
scrapie 173
sea-floor spreading *14–16*
sea level, effect of melting polar ice 20
seaweeds 290–1
second-order reactions 113
secondary bonds 95, 162
  hydrogen bonds 96–8
  in proteins 161, *173–4*
secondary structure, proteins 166, *167*
sedoheptulose-1,7-bisphosphate, chemical
    structure *311*
sedoheptulose-7-phosphate, chemical
    structure *311*
selective breeding 291
  dogs 297
self-organized criticality (SOC) 26
semi-conservative replication of DNA *209*–12
'sense' of polymers *see* directionality of polymers
sensory organs 298
septa, fungi 294–5
serial endosymbiosis theory (SET) *277*
serine
  DIPF inhibition 185
  structure and properties *316*
serine proteases 186

sexual reproduction 200, 203, 283–5
shape of organisms, relationship to gravity 37
shells, atomic 78
SI (Systeme Internationale) units 31
sickle-cell anaemia 219
   malaria resistance 220
sigma (σ) orbitals 86, *87*
sigmoidal curve, reaction rate vs.
      concentration *191*
single bonds, average bond enthalpies *88*
skin
   damage from ultraviolet radiation 41–2
   formation of vitamin D 42
slime moulds 288
Smith, Michael 220
sodium chloride
   crystal lattice structure 95
   dissolution in water 152–*3*
   ionic bonding 94
sodium ions
   concentration gradients 153
   importance in living systems 95
solar energy 3
soluble coenzymes 192
solvation energy, ATP 130
sound energy 5
sp hybridization 92–4
sp² hybridization 90, 92–*3*
sp³ hybridization 88–*9*
space-filling models 102–3, 103, *104*
specialization of cells 278, 289–90
specific heat capacity 71–2
specificity of enzymes 178
spectroscopy 310
sperm 280, 284
spirochaetes, as origin of cilia and
      flagella *277–8*
spontaneity of reactions 58–9, 110
   and entropy 61–2, 65
   and Gibbs energy 65–7
   and redox potentials 227
spontaneous energy flow 70–1, 72
stability numbers 84, 85
standard enthalpy of formation 57
standard free energy change 117
standard reduction potentials 225–7
standard states 117
starch 163, 239, 248, 272
stars, evolution 11–12
state functions 53–4
   additivity 57–8
   and cyclic processes 59
static energy balance 7
statins 184
steady state (dynamic energy balance) 7–8,
      127, 133

stearic acid 156–7
steric strain 90
stomata 291
storage proteins 164
strong nuclear force 35, 43
structural homology 297
structural proteins 163
structure–function relationships,
      biomolecules 101–2
subduction 15, 15–16
substrate-level phosphorylation 129, 237–8, 259,
      260, 263
succinate
   chemical structure *312*
   oxidation to fumarate 269
succinate dehydrogenase 229, *230*, 233
succinyl-CoA
   chemical structure *312*
   conversion to succinate 269
sucrose 239
sugars, water-solubility 151
sulphur cycle 252
Sun
   dying stages 11–12, 29
   energy dissipation 134
   as energy source 69
   nuclear fusion 45–6, 76
sunscreens 41–2
superconductors 6–7
supernovae
   as cause of mass extinctions 19, 26
   formation of heavy elements 76
surroundings 52
Sutton, Walter 201
symbiotes, fungi 295
systems 9, 51–*2*

Tamiflu (Relenza) 103, 185
tectonic plates 14–*15*
temperature 6, 41
   effect on position of equilibrium 118–19
   relationship to molecular kinetic energy
      110–12, *111*
   relationship to reaction rates 112
   as a state function 53–4
   thermal motion 49
temperature regulation 7, 8
tertiary structure, proteins 166, *167*
thermal cycling, PCR 221–2
thermal energy 41
thermal equilibrium 50
thermal motion 49
thermochemistry 58
thermodynamic zero (absolute zero) 6
thermodynamics 4, 48–9
   laws of

first (energy conservation law) 5, 52–3, 55–6
    second law 62, 67–9, 135, 136
    third law 72
  systems and surroundings 51–2
thermogenin 237
Thermoplasma 275
Thermus aquaticus, as source of DNA polymerase 222
thiamine pyrophosphate *196*, 262
thioesters *91*, 259
  in citric acid cycle 261–2
  in glycolysis 260–1
thiols *91*
Thomson, William (Lord Kelvin) 48
three-dimensional structure of molecules 102–3
  proteins 161
three-domain classification of life 274–5
three-site alternating binding site mechanism, ATP synthase 235, *236*
threonine, structure and properties *317*
thylacoids 239, *240*
thymine 206–7
thyroid hormones, effect on metabolism 8
tides 12–13
time, arrow of 136
tissues 200
trace elements 73
trans configuration 92, *93*
transaminase reactions 194–5
transcription 212–13, *216*
transcription factors 289
  energetics 290
transcriptional control of enzymes 190
transcriptome 197
transition state analogues 183
transition states 114
  in enzyme-catalysed reactions 179, 181, 182–3
translation 214–15, *216*
transport across membranes 159, 161
transport proteins 163
transport systems
  cardiovascular system 300
  fungi 294–5
  plants 292
tricarboxylic acid (TCA) cycle *see* citric acid cycle
triglycerides 158
  hydrolysis 101
triple bonds 87
  average bond enthalpies *88*
triplet code 213, *214*
trypsin, action on chymotrypsinogen 186
tryptophan, structure and properties *316*
turnover, molecular 133–4, 135–6
tyrosine
  dissociation constant and pK_a *148*
  structure and properties *317*

ubiquinone (coenzyme Q) 229, 231
ultraviolet (UV) radiation *40*, 41–2, 240, 306, *307*
UMP 211
uncoupling, electron transport chain 237
units 31
  of energy 4
universal ancestor 274
Universal Tree of Life *275*
universality, triplet code 213
universe, changes with time 11
uranium-235, use in power stations 45
uranium-238 44
urinary tract 300

valence 85
valence electrons 83
valine, structure and properties *315*
Van der Waals forces 174
  strength *173*
van't Hoff plots *119*
vaporization, heat of 56
vector quantities 38
velocity 37–8
  definition 33
velocity of a wave 303
vesicle formation 158, 279
visible light *40*, 41, 240, 306, *307*
vision 41
visual spectrum 310
vitamins 196
  vitamin C 169
  vitamin D, formation in skin 42
V_{max}, enzyme-catalysed reactions 181–2
volcanoes, formation 15–16
voltage calculation, redox reactions 226, 227
volume, as a state function 53–4
volvox 278

water 140
  aquatic life 140–1
  in biological processes 141
  boiling 149
  bond dissociation energies 105
  distribution on Earth 140
  entropy of 64
  in enzyme-catalysed reactions 182
  hydrogen bonds 149–50
  ice 149
  mixture with oil 154–5
  origins of 14
  role in protein structure 171
  as a solvent 151–3
  specific heat capacity 71
  structure *150*
water cycle 140, *252*

water vapour, as greenhouse gas 21
waterfalls, energy intensity 69
water-soluble vitamins 196
Watson, James 201
wave equation 303
wavelength 303
wavelike properties of electrons 86
wave-particle duality, electromagnetic radiation 305–6
weak acids 142
weak nuclear force *35*, 43
Wegener, Alfred 14
weight 36, 37
  see also body weight
West Antarctic Ice Sheet (WAIS) collapse, effect on sea levels 20
white dwarf, Sun as 11–12
Wilson, J.T. 14

Woese, Carl 218, 274
work 34, 49–*50*
  as a boundary process 52
  definition 4, 33
  mechanical 36

xanthine oxidase, inhibition by allopurinol 184
xeroderma pigmentosum 217
X-rays *40*, 42, 240, 306, *307*
xylem 292
xylulose-5-phosphate, chemical structure *311*

year analogy, geological timescale 22, 24–6
yeasts 295

zinc, reaction with copper sulphate 225–6
zwitterions 164
zymogens 186